JEFF DENISON
374-1860

C - 700   4
                   10
C - 730.   14
C+ 770   15
                   16
B - 800   17
B 930
B+ 870
A - 900

Engineering
Cost
Analysis

Engineering Cost Analysis

COURTLAND A. COLLIER
University of Florida

WILLIAM B. LEDBETTER
Texas A&M University

HARPER & ROW, PUBLISHERS, New York
Cambridge, Philadelphia, San Francisco,
London, Mexico City, São Paulo, Sydney

1817

Sponsoring Editor: Cliff Robichaud
Project Editor: Jo-Ann Goldfarb
Designer: Michel Craig
Production Manager: Marion Palen
Compositor: Science Typographers, Inc.
Printer and Binder: The Murray Printing Company
Art Studio: Vantage Art, Inc.

Engineering Cost Analysis

Copyright © 1982 by Courtland A. Collier and William B. Ledbetter

Library of Congress Cataloging in Publication Data

Collier, Courtland A.
    Engineering cost analysis.
    Includes index.
    1. Engineering—Costs.   2. Engineering economy.
I. Ledbetter, William Burl. II. Title.
TA177.7.C64   658.4′03   81-6793
ISBN-0-06-041329-8   AACR2

Contents

v

Chapter 3 The Value of One Future Payment Compared to a Uniform Series of Payments (F/A, A/F)   35

Chapter 4 The Value of One Present Payment Compared to a Uniform Series Of Payments (P/A, A/P)   62

Chapter 8 Annual Payments Method of Comparing Alternatives (or, Evaluating Alternatives by Comparing Equivalent Series of Equal Periodic Payments) 152

Chapter 9 Future Worth Method of Comparing Alternatives 174

Preface

The purpose of this book is to provide both students and professional engineers with some basic tools needed in the practice of good engineering decision making. Engineers are basically problem solvers. Their salaries are determined for the most part in a free competitive marketplace by the value of the services they render. As problem solvers, engineers frequently generate a whole range of alternative solutions to the problems confronting them and their clients or employers. These alternatives often require investments of different amounts of resources and bear fruit of varying amounts at different points in time. The methods and techniques presented in this book enable the engineer as decision maker to quantify each alternative—in terms of dollars, where applicable—and select the most productive. In utilizing these tools, the engineer obtains the satisfaction of dealing with a numerical value. The benefits of dealing with hard numbers are that decisions are more easily reached and scarce resources conserved. The value of the engineer's work product is thereby enhanced, and only the most skeptical can doubt that tangible rewards will soon follow. In short, the mastery of the material in this book should prove a satisfying and rewarding experience in many ways to both the student engineer and the practitioner.

Through its order of presentation, careful explanations, and many examples, *Engineering Cost Analysis* is designed to facilitate self-study. This means that classroom time then can be better used to concentrate on applications, problem solving, and clarifying concepts.

The authors express their appreciation to the many pioneers who have labored and published in this field and who are largely responsible for the improved analysis of financial alternatives. Also appreciated are those select students who are courageous and perceptive enough to question unclear presentations. From such questioning come both deeper understanding and improved teaching methods. Since our book is expected to undergo further refinement and revision in the future, questions and suggestions for clarification of the material from all readers, whether student or practitioner, will be sincerely appreciated.

The authors also wish to acknowledge the contribution of the many typists who worked on the manuscript over the years, including Mrs. Beatrice Cullen and Mrs. Carol Laine. And finally we are pleased to dedicate this text to our wives and families. Without their encouragement and support this text would not exist.

COURTLAND A. COLLIER
WILLIAM B. LEDBETTER

Part I
BASIC PRINCIPLES

Chapter 1
Quantifying Alternatives for Easier Decision Making

THE DECISION PROCESS

Life is full of choices. In matters small and large we are constantly confronted with the need to make selections from arrays of mutually exclusive alternative courses of action concerning how best to spend our time, where to live, which job to take, how to determine spending priorities for our money, and so forth. In short, how can we best invest our limited supply of resources such as time and money? Decisions, decisions, decisions. The decision-making process would be greatly simplified if the alternatives could be quantified objectively and some numerical value attached to each. Then an optimum return on our limited resources could be assured by simply selecting the alternative with the highest numerical value from each array.

Not all alternatives can be quantified, but a surprising number can. For instance, the question of whether or not to install a traffic light at an intersection can be reduced to a summation of costs versus benefits to all concerned. The costs of stopping a vehicle, costs of waiting time, costs of damage from accidents, and even of air pollution from idling engines can be assigned some numerical value. Later chapters explain how to equate costs and benefits that occur in different quantities and at different times so that alternatives can be compared directly and the better choice made evident. One of the primary purposes of this text is to enable the reader to make intelligent selections from among a wide variety of alternatives.

As in other areas of life, the practice of engineering requires a considerable amount of decision making. The decision-making process may vary somewhat depending upon the desires of the project sponsor. Most engineering works are sponsored either by the government as public works projects, or by private enterprise, as profit-making ventures.

Public Works Decision Process

Planning and implementation for public works typically involves the following process:

1. List current and future needs of the public that can be met by public works projects.
2. List the alternative proposals for meeting those needs.
3. Compare the alternatives.
   a. Quantify and numerically compare *tangible values*, such as the costs, benefits, effectiveness, or income of each alternative in meeting the need.
   b. Compare the *intangible values*, such as quality, appearance, effect on neighborhood image, pride and respect, the comparative needs of the beneficiaries, and the distribution of the benefits.
4. Establish priorities based on ratios of benefits to cost or equivalent, with proper consideration given to intangibles.
5. Make recommendations for funding, and hold public hearings to obtain approval and funding.
6. Construct and operate the project. Evaluate accomplishments compared to previous estimates.

Since public works planning involves decision making in the spending of the public's tax money, it should be done with the public's sense of priority in mind. The conscientious engineer performs a valuable public service in preparing and presenting comparative numerical data for use by the public when facing decisions between competing alternatives.

Implementing Private Enterprise Projects

Planning for privately owned facilities differs from public works basically in the accountability for funds, both investment and return. In public works, the funds must be fairly distributed to reflect the needs of citizens, their sources of revenue, and other factors of concern to them. In the private sector, the investor has wide discretion concerning where and how much funds are expended, within the broad limits set by law. Usually the private investor is primarily concerned with profitability. As a result, the private investor, like the public works planner, is vitally concerned with the needs of the public, since the return on investment usually depends on the patronage of the consuming public as well as on the satisfaction of individual demands. Therefore, rather than rely on intuition and hope, planners of privately owned projects need factual numerical comparisons in order to make intelligent decisions.

Dollar Value: The Common Denominator

In planning for either public or private projects, all significant aspects of each project are analyzed, and viable alternatives are listed, quantified, and compared. Whatever can be quantified should be, since decision makers find it easier to compare numbers than other qualities. In order to quantify and compare as many aspects of a proposal as possible, a common denominator for most purposes is the dollar value. For this reason, most of the significant attributes of a project are reduced to dollar values wherever possible.

INTEREST: THE RENT PAID FOR USE OF MONEY

Money, like the goods and services it represents, can be owned, or it can be borrowed or owed. When borrowed, owners usually expect some compensation for the inconvenience of doing without their money for the specified period of time, so there is usually an agreed upon rental fee charged for the loan. Of course, the rental fee is more commonly called interest, but the function is the same as other rental fees; that is, the rent (or interest) is paid in order to acquire the use of the owner's money for a stipulated period of time. By the same reasoning, the borrower who has the use of the money expects to pay rent or interest. For instance, a builder often needs short-term money and takes a construction loan to cover the cost of labor and materials put into the project until reimbursed by the owner. An owner usually needs a long-term loan (mortgage), which can be repaid in equal monthly installments out of current income.

Interest Rates

Owners of engineering projects, whether public or private, may finance the project with their own money, or with borrowed money, or with a combination of both owned and borrowed money.

• *Owned Money.* If the owners of the project have money of their own available they may choose between investing the money in the project or they may lend it to others at *interest*. If they invest in the project the money is not available for loan and the interest income is lost. The lost interest income therefore should be charged to the project as one of the costs of the project.

• *Borrowed Money*. If the owners choose to finance the project with borrowed money, the interest paid on the borrowed money is also one of the costs of the project.

• *Combination of Owned and Borrowed Money*. On private projects, most lenders will lend only from about 67 to 80 percent of the value of a project. Therefore, most projects require owners to finance from 20 to 33 percent of the project with their own money. Since most owners expect to make a return on their projects greater than the interest rate charged by money lenders, they usually prefer to borrow as much as possible and use a minimum of their own funds. Borrowing is sometimes called "leveraging" and most owners prefer as much leveraging as possible. On public works projects owned by agencies of the government, 100 percent financing can be obtained commonly since the risk of default is usually small. This type financing is usually done through bond issues.

DETERMINATION OF INTEREST RATES
Interest rates are determined by the following factors:

1. The risk of not getting the loaned money back. A high risk warrants high interest rates. If the prevailing interest rate is 15 percent and there is a 1-in-50 probability of default, add about 2 percent (1/49) to cover this risk, summing to a net interest rate of 17 percent in this case.
2. The supply and demand for funds. A high demand for loans usually drives interest rates up. A typical supply–demand curve is illustrated in Figure 1.1. For the particular money market shown, an amount equal to V dollars will be lent at i interest rate.
3. Overhead costs of bookkeeping, collection fees, and so forth, and length of time of loan. High collection costs, high accounting costs for small weekly payments, low dollar amount of loan, short-term loans with turn-around costs all result in higher interest charges.

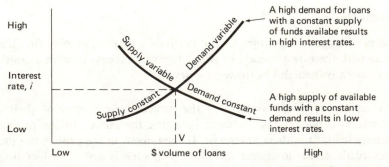

Figure 1.1 Effect on interest rates of supply and demand for loan funds.

4. Government regulations. Usury laws put ceilings on certain types of loans. Also some federally chartered banks are limited in the interest they may pay on deposits, and the discount rate at which banks borrow money is regulated by the Federal Reserve Board.

SIMPLE AND COMPOUND INTEREST

Simple interest is the terminology used to describe interest usually paid to the lender as soon as it is earned. If the simple interest payment is not claimed by the lender when due, no interest accrues on that interest payment no matter how long the lender leaves it unclaimed.

Compound interest on the other hand is interest which if not paid as soon as it is earned is considered as increased principal and thus earns additional interest with time. Compound interest implies that each subsequent interest payment is calculated on the total of principal plus all accumulated interest to date.

SIMPLE INTEREST EXAMPLE:	
Principal (original investment)	$1,000
Interest earned and paid at end of year 1 @ 10%	100
Interest earned and paid at end of year 2 @ 10%	100
New total of principal plus *simple* interest	$1,200
COMPOUND INTEREST EXAMPLE:	
Principal (original investment)	$1,000
Interest earned, end of year 1 @ 10%	100
New principal plus accumulated interest	$1,100
Interest earned, end of year 2 @ 10%	110
New total of principal plus *compounded* interest	$1,210

The percent interest rate is usually expressed as the nominal annual rate, even though interest may actually be paid quarterly, monthly, or daily. For instance, 6 percent nominal interest paid semiannually on a $1,000 investment would yield a check for 3 percent×$1,000=$30, sent once every six months to the lender. The 6 percent refers to the annual nominal interest rate, and the "paid semiannually" indicates that a check for a prorated amount is sent twice a year. If the interest is not paid then—if the interest is compounded—the $30 earned interest (but unpaid) earns 3 percent×$30= $0.90 additional interest the second six months and the total investment plus earned interest would equal $1,060.90 at the close of the first year.

Percent Growth Similar to Interest

The concept of simple interest and compound interest is readily applicable to nonfinancial subjects as well. For instance, population growth is often expressed in terms of simple or compound percentage. If a city that had a population of 100,000 a year ago now has a population of 110,000, the growth rate may be calculated as 10 percent per year. If the same rate of growth continues during this year, the population next year will grow by another 10 percent (or 11,000 people) and the total will reach 121,000. Thus the growth rate would be 10 percent compounding. Sometimes the growth rate is expressed in terms of simple percentage, similar to simple interest. For instance, if a state had a population of 4,000,000 in 1970 and 5,000,000 in 1980, some commentators would calculate the increase of 1,000,000 in 10 years, or an average of 100,000 per year on a base of 4,000,000. The rate of increase therefore is 2.5 percent per year ($100,000/4,000,000 = 0.025$ or 2.5%). The growth rate of many numbers important to all of us are commonly measured in percentages. A partial list includes:

> Consumer Price Index (cost of living)
> Population of our neighborhood, county, state, nation, friendly nations, unfriendly nations, planet earth
> Construction Cost Index
> Unemployment and other sociological indexes such as the crime rate (Index of Crime)
> Increase in traffic count on a given road or at a particular intersection

PROBLEMS FOR CHAPTER 1

Problem Group A

Find or apply the percent increase in several commonly encountered indexes over a one-year period.

A.1 Find a recent issue of *Engineering News Record* (*ENR*) magazine and look up the annual rise in the Construction Cost Index expressed in percent (see the construction scoreboard page). Assume a construction project last year was bid at $10,000,000 but the award was postponed for one year due to high prices. How much would it cost now, one year later, if the Construction Cost Index adequately reflects the change in construction costs? (Note which issue and page numbers of the *ENR* from which the information is derived.)

A.2 Find the current Consumer Price Index and the value of one year ago. Then derive the percent increase over the past year. (See *U.S. News & World Report* under "Trend of American Business" section, or monthly *U.S. Bureau of Labor Statistics Bulletin* [*BLS*], or other.)

A.3 A small retirement village had a population of 1,750 one year ago. During the past year, 14 people died and 52 people moved into the village.
a. What was the percent death rate during the year?
b. What was the percent *net* growth rate during the year?

Problem Group B

Find current interest rates used by credit institutions.

B.1 Call a local savings-and-loan institution or commercial bank and find what current interest rate is being paid on passbook savings. What is the compounding period? Show the name of institution and date.

B.2 Call a local finance company and find what interest rate is charged on a $1,000 personal (unsecured) loan for any time period of your choice. Is the interest simple or compounded? Show name of company and date called.

B.3 Call a local credit union and find what interest rate and compounding period (a) is charged on unsecured loans to members and (b) is paid on savings or share accounts of members.

B.4 What annual and monthly interest rates are charged by the major credit card companies on credit accounts (Visa or Mastercharge)?

Problem Group C

Consider the problem of quantifying some typical tangibles and intangibles.

C.1 When considering whether or not to install a traffic signal at a street intersection, how would you approach the problem of finding dollar-value equivalents to the following:
  a. Waiting time for motorists now trying to cross the intersection without a signal as well as waiting time for motorists stopped by the signal if one is installed.
  b. Safety without the signal compared to safety with the signal.

C.2 When considering whether to specify brick facing for the outside of a proposed school building or whether to leave the outside as painted concrete block, how do you go about evaluating the numerical value of appearance and aesthetics? List the items you would consider. Briefly discuss your approach to evaluating each item.

C.3 Assume you need an automobile, and that you have $4,000 cash to spend either as payment in full or as a cash down payment on a car costing more than $4,000. Describe the features you would consider (including cost) when shopping for this car and describe at least ten factors that would influence your final selection. Indicate beside each factor listed whether it is "tangible" or "intangible."

C.4 List five tangible and five intangible factors influencing your decision to attend the college of your choice.

Chapter 2
The Value of a Single Payment Now Compared to a Single Payment in the Future ($P/F, F/P$)

KEY EXPRESSIONS IN THIS CHAPTER

$P =$ Present value of all payments under consideration.* Present value may be thought of as an equivalent single lump sum cash payment now, at time zero the beginning of the *first* of n time periods. No interest has accrued. The present value differs from an equivalent future value by the amount of interest compounded on the present value during the intervening period.

$F =$ Future value. A single lump sum cash payment occurring in the future at the end of the *last* of n time periods. The future balance of all payments under consideration, together with accrued interest at rate i.

$i =$ Interest rate, assumed to be the annual rate unless otherwise specified. This is the rate at which the balance of the account is charged for use of the funds. Or in nonfinancial problems i is the growth or decline of the base number. It is assumed as annual and compounded unless otherwise noted or obvious from the problem.

*NOTE: The words value, worth, sum, amount, and payment are used interchangeably. For instance, present value, present worth, present sum, and so forth, all mean the same.

$n =$ Number of time periods, frequently years, but may be quarters, months, days, minutes, or as otherwise specified.

Expressions used to describe the time at which payments are made.

BOY = Beginning of the year
BOM = Beginning of the month
EOY = End of the year
EOM = End of the month

INVESTMENT FOR PROFIT

Most investments are made with the hope of some financial gain or profit. When money is invested (deposited) in a savings account, we expect to get more out than we put in. The value of the savings account should increase as interest payments accumulate in the account. In describing the investment in more general terms we can state that the future value (designated as F) of the account is always higher than the present value (designated as P) provided that the interest rate (i) is greater than zero, and that the number of time periods (n) during which interest is paid is equal to or greater than one. The size of the increase in the value of the account will depend upon the interest rate i, and the number of interest payment periods n. A measure of the size of the increase is given by the ratio F/P. This is the ratio of the dollars balance in the account, F, at the end of a series of n compounding periods compared to the original deposit, P, made at the beginning of the first compounding period.

CASH FLOW

As the name implies, cash flow occurs whenever cash or its equivalent "flows" from one party to another. When you pay money for a cup of coffee or the morning newspaper, cash flows from you to the vendor. If the morning paper comes by monthly subscription, the paper carrier has earned a little more with each delivery but cash does not flow until you pay the monthly bill.

The cash flow exchange may occur by using currency, checks, a transfer through bank accounts, or some other means, providing the transaction is readily convertible to cash. Cash flow *in* occurs when you receive payment and cash flow *out* occurs when you pay out. Some other examples of cash flow follow:

1. A savings bond is purchased for $75 that will mature in seven years and can be cashed in at that time for $100. Each year of the seven years, the bond increases in value and could be cashed in at any time. But there is no cash flow until the bond is actually exchanged for money.

2. As an engineer, you agree to design a structure and charge a fee of 7 percent of the cost of the structure payable upon satisfactory completion of the design. When your design work is half-completed you may feel you have earned one-half of your fee (some accountants might even add one-half of the fee to your statement of earned income at this time), but the actual cash flow does not occur until the payment is made upon the completion of the design.
3. As an investor you purchase a choice piece of real estate for $10,000. You pay 20 percent down (cash flow out of $2,000) with the balance due in a lump sum in five years (no cash flow during the five-year period). At the end of five years you pay off the remaining 80 percent due (cash flow out) and sell the property for a $5,000 down payment ($5,000 cash flow in) plus a mortgage that entitles you to $2,000 per year for five years ($2,000 cash flow in at the end of each of the next five years).

Cash Flow Diagram

Cash flow is illustrated graphically by use of a line diagram, a type of graph used extensively because of its simplicity and ease of construction. It consists of two basic parts, (1) the horizontal time line, and (2) the vertical cash flow lines. The horizontal time line is subdivided into n periods, with each period representing whatever time duration is appropriate for the problem under consideration, such as a year, month, day, and so forth. The vertical lines represent cash flow and are placed along the time line at points corresponding to the timing of the cash flow. The vertical lines are not necessarily to scale, although a large cash flow is usually represented by a longer line than a small cash flow. If you borrow $1,000 for one year and agree to pay 8 percent interest upon repayment, the cash flow diagram representing this transaction from your (borrower's) point of view is illustrated in Figure 2.1. Note that n represents the number of completed periods. Therefore, n does not equal 1 in Figure 2.1 until the full period is completed at the right end of the diagram.

In Figure 2.1, the use of the double arrow point designates a derived equivalent value, an unknown value to be found. In this case $1,080 after

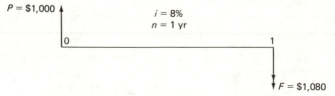

Figure 2.1   Borrower's cash flow line diagram.

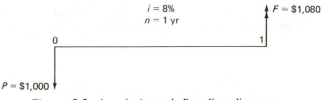

Figure 2.2   Lender's cash flow line diagram.

one year is derived as the equivalent to $1,000 now by simply adding the accrued interest of $80 to the original $1,000 loan.

Normally, receipts or income are represented by upward pointed arrows since they represent cash flow in and an increase in cash available. Conversely, disbursements or expenditures are usually downward pointed arrows indicating a decrease in available cash due to a cash flow out.

Notice that when the cash flow diagram is drawn from the *lender's* point of view, the arrows point in the opposite direction. In loaning you the $1,000, the lender has a cash flow out (arrow downward) and upon repayment the lender has a cash flow in (arrow upward). See Figure 2.2.

If the $1,000 is borrowed for a four-year period ($n=4$) at 8 percent interest per year, the borrower and lender might select one of the following four alternative plans of repayment. These four plans illustrate some of the more common repayment plans but are by no means a complete list of all the alternatives available for repaying $1,000 in four years with 8 percent interest on the balance.

The cash flow diagrams for this example are drawn from the borrower's point of view (borrowed funds are a cash flow in, with arrow upward) and each plan provides for 8 percent interest on the unpaid balance.

Plan A pays interest only on the $1,000 loan at the end of each year. At the end of the fourth year the loan is repaid together with interest due for that year (Figure 2.3).

Plan B repays an equal fraction of principal each year (one-fourth or $250 in this case) plus interest on the amount of principal owed during the year. Thus, at the end of year 1, the interest owed for the use of $1,000 for one year at 8 percent is $80. This $80 interest payment plus repayment of

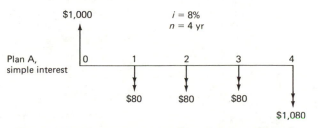

Figure 2.3   Borrower's cash flow diagram, Plan A.

$250 of the principal results in a total cash flow of $330 at the end of year 1. Payments for subsequent years are calculated in a similar manner (Figure 2.4).

Plan C provides for *equal annual* payments, as called for in most standard installment loan contracts. At the end of year 1, $80 is due as interest for the $1,000 owed during the year. The remainder of the payment, $301.92 − $80 = $221.92 pays the balance of principal due, so that only $1,000 − $221.92 = $778.08 is owed during the second year. The interest due at the end of the second year then is only 0.08 × $778.08 = $62.25. The amount of payment is calculated by methods shown in later chapters so that the final payments pays all remaining principal due plus the final interest payment (Figure 2.5).

Plan D accumulates interest until the end of the four years. The balance owed at the end of the first year is $1,080. At the end of the second year the total amount borrowed is considered to be $1,080 instead of the original $1,000, and interest is calculated on this whole amount as 0.08 × $1,080 = $86.40. Interest calculated in this manner is said to be compounded (Figure 2.6).

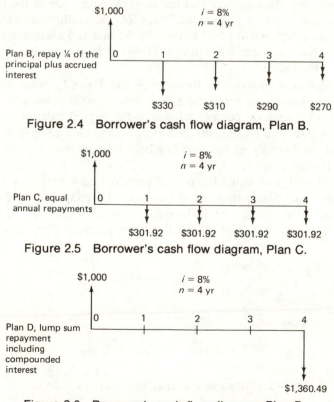

$1,000 $i = 8\%$ $n = 4$ yr

Plan B, repay ¼ of the principal plus accrued interest

0   1   2   3   4

$330 $310 $290 $270

Figure 2.4   Borrower's cash flow diagram, Plan B.

$1,000 $i = 8\%$ $n = 4$ yr

Plan C, equal annual repayments

0   1   2   3   4

$301.92 $301.92 $301.92 $301.92

Figure 2.5   Borrower's cash flow diagram, Plan C.

$1,000 $i = 8\%$ $n = 4$ yr

Plan D, lump sum repayment including compounded interest

0   1   2   3   4

$1,360.49

Figure 2.6   Borrower's cash flow diagram, Plan D.

EQUIVALENCE

The four preceding plans are termed "equivalent" to each other. *Equivalence means that one sum, or series differs from another only by the accrued interest at a stated interest rate i accumulated during the intervening n compounding periods.* In other words, $1,000 today is equivalent to four equal payments of $301.92, or it is equivalent to a lump sum payment of $1,360.49 four years in the future, given that the time value of money (the interest rate i paid for the use of these funds) is 8 percent. If you have a good credit rating and borrow $1,000 from a bank, the bank may tell you that you may repay the $1,000 using *any* one of the preceding four plans because they are equivalent. This equivalence concept is the basis for comparing alternative proposals in monetary terms, and thus your clear understanding of this concept is fundamental to the remainder of this text.

INTEREST USUALLY PAID IN ARREARS

In commercial dealings, some expenditures are customarily paid in advance and some in arrears. For instance, rent payments and insurance premiums are customarily paid in advance at the beginning of the time period, while interest payments and property taxes are usually paid at the end of the period. Sometimes the lender and borrower may ignore custom and agree that interest shall be paid in advance (prepaid interest). The effect is to increase the return to the lender, and consequently increase the effective interest rate, since the lender (rather than the borrower) has the use of the interest payment money during the term of the loan.

SINGLE-PAYMENT COMPOUND AMOUNT FACTOR

If interest is compounded, unpaid interest is treated as principal and this earns further interest. Thus, on a loan of $1,000 at 6 percent per year interest, the interest payment of $60 earned at the end of year 1 may be added to the loan and treated as a loan of $1,060 for the entire second year. Then at the end of the second year, 6 percent interest is due on a loan of $1,060 rather than on just the original $1,000. Thus, the interest due is added to the original loan (or invested capital) and in turn earns interest.

Example 2.1

The sum of $1,000 is borrowed at 6 percent compounded annually, to be repaid in one lump sum at the end of three years. How much should the lump sum repayment amount to at the end of the three-year period?

SOLUTION
At the end of one year the borrower owes the principal ($1,000) plus 6 percent interest $(0.06 \times \$1,000 = \$60)$ which totals $1,060. If the interest

payment is not taken out but simply added to the $1,000 loan, then at the end of two years the amount owed is $1.06 \times \$1,060 = \$1,123.60$. At the end of the full three years, the borrower should repay a total of $\$1,123.60 + 0.06 \times \$1,123.60 = \underline{\underline{\$1,191.02}}$.

In terms of the original principal (P) and interest (i) this sequence of events may be written as shown below, where F is the future sum of all principal and interest accumulated at the end of each time period shown.

end of first year $\qquad F = P + iP = P(1+i)$

end of second year $\quad F = P(1+i) + iP(1+i) = P(1+i)(1+i) = P(1+i)^2$

end of third year $\qquad F = P(1+i)^2 + iP(1+i)^2 = P(1+i)^3$

end of nth year $\qquad F = P(1+i)^n$ $\hfill$ (2.1)

Thus, in Example 2.1, for $1,000 at 6 percent compounded for three years, the total future sum, F, may be found very simply by

$$F = \$1,000(1 + 0.06)^3 = \$1,191.02$$

Note that the factor $(1+i)^n$ is independent of the value of P and represents the ratio of F/P. This factor of $(1+i)^n$ is called the single payment compound amount factor (SPCAF). Note also that in order to find F, this factor is multiplied by P. For convenience this factor may be designated by the notation $(F/P, i, n)$. Using this notation Equation 2.1 would be expressed as

$$F = P(F/P, i, n)$$ $\hfill$ (2.1A)

This notation is *not* a formula, although it is functionally correct in that, when the P values are cancelled, both sides of the equation revert to F.

$$F = \not{P}(F/\not{P}, i, n)$$

When F is known and P is sought, the equivalence factor is simply the reciprocal of Equation 2.1:

$$P = F\frac{1}{(1+i)^n}$$ $\hfill$ (2.2)

The factor $1/(1+i)^n$ is often termed present worth compound amount factor (PWCAF) and can be expressed in its functional form as $(P/F, i, n)$. Using this form, Equation 2.2 becomes

$$P = F(P/F, i, n)$$ $\hfill$ (2.2A)

As an aid to remembering the sequence of characters, note that when the F values are cancelled, both sides of the equation revert to P.

For Equations 2.1 and 2.2 the interest rate, i, is the interest rate per *period* and the time, n, is n compounding *periods*. Thus, i and n are linked to an identical number of compounding periods. If i is expressed as a rate per quarter, then n must be expressed in quarters of years, and so forth. These notations (2.1A and 2.2A) will appear frequently in later chapters.

SOLUTION USING THE TABLES IN APPENDIX A

The tables in Appendix A contain values of $F/P=(1+i)^n$, as well as values of the inverse $P/F=1/(1+i)^n$. For many problems use of the tables saves time. Each table contains values for only one interest rate i. The time periods n are shown in the extreme right- and left-hand columns of each page. Example 2.1 could be solved by proceeding through the following steps:

Repeat Example 2.1   (*Given P, i, n, find F*)

If $1,000 is borrowed for three years at 6 percent per year, how much is due at the end of three years?

SOLUTION

1. Find the table in Appendix A for $i=6$ percent.
2. Proceed down the n column to $n=3$.
3. Go horizontally to the right along the $n=3$ line to the F/P column, finding $F/P=1.1910$.
4. Since $F/P=1.1910$, then $F=P\times1.1910$. Since P is given as $1,000 then $F=\$1,000\times1.1910=\$1,191$, the future amount resulting from $1,000 invested principal plus compound interest at 6 percent after three years.

Note the answer is correct to five significant digits *only*, because the tables in Appendix A are complete to only five significant digits. Thus, some round-off has been introduced through the tables and the reader is cautioned not to include more digits in the answer than are significant.

Examples of Single Payment, (Deposited or Borrowed) Accumulating Interest Compounded, and Accruing to a Larger Future Sum

If any three of the four variables are given, the fourth may be found, as illustrated in the following examples.

Example 2.2   (*Given P, i, n, find F*)

Deposit $1,000 now at 8 percent interest, per period. What is the accumulated sum after ten periods?

SOLUTION

$$F = P(1+i)^n = \$1,000(1+0.08)^{10} = \$1,000 \times 2.1589 = \underline{\underline{\$2,159}}$$

or, \$1,000 deposited now at 8 percent will accumulate to \$2,159 in ten periods. [From Appendix A, the $(F/P, 8\%, 10)$ is seen to be 2.1589.]

Example 2.3 (*Given F, i, n, find P*)

The sum of \$10,000 will be needed in ten periods to meet a certain need. How much could be invested *now* at 8 percent compounded interest to accumulate to \$10,000 in ten periods?

SOLUTION

$$P = \frac{F}{(1+i)^n} = \frac{\$10,000}{(1+0.08)^{10}} = \frac{\$10,000}{2.1589} = \underline{\underline{\$4,632}}$$

or, \$4,632 invested now at 8 percent will accumulate to \$10,000 in ten periods. [From Appendix A the $(F/P, 8\%, 10)$ is seen to be 2.1589.] Note that $(P/F, 8\%, 10)$ is 0.46319. Therefore $P = 10,000(0.46319) = \underline{\underline{\$4,632}}$.

Example 2.4 (*Given P, F, i, find n*)

How many years must \$1,000 stay invested at 6 percent to accumulate to \$2,000?

SOLUTION

$$F = P(1+i)^n$$

$$2,000 = 1,000(1+0.06)^n$$

$$n = \underline{\underline{11.9 \text{ yr}}}$$

ALTERNATE SOLUTION 1
Use the F/P tables in Appendix A. $F/P = 2,000/1,000 = 2.0$. In the $i = 6$ percent tables of Appendix A find values of F/P that bracket the value $F/P = 2.0$. Therefore, find $(F/P, 6\%, 12) = 2.012$, and $(F/P, 6\%, 11) = 1.898$. Then by rectilinear interpolation*

$$\frac{2.0 - 1.898}{2.012 - 1.898} \times 1 \text{ yr} + 11 \text{ yr} = \underline{\underline{11.9 \text{ yr}}}$$

If \$1,000 is invested for 11.9 years at 6 percent it will accumulate to \$2,000.

---

*Caution: The relationship is exponential and thus curvilinear, not rectilinear; therefore, a rectilinear interpolation is an approximation, and interpolating between two points that are too widespread will result in a significant error.

ALTERNATE SOLUTION 2
Use natural logarithms to find $n = 11.9$ years.

Example 2.5   (*Given P, F, n, find i*)

Ten years ago you invested $1,000 in a stamp collection. Today you are offered $2,000 for the collection. What is the equivalent annual compounded interest rate?

SOLUTION

$$2000 \left(1 + .0125\right)^{12}$$

$$F = P(1+i)^n$$
$$2,000 = 1,000(1+i)^{10}$$
$$(1+i)^{10} = 2.0$$
$$1+i = 1.07177$$
$$i = 0.072 = 7.2\%$$

ALTERNATE SOLUTION 1
Using rectilinear interpolation of tables in Appendix A, for $(F/P, i, 10) = 2.000$,

i		F/P
8%	=	2.159
i unknown	=	2.00
7%		1.967

$$i = 7\% + \left(\frac{2.00 - 1.967}{2.159 - 1.967}\right) \times 1\% = 7.17\% = 7.2\%$$

ALTERNATE SOLUTION 2
Using roots,

$$\frac{2,000}{1,000} = (1+i)^{10}$$
$$i = 2^{0.1} - 1$$
$$i = 0.0718 = 7.18\% = 7.2\%$$

ALTERNATE SOLUTION 3
Use natural logarithms to find

$$\ln\left(\frac{2,000}{1,000}\right) = 10\ln(1+i)$$
$$\ln(1+i) = \frac{\ln 2}{10} = 0.06931$$
$$1+i = e^{0.06931}$$
$$i = 0.0717 = 7.2\%$$

NOMINAL AND EFFECTIVE INTEREST

To clarify the relationship between the types of interest rates discussed in this chapter the following nomenclature is used:

i = interest rate per interest period

i_n = nominal *annual* interest rate (often termed r)

i_e = effective interest rate (usually annual but can be for longer or shorter compounding periods)

n = number of compounding periods (often years, but not always)

m = number of compounding periods in *one* year

e = the base of natural logarithms, 2.71828 +

Throughout most of this text and other literature on this subject, the subscripts to i are not used to differentiate between the three types of interest rates. The reader is usually able to determine the type of interest under discussion by the context in which it is used. It is important, therefore, that the distinction between these types of interest be carefully noted, since most discussions regarding interest rates assume readers can discern the differences on their own.

1. The *actual* interest rate, i, is the rate at which interest is compounded at the end of each period, regardless of how many compounding periods occur in one year. Thus, interest compounded monthly at 1 percent per month implies an actual interest rate of 1 percent per month.

2. The *nominal* interest rate, i_n, is the *annual* interest rate *disregarding* the effect of end-of-period compounding where the periods are less than one year. Thus, an actual interest rate of 1 percent per month results in a nominal rate of 12 percent per year. In equation form this appears as

$$i_n = im \tag{2.3}$$

or, a nominal interest rate of

$$12\% = 1\%/\text{month} \times 12 \text{ months/year}$$

Most financial instruments and accountants use the nominal interest rate, followed by the number or type of compounding periods per year. Thus, 6 percent compounded quarterly means $i_n = 6$ percent and $m = 4$ (number of quarters per year). The reader is cautioned to carefully determine the actual interest intended when dealing with any financial calculations as it is possible for one or more of the various parties involved to become confused with this method of specifying interest rate.

3. The *effective* interest rate, i_e, is the *annual* rate *including* the effect of compounding at the end of periods shorter than one year. Thus, an actual rate of 1 percent per month implies an effective rate of $1.01^{12} - 1 = 0.1268$, or 12.68 percent per year. In equation form using i,

$$i_e = (1+i)^m - 1 \tag{2.4}$$

$$i_e = 1.01^{12} - 1 = 0.1268 \quad \text{or} \quad 12.68\%/\text{yr}$$

or using i_n,

$$i_e = \left(1 + \frac{i_n}{m}\right)^m - 1 \tag{2.5}$$

$$i_e = \left(1 + \frac{0.12}{12}\right)^{12} - 1 = 0.1268 \quad \text{or} \quad 12.68\%/\text{yr}$$

Note: Effective interest rate "i_e" is, by convention, usually the *annual* interest rate, although it is sometimes used as the equivalent interest rate per payment period where there are several compounding periods between payments. The actual interest rate "i" may or may not be for a one-year period. Also, note that when $m = 1$, $i = i_n = i_e$.

Alternatively, where only one compounding period separates P and F, and the known data include the amount, P, deposited at the beginning of the period, and the resulting balance, F, at the end of the period, i can be found as

$$i = F/P - 1 \tag{2.6}$$

The following examples illustrate nominal and effective interest rates.

Example 2.6   (*Given P,F,i,i_n, find n*)

How many months must $1,000 stay invested at 6 percent compounded monthly to accumulate to $2,000? (Compare with Example 2.4.)

SOLUTION
The nominal interest rate is given as $i_n = 6$ percent per year, so the actual interest rate is found by Equation 2.3 as

$$i = \frac{i_n}{m} = \frac{6}{12} = \frac{1}{2}\%/\text{month} = 0.005$$

$n =$ number of *monthly* periods

Then the value n in terms of months may be determined from the equation

$$F/P = (1 + i)^n$$

as

$$n = \frac{\ln(F/P)}{\ln(1 + i)} = \frac{\ln 2}{\ln 1.005} = \underline{\underline{139 \text{ months}}}$$

ALTERNATE SOLUTION
An alternate solution involves finding the effective annual interest rate i_e and solving by the same equation for n in terms of years. The i_e is found as

$$i_e = (1 + i)^n - 1 = 1.005^{12} - 1 = 0.0617$$

Then

$$n = \frac{\ln(F/P)}{\ln(1 + i)} = \frac{\ln 2}{\ln 1.0617} = \underline{\underline{11.58 \text{ yr}}}$$

This answer of course coincides with the previous answer since 139 months divided by 12 months per year = 11.58 years.

CONTINUOUS INTEREST

If the number of compounding periods, m, per year is allowed to increase, then the length of each compounding period (expressed as a fraction of a year, $1/m$) becomes shorter and shorter until the duration of each period approaches zero. In mathematical terms this can be written

if $1/m \rightarrow 0$, then $m \rightarrow \infty$

Since $i_e = [1 + (i_n/m)]^m - 1$, as m gets infinitely large, the limiting value of $[1 + (i_n/m)]^m - 1$ can be shown to be

$e^{i_n} - 1$

Thus, for continuous compounding,

$$i_e = e^{i_n} - 1 \tag{2.7}$$

Combining this with Equation 2.5 yields an equation for determining a future value of a present sum with interest compounding continuously.

$$F = Pe^{i_n n}, \quad \text{or} \quad P = F/e^{i_n n} \tag{2.8}$$

If n or i are the unknown variables then the equation is simply transposed to read

$$n = \frac{\ln(F/P)}{i_n} \tag{2.8A}$$

or

$$i_n = \frac{\ln(F/P)}{n} \tag{2.8B}$$

For example, a nominal interest rate of 12 percent per year compounded continuously yields an effective annual interest rate of

$$i_e = e^{0.12} - 1 = 0.1275 \quad \text{or} \quad 12.75\%$$

If the sum of \$1,000 is invested for ten years and compounded continuously at a nominal interest rate of 12 percent per year, the resulting balance will be (from Equation 2.8)

$$F = Pe^{i_n n} = \$1,000 \times e^{0.12 \times 10} = \$3,320$$

Alternatively, how long will it take for a deposit of \$1,000 to double in value if compounded continuously at $i_n = 12$ percent? Find

$$n = \frac{\ln(F/P)}{i_n} = \frac{\ln 2}{0.12} = 5.8 \text{ yr}$$

The concept of continuous compounding is useful as an approximation to compound increases of income, expenditures, population, traffic count, or similar events which are spread more or less evenly throughout the year instead of concentrated at year's end.

OTHER APPLICATIONS OF COMPOUND INTEREST FORMULAS

Growth rates of many kinds follow the same pattern as money drawing interest. For example, consumer demand growth predictions are often estimated by use of compound interest formulas.

Example 2.7 (*Find n, given P, i, F, then find F, given P, i, n*)

A city's sewage treatment plant has capacity enough to serve a population of 70,000. The current population is 50,000 and growing at the rate of 5 percent. (See Figures 2.7 and 2.8.)

1. How many years before the plant reaches capacity (a) if compounding occurs continuously; (b) if compounding is calculated at the end of each year (EOY)?
2. Additions to the plant should be planned to serve for a ten-year period. What increment of population should the next addition serve (a) if compounding occurs continuously; (b) if compounding is calculated at EOY?

SOLUTION 1
Find n, given $P = 50,000$, $F = 70,000$, $i = 5\%$.

(a) Continuous compounding.

$$n = \frac{\ln(F/P)}{i} = \frac{\ln(70,000/50,000)}{0.05} = 6.7 \text{ yr}$$

(b) EOY compounding.

$$F = P(1+i)^n$$

$$70,000 = 50,000(1+0.05)^n$$

$$1.4 = 1.05^n$$

$$\ln 1.4 = (n)\ln 1.05$$

$$n = \frac{\ln 1.4}{\ln 1.05} = 6.9 \text{ yr}$$

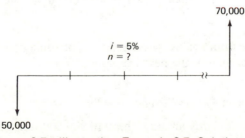

Figure 2.7 Illustrating Example 2.7, Solution 1.

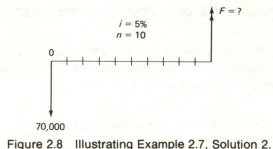

Figure 2.8   Illustrating Example 2.7, Solution 2.

In either case the population will reach 70,000 about seven years from now if growth continues at the same rate. For relatively small values of n the difference between continuous and end-of-period compounding is usually small. For large values of n the results of continuous compounding can easily double the value for end-of-period compounding.

SOLUTION 2
Find F, given $P = 70,000$, $i = 5\%$, $n = 10$.
    (a) Continuous compounding.

$$F = Pe^{i \times n} = 70,000 e^{0.05 \times 10}$$

  $=$ 115,410 future population

    $-$70,000 present capacity

    45,410 capacity of new addition

    (b) EOY compounding.

$$F = 70,000(1 + 0.05)^{10}$$

  $=$ 114,022 population 10 years after city
                reaches 70,000

  $-$ 70,000 population served by present plant

    44,022 increment of population served by
              addition to sewage plant

A population of about 45,000 needs to be served by the addition to the sewage plant.

Comparative future values sometimes are of some concern and need to be found, as illustrated in the next example.

Example 2.8   (*Given P_1, i_1, P_2, i_2, find n so that $F_1 = F_2$*)

An older port city "J" in your state ships 10,000,000 tons of cargo per year, but shipments are diminishing at a rate of -0.7 percent compounded annually. Another port "T," ships 8,000,000 tons per year but is increasing at 5.2 percent annually. How long before the two ports ship equal amounts, if the percent changes in shipping tonnages continue at the same rate (1) compounding annually, and (2) compounding continuously?

SOLUTION 1

Let $F_1 = F_2$, so $P_1(1+i_1)^n = P_2(1+i_2)^n$, then

$$\ln\left(\frac{P_1}{P_2}\right) = n \times \ln\left(\frac{1+i_2}{1+i_1}\right)$$

or

$$\ln\left(\frac{10}{8}\right) = n \times \ln\left(\frac{1.052}{0.993}\right)$$

$$n = \frac{0.223}{0.0577} = 3.9 \text{ yr}$$

At that time the amount shipped would be 9.7 million tons.

SOLUTION 2

This brings up a dilemma. If i_2 is taken as -0.007, you quickly discover you cannot take the logarithm of a negative number, so simply substituting values into the formulas and solving does not yield an answer! The problem can be solved two ways.

ALTERNATE SOLUTION 2A

$$F_1 = F_2, \qquad \text{so} \quad P_1 e^{ni_1} = P_2 e^{ni_2}$$

so,

$$P_1/P_2 = e^{ni_2}/e^{ni_1}$$

then

$$\ln(P_1/P_2) = (n \times i_2) - (n \times i_1) = n(i_2 - i_1)$$

and

$$n = \frac{\ln(P_1/P_2)}{i_2 - i_1} = \frac{\ln(10/8)}{0.052 - (-0.007)}$$

$$n = 3.78 = 3.8 \text{ yr}$$

ALTERNATE SOLUTION 2B

Convert continuous interest to annual effective interest and solve the same as Solution 1, that is,

$$i_{1e} = e^{0.052} - 1 = 0.05338$$

$$-i_{2e} = -(e^{0.007} - 1) = -0.00702$$

then,

$$\ln \frac{P_1}{P_2} = n \times \ln \left(\frac{1+i_2}{1+i_1} \right)$$

$$\ln \frac{10}{8} = n \times \ln \left(\frac{1.05338}{1-0.00702} \right)$$

$$n = 3.78 = \underline{\underline{3.8 \text{ yr}}}$$

Thus, the continuous compounding slightly shortens the time required to achieve equal tonnages in this example. As values of n increase, the differences between the results of continuous compounding and periodic compounding can increase to sizable amounts.

RESALE OR SALVAGE VALUE

Often the incomes or costs of a facility will be separated by a period of time. For example some income may be anticipated due to resale of a facility at some future date. The present value of this future income should be considered when calculating the net present value of any investment, as illustrated in the following example:

Example 2.9

A public works facility has an initial estimated cost of $100,000. At the end of ten years the facility will be replaced and have a resale or salvage value estimated at $50,000. If $i = 6$ percent, what is the net present value of the facility, considering both the present cost and the anticipated income from resale of the facility? (See Figure 2.9.)

SOLUTION
The present value of the facility consists of the initial cost ($100,000) less the present value of the $50,000 income from salvage ten years from now.

$$P_1 = -\$100,000$$

$$P_2 = F(P/F, i, n) = \$50,000 \underset{0.5584}{(P/F, 6\%, 10)} = \$27,920$$

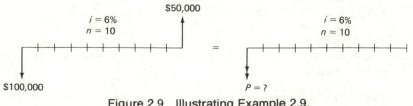

Figure 2.9   Illustrating Example 2.9.

Since P_2 is an income, it has an opposite sign from the cost. Therefore, the net present value is the difference between P_1 and P_2. This yields $(-\$100,000 + 27,920) = -\$72,080$ net present value (cost in this case). Another way of viewing the proposal is:

1. The city raises $72,080 cash to pay for the facility.
2. In addition the city gets a loan for the remaining $27,920 at 6 percent, with interest compounded for ten years with repayment due in a single lump sum at the end of that time.
3. At the end of ten years, the loan has now accumulated principal and interest to total $50,000, and the facility is sold for $50,000 to pay off the loan.
4. Total cash outlay (present value) to city at the start of the project is $72,080.

COMPARISON OF ALTERNATIVES

When confronted with alternatives involving costs or income, the usual approach is to compare the alternatives on a common basis and select the least costly, or most profitable. The same approach is used when the alternatives involve a choice between building a project all at one time or in stages. The following example illustrates one logical method of solution.

Example 2.10   (*Given P_1, i, n, P_2, F_2, compare PW_1 versus PW_2*)

A subdivision developer asks your opinion on whether to construct roads all at once or in stages. He finds he can put in the base course and pavement complete now for $210,000. As an alternative, the county engineer will permit him to install only the base course now (cost estimated at $120,000), with the paving installed two years from now (cost estimated at $100,000). The developer lends and borrows at 10 percent (so use $i = 10\%$ for this example). Which do you recommend as the more economical alternative?

SOLUTION
The present worth (cost in this case) of each alternative may be determined, and the alternative with the lowest cost selected. (See Figure 2.10.)

present cost of future pavement cost, $P_3 = F(P/F, i, n)$
$P_3 = \$100,000 \, (P/F, 10\%, 2)$ $= -82,640$
$\qquad\qquad\qquad 0.8264$
present cost of base course $= -120,000$

total present cost of alternative $B = -\$202,640$

In other words, the developer can either allocate $210,000 (alternative A) for the base and pavement installation all at once now, or allocate

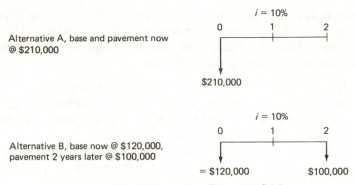

Figure 2.10   Illustrating Example 2.10.

$202,640 (alternative B) for stage construction. If stage construction is selected (alternative B), $120,000 is spent now for the base, while $82,640 is invested at 10 percent to accumulate to $100,000 in two years, enough to pay for the pavement installation at that time.

Conclusion. The developer saves ($210,000−$202,640)=$7,360 by utilizing stage construction, if all goes according to plan. This concept is fully developed in Chapter 7.

Note: As a practical matter the developer may also consider other factors such as:

1. The worry and risk of future price increases may not be worth the $7,360 saving. This consideration favors alternative A.
2. The extra two years of pavement life due to the two-year delay in installation favors alternative B.

DOUBLING TIME AND RATE, RULE OF 70

The "Rule of 70" is a useful method for approximating the time or rate required to double the value of a present sum which is increasing at a compound interest rate. Given the compound interest rate in percent, divide that percent into 70 and find the approximate time required to double the initial value. Conversely, given the doubling time, the interest rate may be determined by dividing the doubling time into 70. For example,

> QUESTION: How long does it take for the population of a country to double if the growth rate is 3.5 percent compounded?
> ANSWER: $70/3.5=20$ years (actually, $n=20.15$ yr).
> QUESTION: If a $1,000 investment made 5 years ago is worth $2,000 today, what has been the rate of return?
> ANSWER: $70/5$ years$=14$ percent (actually, $i=14.87\%$).

SUMMARY

In this chapter the following equivalence factors were developed between P and F:

$$F=P(1+i)^n \quad \text{or} \quad F=P(F/P,i,n)$$

$$P=\frac{F}{(1+i)^n} \quad \text{or} \quad P=F(P/F,i,n)$$

These factors are dependent only upon i and n and thus are independent of the particular values of F or P. Knowing any three of the four variables (F,P,i,n) the fourth can be easily determined. Note that P occurs at the beginning of the first of n compounding periods, while F occurs at the end of the last of n compounding periods, while i is the interest rate per period.

Also introduced in this chapter are the concepts of nominal and effective interest. The nominal annual interest, i_n, is

$$i_n=im$$

while the effective annual interest, i_e, is

$$i_e=(1+i)^m-1=\left(1+\frac{i_n}{m}\right)^m-1$$

For the special case when interest is compounded continuously, the effective annual interest, i_e, is

$$i_e=e^{i_n}-1$$

F and P may be found by

$$F=Pe^{i_n n}$$

$$P=\frac{F}{e^{i_n n}}$$

PROBLEMS FOR CHAPTER 2 ($F/P, P/F$)

Note: Assume all payments are made at the end of the period unless otherwise noted in the problem. EOY is an abbreviation for end of year.

Problem Group A

Simple and compound interest. Construct cash flow diagrams.

A.1 Which of the following are examples of simple interest and which are examples of compound interest?
   a. Jones buys a bond for $1,000, receives a check for $60 every year until the bond's maturity, then receives $1,000.
   b. Jones deposits $1,000 in a savings account bearing 6 percent interest. He makes no withdrawals for a period of ten years. At the end of this period he closes the account and receives $1,791.

A.2 Construct a cash flow diagram illustrating the following cash flows, from (a) borrower's viewpoint; (b) lender's viewpoint.

EOY	
0	Loan of $5,000 @ 10% interest
1	Pay $500 interest
2	Repay $5,000 plus $500 interest

A.3 Construct a cash flow diagram illustrating the cash flows involved in the following transaction from the borrower's viewpoint. The amount borrowed is $2,000 at 10 percent for five years.
   a. Year end payment of interest only. Repayment of principal at the end of the five years.
   b. Year end repayment of one-fifth of the principal ($400) plus interest on the unpaid balance.
   c. Lump sum repayment at EOY 5 of principal plus accrued interest compounded annually.
   d. Year end payments of equal size installments as in a standard installment loan contract (by methods developed in later chapters the amount of each installment will be $527.60).

Problem Group B

Given three of four variables, find P, F, or i.

B.1 A contractor offers to purchase your old D-9 tractor dozer for $10,000 but cannot pay you the money for 12 months. If you feel $i=1$ percent per month is a fair interest rate,
   a. What is the present worth to you today of the $10,000 if it is paid 12 months from now? (How much should be deposited today into an account bearing interest at 1 percent per month in order to accumulate to $10,000 in 12 months from now?)
   b. What annual effective interest rate is involved?
B.2 If the starting salaries of engineering graduates are expected to increase at the rate of 12 percent per year, using the most recent average starting salary, what will be the average starting salary (a) next year, (b) 5 years from now, and (c) 25 years from now?
B.3 A firm wants to lease some land from you on a 20-year lease and build a warehouse on it. As your payment for the lease, you will own the warehouse at the end of the 20 years, estimated to be worth $20,000 at that time.
   a. If $i=8$ percent, what is the present worth of the deal to you?
   b. If $i=2$ percent per quarter, what is the present worth of the deal to you?
B.4 A new branch of the Interstate is expected to open a new area for tourist trade five years from now. Your client expects to construct a $1,000,000 facility at that time. She has funds available now which can be invested at 12 percent compounded annually. How much should she so invest in order to have the $1,000,000 ready at the end of five years?
B.5 In payment for engineering services a client offers your choice between (a) $10,000 now and (b) a share in the project which you are fairly certain you can

cash in for $15,000 five years from now. With $i=10$ percent, which is the most profitable choice?

B.6 The increase in new purchase price of a certain model grader has averaged an annual rate of 9 percent compounded. If this model can be purchased right now for $12,000 new, what is the expected price for a replacement in five years if the price increase curve continues on the same trend?

B.7 Your client has 1,000 acres of "average" residential-development-type land now worth $3,000 per acre. What should it be worth in five years if it appreciates at the rate of 8 percent compounded (a) annually? (b) quarterly? (c) monthly? (d) continuously?

B.8 A reliable client wants to pay you for $10,000 worth of work with an IOU payable five years from now. If you can borrow money from the bank at 15 percent to meet your cash needs between now and then, for how much should he make out the note in order to cover principal plus interest in one lump sum?

Problem Group C

Find i. (F may also be requested.)

C.1 Assume a U.S. Indian head nickel put aside when new 60 years ago will sell now for $2.00. What compound annual interest rate has been earned?

C.2 In 1967 the Consumer Price Index (a commonly used measure of inflation) was 100. Eleven years later it was 193.
a. What was the annual compound growth rate?
b. If it continues compounding at the same annual rate, what will it be in 11 more years?
c. Why can't the increase for the first 11 years simply be doubled to find the increase for the 22-year period?

C.3 Twenty years ago laborers were getting $1.00 per hour in a certain city. Now they receive $10.00 per hour.
a. What is the compound annual growth rate?
b. Assuming the same compounded annual rate, what will they get 20 years from now.

C.4 As a contractor you need to buy a dragline. The equipment dealer knows you well and offers to take either $20,000 cash now, or a note for $30,000 payable in one lump sum five years from now. What annual compound interest rate would the note be equivalent to?

C.5 Assume that 22 years ago the sum of $10,000 was invested in a broad range of NYSE stocks representative of the Dow Jones Industrial Average, and all dividends were reinvested, and that the total investment is worth $104,153 today. What compound rate of increase is this?

C.6 A city with a current population of 50,000 wishes to control its growth rate so that it will reach 100,000 population in 20 years. The regulation will occur by limiting the number of building permits for dwelling units. They now have 15,625 dwelling units which yield 50,000/15,625, an average of 3.2 persons per dwelling unit.
a. What should be the annual compound rate of growth expressed in percent?
b. How many building permits should be issued this year?

    c. At the end of the tenth year what will the population be?

    d. How many permits should be issued for the following year (year 11)?

C.7 A calculator selling for $130 two years ago sells for $59.95 today. What rate of negative interest per year does this change in price represent?

Problem Group D

Find n. (F may also be requested.)

D.1 Assume a medium sized town now has a peak electrical demand of 105 megawatts increasing at an annually compounded rate of 15 percent. Assume the generating capacity is now 240 megawatts.

    a. How soon will additional generating capacity be needed on-line?

    b. If the new generator is designed to take care of needs five years past the on-line date, what size should it be? Assume the present generators continue in service.

D.2 Assume that the U.S. population is now 218,000,000 and will increase this coming year by 1,700,000.

    a. What is the current rate of increase?

    b. If the increase continues to compound annually at that rate, how long will it be before we reach 300,000,000?

D.3 If the birthrate in a certain country is 35 per 1,000 population, and the death rate is 12 per 1,000, how many years does it take for the population to double, assuming no net loss to emigration?

D.4 A city now has a water treatment plant with a 10 million gallons per day (MGD) capacity. The peak actual consumption this year is recorded as 8.7 MGD. Each year the peak consumption is expected to increase by 7.2 percent compounded annually. A consultant recommends construction of an addition to the existing plant which would increase capacity to a total of 30 MGD. Find how many years it will be before this 30-MGD capacity is fully utilized if growth in peak demand continues at this annual rate.

D.5 A county school district now has two high schools with a total student enrollment capacity of 2,500 students. The actual enrollment this year is 2,200 students. Due to general population growth in the area, the school enrollment is expected to increase by 3.8 percent compounded annually. If another high school is added to the system, the total capacity will be raised to 4,000 students. How long will it be before this total capacity is fully utilized?

Problem Group E

Find i and n.

E.1 The population of Florida is estimated at 9,540,000 this year with 4,500 new residents added each week.

    a. What nominal annual growth rate in percent is this?

    b. What effective annual growth rate in percent is this?

    c. How long will it take for the population to reach 10.5 million if the growth continues at the same rate?

    d. How long will it take for the population to reach 15 million if the nominal annual growth rate is used and compounded annually?

E.2 In 1910 the world's population was estimated as 1 billion people. In 1970 the population was estimated at 3.6 billion. In 1980 the population was estimated at 4.2 billion.

    a. Compare the rate of increase for the 60-year period to the 10-year period. (Use continual compounding.)

    b. If the *i* found in part (a) for the most recent period remains constant, how long will it take for the population to double?

    c. There are about 52 million square miles of land area on the earth. Assume that about one-half of this area is arable and that an average of 1 acre is required to grow food for each person. How long will it take for the population to reach the limit that can be sustained by the available food, if these figures are valid, and the rate of increase continues as found in part (a)? (640 acres = 1 square mile.)

Problem Group F

Nominal and effective interest, continuous compounding.

F.1 Assume a certain two-lane road has a traffic volume of 10,000 cars per day increasing at a compound rate of 10 percent per year. It will be two years before a new road can be built and the new facility should be designed to be adequate for an additional 15 years.

    a. If the simplifying assumption is made that each two-lane pair (one lane in each direction) can carry 15,000 cars per day, how many lanes should be provided?

    b. How many lanes are needed if the compounding is continuous rather than annual?

F.2 In 1626 Peter Minuit, governor of the Dutch West India Company bought Manhattan Island from the Indians for $24 worth of beads, cloth, and trinkets.

    a. If the Indians had insisted on $24 cash and invested the $24 at 6 percent, what would the compounded amount be worth today? (Find F.)

    b. If the compounding were continuous instead of annual, how much would the $24 investment be worth today?

    c. The actual assessed value of real estate in Manhattan, land only, was $8,750,000,000 in 1980. What has been the average percent increase in value compounded each year?

F.3 A city's current population is 65,000 and compounding at the rate of 5 percent per year.

    a. What will the projected population be in ten years?

    b. How much time will it take to reach 200,000 at this rate?

    c. What will the projected population be in ten years if the compounding is continuous rather than annual?

F.4 A $1,000 investment earns $20 in interest each quarter.

    a. What actual rate of interest is earned?

    b. What nominal annual rate of interest is earned?

    c. What annual effective rate of interest is earned?

F.5 A major credit card company charges $1\frac{1}{2}$ percent interest on the unpaid balance each month.
  a. What nominal annual interest rate is charged?
  b. What effective annual interest rate is charged?
F.6 A depositor at a savings and loan association deposits $1,000 in a savings account.
  a. If the account earns 8 percent nominal interest compounded quarterly, how much interest would the $1,000 earn after one year? After five years?
  b. If the account earns 9 percent interest compounded continuously, how much interest would be earned after one year? After five years?

Chapter 3
The Value of One Future Payment Compared to a Uniform Series of Payments (F/A, A/F)

KEY EXPRESSIONS IN THIS CHAPTER

A = Uniform series of n end-of-period payments or receipts, with interest compounded at rate, i, on the balance in the account at the end of each period.

$\bar{i}$ = Interest compounded continuously.

ETZ = Equation time zero, with reference to a particular payment series, the time at which $n = 0$.

PTZ = Problem time zero, the present time according to the statement of the problem narrative.

THE FUTURE SUM OF PERIODIC DEPOSITS

In planning the future of a city, a private company, or the family finances, the need for periodic deposits into savings in anticipation of large future lump sum cash outlays can often be foreseen. A city may need new buildings, parks, or transportation facilities, a company may need a new plant or store, and a family may save toward a down payment for a new car or home. One method of saving for these future lump sum outlays is to periodically deposit uniform amounts into an interest bearing account. Many families have savings accounts to serve this purpose, while businesses and governments sometimes use interest bearing sinking funds to accumulate capital.

A simple equation may be derived relating the future sum, F, that will accumulate from a uniform series of n end-of-period deposits, A, with interest compounding at the end of every period at rate i on all sums on deposit. (As a memory aid, the A used to designate periodic deposits can be associated with Annual deposits even though periods of other durations are sometimes used.)

Generalized Problem

Find the total future sum, F, at the end of the nth period from n end-of-period deposits each of amount A, plus interest compounded at the end of every period at rate i.

Generalized Solution, Development of F/A Equation

If *no* money is deposited at the *beginning* of the first period, and A_1 dollars are deposited at the *end* of the first period, the future sum F accumulated at the end of that one period is simply $F=A_1$, since A_1 has not had time to earn any interest.

If A_1 dollars are deposited at the end of the first period, and the same amount of A_2 dollars are deposited at the end of the second period, the future sum F accumulated at the end of the two periods is $F=A_1+A_1i+A_2$. Since $A_1=A_2$ this may be written as $F=A[1+(1+i)]$. At the end of each period the amount remaining on deposit is multiplied by $(1+i)$ representing the principal plus accruing interest. In addition, at the same time the interest is paid, another periodic deposit is made. So, if A dollars are deposited at the end of each period for n periods, accumulating i interest at the end of each period, the future amount F accumulated at the end of each period for up to n periods is:

END OF PERIOD	ACCUMULATED AMOUNT AT THE END OF EACH PERIOD
1	$F=A$
2	$F=A(1+i)+A=A[1+(1+i)]$
3	$F=A[1+(1+i)](1+i)+A=A[1+(1+i)+(1+i)^2]$
⋮	
n	$F=A[1+(1+i)+(1+i)^2+\cdots+(1+i)^{n-1}]$

The general form of this equation for n periods is

$$F=A\left[1+(1+i)+(1+i)^2+\cdots+(1+i)^{n-1}\right]$$

(3.1)

The reduction of the general form to a simpler more practical form is performed by the following sequence:

1. Multiply both sides by $(1+i)$.

$$F(1+i)= A\left[(1+i)+(1+i)^{2}+ \cdots +(1+i)^{n}\right] \tag{3.2}$$

2. Subtract Equation 3.1 from Equation 3.2.

$$Fi = A\left[-1+(1+i)^{n}\right]=\left[(1+i)^{n}-1\right]$$

3. Multiply both sides by $1/i$ to get the equation relating F, A, n, and i.

$$F= A\left[\frac{(1+i)^{n}-1}{i}\right] \quad \text{or} \quad F/A=\frac{(1+i)^{n}-1}{i} \tag{3.3}$$

Note once again that the factor $[(1+i)^{n}-1]/i$ is independent of the amount of either F or A. This factor is sometimes termed the uniform-series, compound-amount factor (USCAF).

Graph of F/A Versus n

A graph of the equation $F/A=[(1+i)^{n}-1]/i$ is shown in Figure 3.1. Note the rapid increases in future value, F, with increases in either n or i.

Graph of the Balance in the Account, F Versus n

The balance in the account graph is shown in Figure 3.2, which illustrates the timing of the cash flows of A related to F. Note that the first deposit of A occurs at the *end* of the first time period. Therefore the first interest is not earned until the end of the second time period since interest is paid in arrears. At the end of the last (nth) period there occurs both a final deposit A and a final interest payment. It is important to remember that the F/A

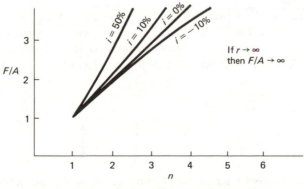

Figure 3.1   Graph of the equation of F/A.

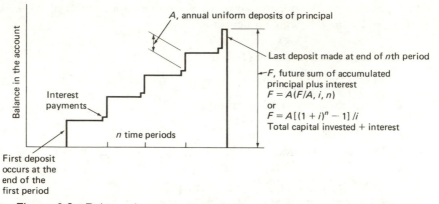

Figure 3.2 Balance-in-account graph showing the timing of cash flows.

series begins at a point in time that is one time period before the first periodic payment A is made. In other words, equation time zero for the F/A series occurs one time period *before* the first periodic payment A occurs.

Notation

Rather than write out the equation each time, a briefer notation format is frequently used:

$$F = A(F/A, i, n) \qquad (3.3A)$$

Cash Flow Diagrams

The general form of the cash flow diagram illustrating the F/A relationship is shown in Figure 3.3. The double arrow notation indicates an unknown value to be found (unknown dependent variable). If any three of the four variables are known or determinable, the fourth can be found. Therefore any of the four variables may at some time appear as the unknown dependent variable.

Note that the first periodic deposit, A, occurs at the *end* of the first period. Also note that the future value, F, occurs *simultaneously* with the last periodic deposit, A. You are cautioned against common errors illustrated in the incorrect USCAF cash flows which follow (Figures 3.4 and 3.5).

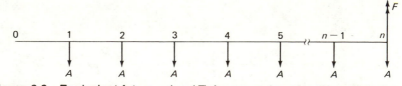

Figure 3.3 Equivalent future value (F) from a series of uniform periodic amounts (A).

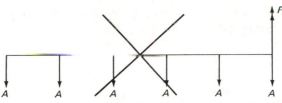

Figure 3.4   Incorrect representation of F/A relationship, with extra (initial) deposit.

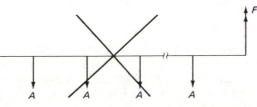

Figure 3.5   Incorrect representation of F/A relationship, with missing (final) deposit.

Uniform Series Sinking Fund Factor, USSFF

The reciprocal of the USCAF is sometimes referred to as the uniform series sinking fund factor (USSFF) and written as:

$$A = F\left[\frac{i}{(1+i)^n - 1}\right] \tag{3.4}$$

In functional notation

$$A = F(A/F, i, n) \tag{3.4A}$$

Note that the $(F/A, i, n)$ factor in Equation 3.3 and its reciprocal factor $(A/F, i, n)$ in Equation 3.4 are dependent on the values of i and n *only*, and are independent of the values for F or A. For convenience, values of A/F and F/A have been tabulated in Appendix A for commonly used combinations of i and n.

APPLICATION OF THE F/A AND A/F EQUATIONS

F/A problems involve the four variables, F, A, i, n. Whenever any three of the variables are known, the fourth may be determined. The following four examples illustrate typical situations involving each of the four variables, in turn, as the unknown.

Example 3.1   (*Find F, given A, i, n*)

If five equal annual deposits of $100 are made at the end of each year into an investment that earns interest compounded at 10 percent annually, what is the value of the investment (including deposits and accumulated interest) at the end of five years?

SOLUTION USING FORMULA

$$F = A[(1+i)^n - 1]/i$$

$$F = \$100[(1.10)^5 - 1]/0.10$$

$$F = \$100 \times 6.1051 = \underline{\underline{\$611}}$$

SOLUTION USING TABLES

For some problems a solution using the tables in Appendix A for F/A and A/F saves time. The value of $[(1+i)^n - 1]/i$ is represented by F/A, and the inverse is represented by A/F. Each table contains values for only one interest rate i. The time periods n are shown in the left and right outer columns of each table.

Example 3.1 could be solved by the following step-by-step procedure:

1. Find the table in Appendix A for $i = 10$ percent.
2. Proceed down the n column to find $n = 5$.
3. Go to the right along the $n = 5$ line to the F/A column finding $F/A = 6.1051$.
4. Since $F/A = 6.1051$ then $F = A \times 6.1051$. Since $A = \$100$ then $F = \$100 \times 6.1051 = \underline{\underline{\$611}}$.

Example 3.2 (*Find A, given F, i, n*)

A person, now 22, expects to retire in 40 years at age 62. He anticipates that a lump-sum retirement fund of $400,000 will see him nicely through his sunset years. How much should be paid annually at the end of each year for the next 40 years to accumulate a $400,000 retirement fund if interest accrues at 6 percent compounded on the amount paid in?

SOLUTION

$$A = F(A/F, i, n)$$

$$A = \$400,000 \; (A/F, 6\%, 40)$$

$$A = \$400,000 \; (0.0065)$$

$$A = \underline{\underline{\$2,600/\text{yr}}}$$

Example 3.3 (*Find n, given A, F, i*)

A city finances its park acquisitions from a special tax of $0.02 a bottle, can, or glass of all beverages sold in the city. It derives $100,000 a year from this tax. The city has an option to buy a 40-acre lakefront park site for

$1,000,000 as soon as the money is available. How soon will this be if the tax receipts are invested annually at the end of each year at 6 percent?

SOLUTION BY LOGARITHMS

$$F/A = \left[(1+i)^n - 1\right]/i$$

$$\frac{\$1,000,000}{100,000} = \frac{\left[(1.06)^n - 1\right]}{0.06}$$

$$10 \times 0.06 + 1 = 1.06^n$$

$$\frac{\ln 1.6}{\ln 1.06} = n = 8.07 \text{ yr} = 8.1 \text{ yr}$$

SOLUTION BY TABLES

The same result may be obtained by an alternative method, by interpolation of the tables in Appendix A, as follows:

$$A = F(A/F, i, n)$$

$$100,000 = 1,000,000 \, (A/F, 6\%, n)$$

Then

$$(A/F, 6\%, n) = \frac{100,000}{1,000,000} = 0.1000$$

Find in the 6% table

$$(A/F, 6\%, 8) = 0.1010 \quad 0.1010$$

$$(A/F, 6\%, n) = 0.1000$$

$$(A/F, 6\%, 9) = \qquad\qquad 0.0870$$

$$\overline{\qquad\qquad 0.0010 \quad 0.0140}$$

Interpolation yields:

$$n = 8 + \frac{0.0010}{0.0140} \times 1 = 8.07 \text{ yr} = 8.1 \text{ yr}$$

Example 3.4   *(Find i, given A, F, n)*

Referring to Example 3.3, the Recreational Advisory Board feels that the park land should be acquired at no later date than seven years from now and that a financial consultant should be used to increase the interest income on invested special tax funds. How much would the interest have to increase in order to reach $1,000,000 in seven years?

SOLUTION

Since i appears twice in the equation the answer is easier to find by interpolation of the tables than by solution of the equation.

$$A = F(A/F, i, n)$$

$$100,000 = 1,000,000(A/F, i, 7), \qquad (A/F, i, 7) = 0.100$$

Find by thumbing through the tables where $(A/F, i, 7)$ brackets 0.100,

$$(A/F, 11\%, 7) = 0.1022 \qquad 0.1022$$

$$(A/F, i, 7) = 0.100000$$

$$(A/F, 12\%, 7) = \qquad\qquad 0.0991$$

$$\overline{0.0022} \quad \overline{0.0031}$$

Interpolation yields:

$$i = 11\% + \frac{0.0022}{0.0031} \times 1\% = 11.71\% = \underline{\underline{11.7\%}}$$

WHEN PROBLEM TIME ZERO DOES NOT CORRESPOND WITH EQUATION TIME ZERO

As explained previously, the beginning point in time for the F/A equation is one period of time before the first payment in the series, as illustrated in Figure 3.1. Many problem situations occur where the beginning time for the problem (problem time zero or PTZ) does not coincide with the beginning time for the equation (equation time zero or ETZ). The general approach is to subdivide the problem into simple components and sum the results. The following examples illustrate the method.

Example 3.5 *(PTZ≠ETZ, series n=20, problem n=30)*

A young engineer decides to save for a down payment on a new car to be purchased $2\frac{1}{2}$ years from now. She decides to deposit $50 per month into a savings account paying 6 percent nominal interest compounded monthly (use $i = 6\%/12 = 0.5\%/\text{month}$, $n =$ the number of months). She plans to make a series of 20 regular monthly deposits with the first deposit scheduled for one month from today. The account will then receive no more deposits but will receive monthly compound interest on the balance for another 10 months. How much will be in the account at the end of the 30-month period?

SOLUTION

A cash flow line diagram illustrates the cash flow of the deposits. The problem can be subdivided and solved in two parts and the results summed (see Figure 3.6).

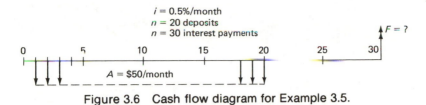

Figure 3.6   Cash flow diagram for Example 3.5.

1. First find the lump sum future amount, F_1, in the account at the end of the first 20-month period using Equation 3.1,

$$F_1 = A\left[\frac{(1+i)^n - 1}{i}\right] = 50\left[\frac{(1.005)^{20} - 1}{0.005}\right] = \$1,049$$

2. Second, find the future amount, F_2, that will accumulate as a result of just the lump sum F_1 balance plus monthly interest payments, without any additional deposits for the final ten-month period. F_1 becomes the original deposit P for this equation.

$$F_2 = P(1+i)^n = \$1,049(1.005)^{10} = \$1,103$$

Another frequently encountered case of problem time zero *not* corresponding with equation time zero occurs when the initial payment in a series occurs at the *beginning* of the first period rather than the *end* of the first period. These are easily handled by either of two methods:

1. Consider the beginning payment as a one-time payment and add its future value to the value of the remaining part of the series.
2. Move equation time zero (*ETZ*) to the left of problem time zero (*PTZ*) by one year, that is, assume time zero for the F/A equation occurs one year before the first deposit is made and count n from the new time zero.

Example 3.6   (*PTZ≠ETZ first payment in series occurs at PTZ*)

Assume a proud grandfather bought his grandson a $100 bond on the day he was born and continued buying an additional one on every birthday. How much is the total value of the investment on his twenty-first birthday after the twenty-second bond has been received? Assume all the bonds accumulate 6 percent interest compounded annually.

SOLUTION
First draw a cash flow diagram as shown in Figure 3.7. The initial $100 bond at time zero does not fit in the standard F/A series and so will be

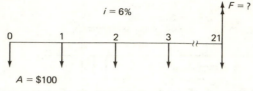

Figure 3.7 Cash flow diagram for Example 3.6.

treated separately as

$$F_1 = P(F/P, i, n)$$

$$F_1 = \$100(F/P, 6\%, 21) = \$340$$
$$\qquad\qquad 3.3995$$

The remaining 21 bonds are a standard F/A series

$$F_2 = A(F/A, i, n)$$

$$F_2 = \$100\,(F/A, 6\%, 21) = \underline{\$3,999}$$
$$\qquad\qquad 39.992$$

total $(F_1 + F_2) = \underline{\underline{4,339}}$

ALTERNATE SOLUTION

This problem could also be solved by treating the initial deposit as if it had occurred at the end of period one. Then time zero for the F/A equation is moved back to one year before the baby is born. The corrected cash flow diagram is shown in Figure 3.8. Counting from the new time zero, $n = 22$ when the twenty-first birthday occurs. The equation becomes

$$F = A(F/A, i, n) = 100(F/A, 6\%, 22) = \underline{\underline{\$4,339}}$$
$$\qquad\qquad\qquad 43.392$$

TWO OR MORE SEQUENTIAL SERIES

Most problems that at first seem complex can be broken down into relatively simple components. For instance, a problem may be composed of more than one series of periodic payments. These may be treated very

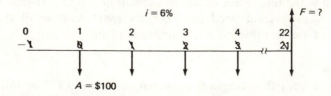

Figure 3.8 Cash flow diagram for Example 3.6 Alternate Solution.

simply as two or more separate series (split series). Each series is solved independently of the other and the results summed at the same time.

Example 3.7 (*Two sequential series with different i values*)

The stadium seats in a certain high school football stadium are expected to need replacement in another nine years from today. The school board has been anticipating this and five years ago began depositing $1,000 per year into an account bearing interest at 6 percent per year. It is now the end of the fifth year and the fifth annual deposit has just been made. The bank announces that beginning today, all funds on deposit will bear interest at 7 percent. The school board, realizing that costs are increasing, decides to raise their deposits to $2,000 per year, beginning with the deposit due one year from today. As consulting engineer to the school board, you are requested to quickly calculate how much the account will total at the end of nine more years if the deposits remain at $2,000 per year and interest remains at 7 percent.

SOLUTION
The cash flow line diagram illustrating this example is shown in Figure 3.9. The problem may be separated into three different parts:

1. The amount on deposit at the end of five years is F_1.

$$F_1 = \$1,000(F/A, 6\%, 5) = \$5,637$$
$$5.637$$

2. This amount, $5,637 may now be considered as the present value, P_1, of a lump sum deposit that begins drawing interest at 7 percent from the end of the fifth year until the end of the fourteenth, or $n = 9$. The future amount that accumulates at the end of nine years from this one source is

$$F_2 = \$5,637(F/P, 7\%, 9) = \$10,360$$
$$1.8384$$

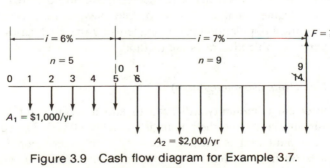

Figure 3.9 Cash flow diagram for Example 3.7.

3. From the end of the sixth year until the end of the fourteenth year deposits of $2,000 each are expected with $i=7$ percent on the accumulated balance. This results in the accumulation of

$$F_3 = \$2,000(\,F/A,7\%,9) = \$23,956$$
$$11.978$$

The total amount on hand at the end of 14 years is

$$F_2 + F_3 = \$10,360 + \$23,956 = \$34,316$$

Example 3.7 was separated into three different parts, and the final answer summed the results at the same time (in this case F at 14 years). To sum F values (or P values) they must all be transferred to the same point on the cash flow diagram.

NEGATIVE INTEREST RATE

Problems sometimes occur where the interest rate (or equivalent growth rate) is negative. For instance, an investment that turns out poorly may have a negative net cash flow (you put in more than you get out) and thus yields a negative rate of return. Investors who take risks usually have experienced this type investment at one time or another on at least some of their ventures.

Population growth patterns can also yield a negative rate of increase. For instance, consider a retirement area with a large number of older people. Due to the large proportion of older residents, the birth rate may be lower than the death rate, leading to a negative natural increase. The natural rate of increase is often simulated by compound interest, so a negative increase may be simulated by a negative interest rate. For instance, if the annual birth rate is 8 per 1000 population and the annual death rate is 14 per 1000 population, then the natural increase is a negative 6 per 1000 population. Thus, *if there were no immigration* into the area from outside, the population would decline at the rate of 0.6 percent per year ($+0.8\%-1.4\%$) compounded. However, if there *were* immigration into the area, and if the characteristics of the incoming residents were similar to those already there, and if these new residents continued to move into the area at a fairly constant rate (same number of new people moving in per year), then estimates of future population may be made using the F/A series equation. The following example illustrates the method.

Example 3.8 (*Negative interest rate*)

A new retirement community will need to add a second unit to the new water treatment plant when the population reaches 40,000. The community has just been built and is starting out now with a zero population. An average of 6,000 new residents are expected to move into the community each year. The estimated birthrate for the community is 8 per 1,000 and the death rate is 14 per 1,000 resulting in a negative annual increase of -0.6 percent (so $i = -0.006$). How long will it be before the second unit is needed for the new water treatment plant?

SOLUTION
The population must reach 40,000 before the second unit is needed, so that $F = 40,000$. The number of new residents expected each year is 6,000, so $A = 6,000$. The residents that are there should have a natural rate of increase of -0.6 percent, so $i = -0.006$.

A diagram similar to the cash flow line diagram may be drawn for this problem as shown in Figure 3.10. The similarity to previous problems involving deposits into an interest bearing account should be apparent. In this case, the accumulated amount "on deposit" loses at rate i instead of gaining at rate i.

To solve for the number of years, n, that will yield a net population increase of 40,000, use the F/A equation, as follows.

$$F/A = \frac{[(1+i)^n - 1]}{i}$$

$$\frac{40,000}{6,000} = \frac{(0.994)^n - 1}{-0.006}$$

$$6.6667 \times (-0.006) + 1 = (0.994)^n$$

$$0.96 = (0.994)^n$$

$$\frac{\ln 0.96}{\ln 0.994} = n = \underline{\underline{6.8 \text{ yr}}}$$

Note that since the negative i is small, and n is not large either, the number of years required to reach 40,000 is just a little more than if i were zero ($40,000/6,000 = 6.7$ yr).

If the retirement community in Example 3.8 already has an existing population at PTZ, the solution involves one more equation and a slightly different logarithmic equation, as follows.

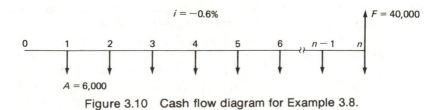

Figure 3.10 Cash flow diagram for Example 3.8.

Example 3.8a (*Find n, given F, P, A, i*)

Assume the retirement community already has a base population of 10,000 persons. Now how long will it take to reach a population of 40,000 with all other factors the same as Example 3.8?

SOLUTION
Given

$$F = 40,000 \text{ pop.}$$
$$P = 10,000 \text{ pop.}$$
$$A = 6,000/\text{yr}$$
$$i = -0.006$$

Find *n*.

$$F_1 = P(1+i)^n$$

$$F_2 = A\left[\frac{(1+i)^n - 1}{i}\right]$$

$$F = F_{\text{total}} = F_1 + F_2 = P(1+i)^n + A\left[\frac{(1+i)^n - 1}{i}\right]$$

$$F = P(1+i)^n + \frac{A(1+i)^n}{i} - \frac{A}{i}$$

$$(1+i)^n = \frac{F + A/i}{P + A/i}$$

$$n = \frac{\ln\left(\dfrac{F + A/i}{P + A/I}\right)}{\ln(1+i)}$$

$$n = \frac{\ln\left[\dfrac{40,000 + 6,000/(-0.006)}{10,000 + 6,000/(-0.006)}\right]}{\ln(1-0.006)} = \underline{\underline{5.1 \text{ yr}}}$$

RATES, FARES, AND TOLLS

Frequently the cost of financing capital investments are met by charging each individual user a certain amount for each use or unit consumed. The amount charged per use is usually called the fare or toll, while the amount charged per unit consumed is commonly called the rate. For instance, to accumulate funds to build a new water treatment plant the current water rate charge may be increased by an extra $0.10 per thousand gallons of water, or a sinking fund for a proposed toll bridge may be funded by charging a toll of $0.25 per vehicle.

When the total cost of the facility is known and the rate is unknown, the customary approach is to substitute an unknown "y" (any letter will do) for the rate and set up an equation. The following example illustrates the procedure.

Example 3.9 (*Find the required fare increase, y, with sequential series*)

A city transit authority is planning on replacing their aging bus fleet six years from now at an estimated lump sum cost of $1,000,000. To accumulate this amount they decided to raise bus fares by whatever amount is required, with the raised fare going into effect three months from today. The increased fare increment will be deposited monthly at the end of each month of collections (the first deposit will be made four months from today) into a fund bearing interest at 0.5 percent per month, compounded monthly. The number of fares per month for the next six years is estimated as

DURING YEARS	FARES PER MONTH
1 through 2	140,000
3 through 6	160,000

How much of a fare increase per passenger is necessary in order to reach the desired lump sum? The fare increase per passenger will remain constant until the end of the sixth year.

SOLUTION
The amount collected each month equals the number of fares per month times the fare per passenger designated by "y." For the first 21 months of collections the monthly receipts should be

$$A_1 = 140{,}000\,y$$

The monthly receipts for the final 48 months are designated as

$$A_2 = 160{,}000\,y$$

The cash flow diagram for this problem is shown in Figure 3.11.

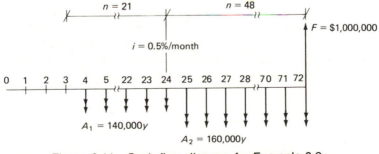

Figure 3.11 Cash flow diagram for Example 3.9.

The first series involves 21 payments $(24-3=21)$. A future lump sum F_1 equivalent to this first series is found at EOM 24 as

$$F_1 = 140,000\,y\,(F/A,0.5\%,21) = 3,092,000\,y$$
$$22.084$$

Then this lump sum F_1 is treated as P_1 and transferred another 48 months $(72-24=48)$ to an equivalent future lump sum F_2 at EOM 72, as

$$F_2 = 3,092,000\,y\,(F/P,0.5\%,48) = 3,928,000\,y$$
$$1.12705$$

The future value F_2 of the second series of payments A_2 is found at EOM 72 as:

$$F_3 = 160,000\,y\,(F/A,0.5\%,48) = 8,656,000\,y$$
$$54.0978$$
$$F_2 + F_3 = 12,584,000\,y$$

The total amount needed at the end of the six-year period is $1,000,000, so $F_2 + F_3$ are equated to this amount and solved for y. Thus,

$$y = \frac{\$1,000,000}{12,584,000} = \$0.079 \quad \text{or} \quad \$0.08 \text{ per passenger}$$

SIGN OF F VERSUS SIGN OF A

In most cash flow diagrams the equivalent future value arrow for F is drawn on the opposite side of the line as the arrows representing uniform periodic deposits, A. Since A represents the deposits and F the lump-sum withdrawal, this opposite sign is correct. However, sometimes a lump-sum future value is desired which depicts a lump-sum deposit equivalent to the A series of deposits, and has the same sign as the periodic deposits, A. In such cases it is correct to show all arrows on the same side of the cash flow diagram base line. Example 3.10 illustrates this difference.

Example 3.10   (*Find F, given A, i, n*)

Assume your banker is willing to accept equivalent quarterly payments (one payment every three months) on a home mortgage instead of the usually stipulated monthly payments of $500 per month. If the interest rate is 1 percent per month, what would be the equivalent quarterly payments? The cash flow diagram for this problem is shown in Figure 3.12.

SOLUTION USING FORMULA

$$F = A\left[(1+i)^n - 1\right]/i$$
$$F = 500\left[(1.01)^3 - 1\right]/0.01 = 500 \times 3.0301 = \$1,515$$

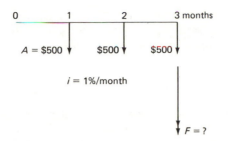

Figure 3.12   Cash flow diagram for Example 3.10. Note the arrow for the equivalent future value *F* points in the same direction as the periodic values *A*.

Conclusion. The payment of $1,515 per quarter gives the banker the equivalent of three monthly payments of $500 per month.

NOMINAL AND EFFECTIVE INTEREST RATES

Example 3.11   *(Find i_n and i_e, given i, n)*

In Example 3.10 the actual interest rate was 1 percent per month. What annual *nominal* interest rate and annual *effective* interest rate does this represent?

SOLUTION
From Equation 2.3, the annual nominal interest rate, i_n, is

$$i_n = im = 1 \times 12 = \underline{\underline{12\%}}$$

From Equation 2.4, the annual effective interest rate, i_e, is

$$i_e = (1+i)^n - 1 = (1.01)^{12} - 1 = 0.1268 = \underline{\underline{12.7\%}}$$

FUNDS FLOW PROBLEMS

In most business transactions and economy studies interest is compounded at the end of each discrete compounding period of time and cash flows are assumed to occur either as lump sums at the beginning or end of such periods or as series of end-of-period payments. This will be the case for most situations in this text. However, occasionally different flows occur, as illustrated in the four types of cases described in the following paragraphs.

• *Case 1. Long intervals between payments in a series are some multiple of the shorter interest compounding periods (interest is compounded more than once between payments in a series).* The cash flow diagram is shown in Figure 3.13 for a typical situation involving more than one compounding period between payment deposits (e.g., monthly compounding, annual payments).

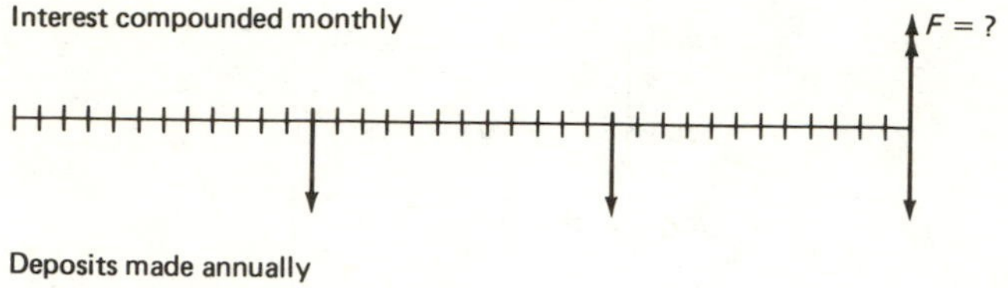

Interest compounded monthly

$F = ?$

Deposits made annually

Figure 3.13 Cash flow diagram for Case 1: Interval between payments consists of several compounding periods.

Where a regular series of payments are made at intervals of more than one interest-compounding period, then two common methods are available to find the future sum, the sinking fund method, and the equivalent interest method.

• *Method 1. Sinking Fund Method.* Use the A/F equation (with $n=m$, the number of compounding periods between payments) to find the periodic payment at the end of every interest compounding period that is equivalent to the less frequent payment given. Then use the F/A equation (with $n=n$, the number of payments) to determine the future value of the equivalent payments.

Example 3.12 (*Interval between payments consists of several compounding periods.*)

Assume the $1,200 payments in this case occur annually, while the compounding is monthly at 1 percent per month rate. Find the equivalent future lump sum at EOM (end of month) 36.

SOLUTION
The cash flow diagram appears as shown in Figure 3.14. Find the monthly payment that is equivalent to the annual payment of $1,200.

$$A = F(A/F, i, n) = \$1,200(A/F, 1\%, 12) = \$94.62/\text{month}$$
$$0.0785$$

Then the equivalent future lump sum at EOM 36 is simply

$$F = A(F/A, i, n) = \$94.62(F/A, 1\%, 36) = \$4,075$$
$$43.076$$

$F = ?$

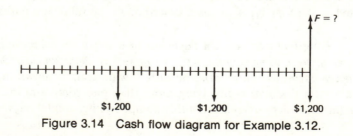

$1,200 $1,200 $1,200

Figure 3.14 Cash flow diagram for Example 3.12.

Thus if \$1,200 is deposited at the end of every 12 months for 36 months, and if interest is credited every month compounded at 1 percent per month on the balance in the account, then the balance in the account at EOM 36 is \$4,075.

- *Method 2. Effective Interest Method.* Instead of compounding the interest every month at 1 percent, compound the interest every m months (corresponding to the payment interval) at a higher interest rate. This rate will be a little higher than m times i to account for the compounding. The effective interest is actually $i_e = (1+i)^m - 1$.

Example 3.13 *(Equivalent interest rate to find F given A)*

Assume the same \$1,200 annual payment and $i = 1$ percent per month as in Example 3.12. Find F at EOM 36.

SOLUTION
The effective interest rate for compounding every 12 months equivalent to compounding at 1 percent per month is

$$i_e = (1+i)^m - 1 = 1.01^{12} - 1 = 0.12683 = 12.68\%$$

Thus the future lump sum equivalent is found as

$$F = A \frac{(1+i)^n - 1}{i} = \$1,200 \frac{1.12683^3 - 1}{0.12683} = \$4,076$$

This indicates that if deposits of \$1,200 are made annually and compounded annually at the 12.68% effective interest rate, the balance at EOM 36 will total the same as \$1,200 deposited annually and compounded monthly at 1 percent per month.

- *Case 2. Interest is compounded continuously, while funds are received or dispersed discretely at the beginning or end of any period.* The continuous compound interest rate may be converted to its equivalent *annual* effective interest rate by use of Equation 2.7 (Chapter 2), as follows:

$$\bar{i}_e = e^{i_n} - 1$$

Then the problem is solved by use of the previously derived factors for discrete compounding (A/F, F/P, etc.). Example 3.14 illustrates this approach.

Example 3.14 *(Find F, given A, n, $\bar{i}$)*

One thousand dollars per year is deposited now and at the end of every year for the next seven years into a savings account earning 5 percent interest compounded continuously. How much is the balance in the savings account at the end of the seventh year?

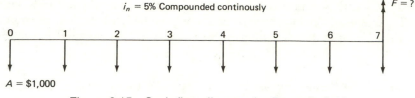

Figure 3.15 Cash flow diagram for Example 3.14.

SOLUTION
The cash flow diagram for this problem is shown in Figure 3.15. Since interest is compounded continuously, the effective continuous annual interest rate, $\bar{i}_e$, can be calculated from Equation 2.7.

$$\bar{i}_e = e^{i_n} - 1 = e^{0.05} - 1 = 0.05127$$

Then the problem can be solved by separating the annual cash flows into two parts (to fit the factors), and substituting $i = 0.05127$ in the appropriate equations to wit

$$F_1 = 1,000(F/P, i, n) = 1,000(1 + 0.05127)^7 \qquad = \$1,419.06$$

$$F_2 = 1,000(F/A, i, n) = 1,000\left[\frac{(1 + 0.5127)^7 - 1}{0.05127}\right]$$

$$= 1,000(8.17354) \qquad\qquad = \underline{\$8,173.54}$$

$$F = F_1 + F_2 \qquad\qquad = \underline{\underline{\$9,592.60}}$$

In Example 3.14, notice that the difference in continuous compounding and discrete compounding is rather small (in this case 5.127% versus 5%). As long as interest rates are low the differences will be small, but as interest rates increase the differences will also increase.

The problem is only slightly more complicated if the periods are *not* yearly. In such cases an *actual* interest rate per period must be determined and then used. Example 3.15 illustrates this approach.

Example 3.15 (*Find F, given A, i_n, n*)

John deposits \$200 a month in a savings account earning 12 percent compounded continuously. How much money will he have in his account after one year (12 deposits), assuming his first deposit occurs at the end of the first month?

SOLUTION
The cash flow diagram for this problem is shown in Figure 3.16. In order to solve this, first a discrete interest rate per period must be determined using Equation 2.3 from Chapter 2.

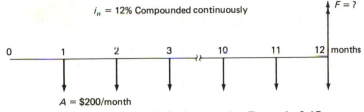

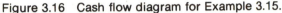

Figure 3.16   Cash flow diagram for Example 3.15.

$$i_n = im \quad \text{or} \quad i = \frac{i_n}{m}$$

Thus,

$$i = 12\%/12 \text{ months} = 1\%/\text{month}$$

But, this i is *not* the actual interest rate because this problem specifies that interest is compounded continuously. The effective continuous interest rate per period (month) may be calculated from Equation 2.7, to wit

$$\bar{i}_e = e^i - 1$$

where $\bar{i}_e$ is the *effective* continuous interest rate *per period* (month in this case). Therefore,

$$\bar{i}_e = e^{0.01} - 1 = 0.0100502$$

Now that an equivalent effective (in this case actual) continuous interest rate per month has been determined, the problem may be solved using Equation 3.3.

$$F = A\left[\frac{(1+i)^n - 1}{i}\right]$$

$$F = \$200\left[\frac{(1+0.0100502)^{12} - 1}{0.0100502}\right] = \underline{\underline{\$2,537}}$$

· *Case 3. Interest is compounded periodically, while funds are received or disbursed discretely within the periods, not necessarily at the beginning or end of any period.* In this case, since interest is earned or payed on the balance at the close of each period, most engineering cost analyses omit any interest attributable to the time between the receipt of the funds and the end of the compounding period as well as interest attributable to the time between the payment of the funds and the beginning of the compounding period. Most banks do not pay interest on funds withdrawn from a savings account prior to the close of an interest bearing period. For example, if you place $1,000 in a savings account on January 1 earning 1.5 percent per quarter (6% nominal/yr) and withdraw the $1,000 two months later on March 1, you may earn zero interest because you did not keep the money in the account for the full quarter. In actual practice the reverse is not the case, as $1,000 paid into such an account generally earns interest from the day it enters. But, in engineering calculations it generally is not necessary to consider intermediate compounding effects, and cash flows may usually be simplified to coincide with interest periods. The following illustrates this approach.

Example 3.16 (*Find F, given several A sequences, i, n*)

Given the cash flows shown in Figure 3.17, with $i=5$ percent per quarter, find the equivalent future lump sum F.

SOLUTION

Since interest is earned at the close of each quarter, the simplified cash flow diagram is shown in Figure 3.18. Notice the receipts (positive cash flows) can be grouped at the beginning of each quarter in which they are received, while the disbursements (negative cash flows) can be grouped at the end of each quarter. And the cash flow diagram can be further simplified by summing each quarter's totals as shown in Figure 3.19.

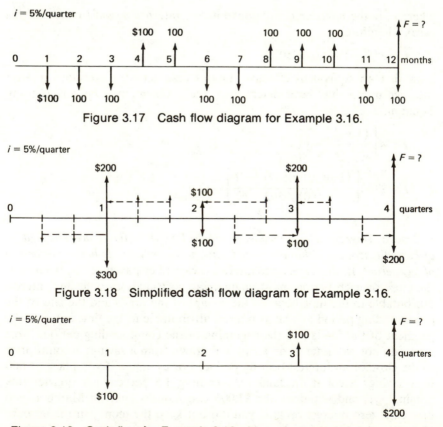

Figure 3.17   Cash flow diagram for Example 3.16.

Figure 3.18   Simplified cash flow diagram for Example 3.16.

Figure 3.19   Cash flow for Example 3.16 with each quarter's totals summed.

• *Case 4. Interest is compounded either periodically or continuously while funds are received or disbursed more or less continuously.* This case involves such revenue generating endeavors as toll bridges, cash retailers, and the like. Depending upon how often receipts are deposited, the cash flow may be considered continuous. When large sums of money are involved, it may be necessary to more precisely calculate the true equivalence factors to be used in such situations. For those cases the reader is referred to such texts as *Engineering Economy* by DeGarmo, Canada, and Sullivan.* However, for most engineering cost analysis studies, sufficiently close approximations may be obtained by lumping cash flows at discrete time periods and treating them as either Case 2 or Case 3.

SUMMARY

In this chapter the relationship between a future value, F, and a uniform series of payments (or receipts) are described. The factors are

$$F = A \left[\frac{(1+i)^n - 1}{i} \right] = A(F/A, i, n)$$

or

$$A = F \left[\frac{i}{(1+i)^n - 1} \right] = F(A/F, i, n)$$

The $(F/A, i, n)$ factor is sometimes termed the uniform series compound amount factor (USCAF), while the $(A/F, i, n)$ factor is often termed the uniform series sinking fund factor (USSFF). Note that one is the reciprocal of the other.

The cash flow diagram for these two factors is shown in Figure 3.20. Note that the *first* time A occurs is at the *end* of the first year and that F and the last value of A occur simultaneously.

Values for these two factors for common combinations of i and n can be found in Appendix A.

Finally, funds flow problems were discussed for four cases, to wit:

Case 1. Interest is compounded more than once between payments in a series of payments. Either the sinking fund method or an equivalent effective interest rate may be used to solve this type problem.

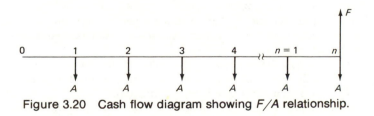

Figure 3.20   Cash flow diagram showing F/A relationship.

*E. Paul DeGarmo et al., *Engineering Economy*, 6th edition. New York: Macmillan, 1979.

Case 2. Interest is compounded continuously while funds are received or dispersed discretely at the beginning or end of any period. In this case, calculate the effective annual interest rate and use the previously derived equivalence factors.

Case 3. Interest is compounded periodically while funds are received or disbursed discretely within the periods not necessarily at the beginning or end of any period. Here, for most cases the funds may be summed together at the end of each compounding period for disbursements, or at the beginning of the period for receipts.

Case 4. Interest is compounded either periodically or continuously, while funds are received or disbursed more or less continuously. For most engineering applications this situation may be satisfactorily approximated by either Case 2 or Case 3.

PROBLEMS FOR CHAPTER 3 (F/A, A/F)

Note: Assume all payments are made at the end of the period (year in most cases) unless otherwise noted in the problem.

Problem Group A

Given three of the four variables, find A and/or F.

A.1 A young engineer decides to save $240 per year toward retirement in 40 years.
 a. If he invests this sum annually at the end of every year at 9 percent, how much will be accumulated by retirement time?
 b. If by astute investing the interest rate could be raised to 12 percent, what sum could be saved?
 c. If he deposits one-fourth of this annual amount each quarter ($60/quarter) in an interest bearing account earning a nominal annual interest rate of 12 percent, compounded quarterly, how much could be saved by retirement?
 d. In part (c) above, what annual effective interest rate is being earned?

A.2 A municipality wishes to set aside enough of its annual gasoline tax funds to finance a $1,000,000 road project scheduled for contract three years from now. If the annual deposit can be invested at 8 percent, how much should be deposited at the end of each of the three years to accumulate $1,000,000 at the end of the third year?

A.3 Your client is building a small medical office complex with an anticipated economic life span of 20 years. Estimated replacement cost in 20 years is $300,000. Assuming zero salvage value
 a. How much rent should be charged annually just to cover this one item if the rent is deposited at the end of each year and draws 10 percent interest while on deposit?
 b. How much rent should be charged monthly if the rent is deposited at the end of each month in an account drawing 8 percent nominal interest compounded monthly? Compounded continuously?

A.4 A town with a population of 13,700 decides to set up a trust fund with which to finance new parks and recreation facilities. They tax each bottle or can of beverage sold in the town $0.01, and since they consume an average of one beverage per day per person they raise $50,000 per year. They hope to finance one major project every five years. How much will be available to spend at the end of each five year period if

a. The $50,000 is deposited at the end of each year and earns annual interest of 8% compounded quarterly?

b. The $50,000 is deposited in 52 weekly installments and the 8% interest is compounded weekly?

A.5 An insurance company uses 40 percent of its premium income for overhead (OH) (salaries, rent, commissions, profit, etc.). Premium income not used for claims or OH is deposited in an account bearing 12 percent interest per year. The remaining 60 percent of premium income is used to pay death benefits to policy holders. Assume that you are doing a rough estimate of insurance charges for the policy holders that are in the 30 to 35 year age bracket. (The rough estimate does not take into account statistical variations within the group.) The tables show a remaining life expectancy for this group of an average of 44 more years.

a. How much should the insurance company charge per year over the remaining life of each policy holder just to cover each $1,000 worth of death benefits?

b. How much should the total annual charge be to cover overhead as well as death benefits?

A.6 A newly opened toll bridge has a life expectancy of 25 years. Considering rising construction and other costs, a replacement bridge at the end of that time is expected to cost $5,000,000. The bridge is expected to have an average of 300,000 toll paying vehicles per month for the full 25 years. At the end of every month the tolls will be deposited into an account bearing annual interest at 7.5 percent compounded monthly. The operating and maintenance expenses for the toll bridge are estimated at $0.15 per vehicle.

a. How much toll must be collected per vehicle in order to accumulate $5,000,000 by the end of 25 years?

b. What should be the total toll per vehicle in order to include both overhead and maintenance costs as well as replacement costs?

c. What annual effective interest rate is being earned?

Problem Group B

Find A and/or F. Two or more steps may be required.

B.1 Your client is collecting rent of $10,000 per year on a commercial building. Her operating expenses use up $5,000 per year, and she deposits the remaining $5,000 per year into an account that draws 10 percent interest compounded. The accumulated balance in the account including principal and interest at the present time is $42,312, including a very recent year-end deposit of $5,000. Your client plans to demolish the existing structure five years from now (she will make five more deposits at $5,000 each) and replace it with a structure estimated to cost $150,000. How much cash will she need five years from now in addition to what is expected to be in the account?

B.2 A young engineer gets a $50-a-month raise and decides to invest one-half of it in the credit union every month toward a down payment on a car. Assume he deposits $25 now and $25 a month for 20 months and gets 12 percent interest compounded monthly on his deposit. How much will he have in 20 months? (There are 21 deposits in all. The first deposit is at the beginning of the first period and the last at the end of the twentieth period.)

B.3 The central air conditioner compressor and air handling in the local municipal auditorium has an estimated life expectancy of 12 more years. A replacement at that time is expected to cost $200,000. The city manager is considering a plan to set aside equal annual deposits into a fund bearing 8 percent interest so that the $200,000 will be on hand when needed 12 years from now. He is ready to make a deposit *right now* and requests that you, as the consultant, determine how much the deposit should be. (Assume 13 equal annual deposits, including one at the beginning of the first period and one at the end of the last period.)

B.4 If $1,000 is deposited at the end of every month, how much will be contained in the account after five years (60 deposits) if the nominal annual interest rate is 12 percent and compounded (a) monthly? (b) quarterly? (c) annually? (d) continuously?

Problem Group C

Find *i*.

C.1 A young man of 25 decides he will retire at 50 with a sum of $200,000. He feels that by careful budgeting he can invest $2,000 per year toward this sum.
   a. What annual interest rate does his investment need to earn in order to attain his goal?
   b. What actual quarterly interest rate does his investment have to earn if he deposits $500 quarterly?
   c. What effective annual interest rate is earned in part (b)?

C.2 A young engineer decides to save her tax rebate each year and invest it for retirement. She arranges to over-deduct and expects a $1,000 rebate at the end of each year. She is 23 years old now and would like to retire at age 58 with a $300,000 nest egg. How much annual interest must her retirement investments yield in order to reach her goal?

Problem Group D

Find *n*.

D.1 To pay for replacing his dragline, a contractor is setting aside $0.10 a cubic yard moved by the dragline. The dragline is averaging 200 yd^3 per hour and 170 hours per month. The money is invested at the end of every month at the nominal annual rate of 10 percent per year. How many months will pass before he accumulates the $150,000 needed to purchase a new dragline? (Assume monthly compounding of interest.)

D.2 A new city in a retirement area is expected to have an influx of 10,000 new residents per year on the average. Due to the large number of older people attracted to the area, the anticipated birthrate is only 5 per 1,000 per year, whereas the death rate is 15 per 1,000 residents. Therefore, there will be a negative natural annual increase of 10 per 1,000. The city is just getting started and population of the area now is essentially zero. How long will it be before the population reaches 60,000 if growth continues to occur at the estimated rates?

Problem Group E

Find the financing cost.

E.1 A contract calls for a lump sum payment of $1,000,000 to you, the contractor, when a job is completed. Estimated time to complete the job is ten months. Your expenditures will be very close to $95,000 per month. You pay your banker 2 percent per month on the amount loaned out to you and borrow the full $95,000 per month from her at the end of every month. Find the financing cost. (The financing cost is the total amount of interest charges.)

Chapter 4
The Value of One Present Payment Compared to a Uniform Series of Payments ($P/A, A/P$)

INSTALLMENT LOAN

A uniform series of payments such as installment loan payments are familiar to everyone in today's credit-oriented society. Appliances, cars, houses, even vacations, may be purchased on installment plans. Many engineering projects are financed in essentially the same way. Funds to construct the project are borrowed and repaid in annual (or other end-of-period) payments. The value of the periodic repayments (A) must equal the value of the amount borrowed (P) plus the interest compounded periodically at rate (i) on the unpaid balance over n periods of repayment. An equation may be derived for this equivalence relationship as follows.

   P dollars are borrowed now and a uniform series of repayments, A, will be made at the *end* of each of n periods. Each payment A will include two parts: payment of interest on the unpaid balance, plus a repayment of part of the principal.

P/A EQUATION

To derive the P/A equation simply combine the two equations,

$$P = F\left[\frac{1}{(1+i)^n}\right] \qquad (2.2)$$

and

$$F = A\left[\frac{(1+i)^n - 1}{i}\right] \qquad (3.3)$$

Substitute the right side of the Equation 3.3 for F in Equation 2.2 to obtain

$$P = A\left[\frac{(1+i)^n - 1}{i}\right]\left[\frac{1}{(1+i)^n}\right]$$

This may be combined and simplified to

$$P = A\left[\frac{(1+i)^n - 1}{i(1+i)^n}\right] \qquad (4.1)$$

The factor in brackets in Equation 4.1 is termed uniform series present worth factor (USPWF) and its functional form is

$$P = A(P/A, i, n) \qquad (4.1A)$$

The inverse of the USPWF is termed the uniform series capital recovery factor (USCRF) and is written as

$$A = P\left[\frac{(i(1+i)^n}{(1+i)^n - 1}\right] \qquad (4.2)$$

or, functionally, as

$$A = P(A/P, i, n) \qquad (4.2A)$$

As pointed out in Chapters 2 and 3 these factors depend only upon i and n. Values for these two factors are tabulated in Appendix A for commonly used combinations of i and n.

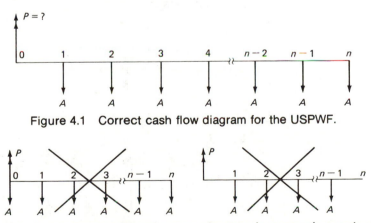

Figure 4.1   Correct cash flow diagram for the USPWF.

Figure 4.2   Illustrations of several commonly occurring errors in constructing cash flow diagrams for the USPWF.

The cash flow diagram for Equation 4.1 (USPWF) is shown in Figure 4.1. Beginners sometimes make errors in constructing the cash flow diagrams and some of the more common errors for this relationship are shown in Figure 4.2.

Graph of P/A

A graph of the equation

$$P/A = \frac{(1+i)^n - 1}{i(1+i)^n}$$

is shown in Figure 4.3. The graph illustrates the response of P/A to variations in i and n. This information is helpful in determining the effect of changes in input variables on problem solutions.

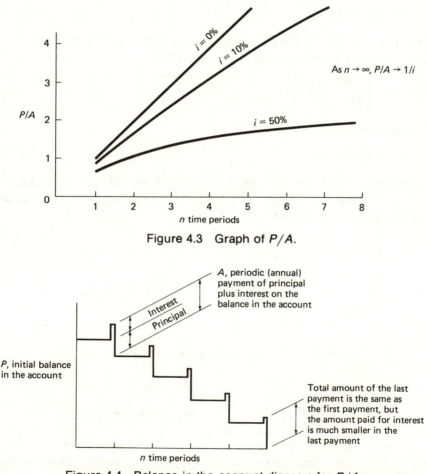

Figure 4.3 Graph of P/A.

Figure 4.4 Balance-in-the-account diagram for P/A.

Graph of Decrease in Principal Balance with Periodic Payments

Figure 4.4 represents the balance of an account that begins with a present amount P on the left side. At the end of each of n time periods, a payment A is made until the account is reduced to zero at the right side. Each periodic payment A consists first of interest on whatever current balance is left in the account. The remaining portion of the periodic payment A pays off another part of the balance, until finally the balance is reduced to zero. Note that the first periodic payment A is made at the *end* of the first period.

It is important to remember that equation time zero (beginning time for the equation) for the P/A series occurs *one period before* the first periodic payment, A, occurs. In analyzing problems, this information is used to find the correct value of n, the number of periods.

As in the equations discussed in earlier chapters, the P/A (or A/P) equation contains four variables, P, A, i, and n. If any three of the variables are given (or can be derived from other data in the problem) then the fourth variable can be found. The next four examples demonstrate how each of the four variables in turn may be found.

Example 4.1 (*Find P, given A, i, n*)

A young contractor decides he can budget $100 per month as payments on a new pickup. The bank offers a 60-month repayment period on new vehicle loans at 1 percent per month on the unpaid balance. How much can be financed under these terms?

SOLUTION

$$P = \frac{A\left[(1+i)^n - 1\right]}{i(1+i)^n}$$

$$P = \frac{\$100\left[(1+0.01)^{60} - 1\right]}{0.01(1+0.01)^{60}}$$

$$P = \$100/\text{month} \times 44.955 = \underline{\underline{\$4,495.50}}$$

Example 4.2 (*Find A, given P, i, n*)

The contractor finds the pickup he likes but the price tag is $6,000. Under the same terms, 1 percent per month and 60 months, what would the monthly payment be, with the first payment due one month from now?

SOLUTION
Using the formula

$$P = \frac{A[(1+i)^n - 1]}{i(1+i)^n}$$

$$\$6,000 = \frac{A[(1+0.01)^{60} - 1]}{0.01(1+0.01)^{60}}$$

$$\$6,000 = A \times 44.955$$

$$A = \frac{\$6,000}{44.955} = \underline{\underline{\$133.47/\text{month}}}$$

Rather than laboriously evaluating the equation each time, the tabulation of P/A and A/P values as provided in Appendix A can be used.

Example 4.3 (*Find i, given P, A, n*)

A contractor buys a $10,000 truck for nothing down and $2,600 per year for five years. What is the interest rate she is paying?

SOLUTION

$$A = P(A/P, i, n)$$

$$\$2,600 = \$10,000(A/P, i, 5)$$

$$0.2600 = (A/P, i, 5)$$

The equation with these data included appears as

$$A/P = 0.26 = \frac{i(1+i)^5}{(1+i)^5 - 1}$$

Since the i appears three times in the equation, it is usually easier to interpolate from the tables. (For values outside the tables, trial and error to bracket the solution, followed by interpolation, is often the simplest method.)

For a solution using the tables, look through the various interest rate tables of Appendix A, each time under the A/P column and across the $n=5$ line until the A/P values are found which bracket 0.2600, as follows:

i	A/P	A/P
10%	0.2638	
i		0.2600
9%	0.2571	0.2571
Subtract	0.0067	0.0029

Then interpolate

$$i = 9\% + \frac{0.0029}{0.0067} \times 1\% = 9.43 = \underline{\underline{9.4\%}}$$

Example 4.4 (*Find n, given P, A, i*)

A new civic auditorium is estimated to cost $5,000,000 and produce a net annual revenue of $500,000. If funds to construct the auditorium are borrowed at $i=8$ percent, how many annual payments of $500,000 are required to retire the debt?

SOLUTION

The equation for P/A is $P/A=[(1+i)^n-1]/[i(1+i)^n]$. Since the n appears twice in the equation it is usually simpler to interpolate from the interest tables in Appendix A. Using the tables, the answer is found as follows:

$$P=A(P/A, i, n)$$

$$\$5,000,000=\$500,000(P/A, 8\%, n)$$

$$(P/A, 8\%, n)=\frac{\$5,000,000}{\$500,000}=10$$

From the interest tables in Appendix A, bracket the P/A factor of 10 and interpolate to find n.

$$(P/A, 8\%, 21)=10.0168$$

$$(P/A, 8\%, n)= \qquad\qquad 10.000$$

$$\underline{(P/A, 8\%, 20)=\ 9.8181} \qquad \underline{9.8181}$$

$$\text{subtract} \qquad\ 0.1987 \qquad\quad 0.1819$$

$$n=20 \text{ yr}+\frac{0.1819}{0.1987}\times 1 \text{ yr}=\underline{20.9 \text{ yr}} \text{ (21 yr)}$$

Of course this does *not* mean that 20 payments of $500,000 each will be made at the end of 20 successive years and then another $500,000 will be paid in another 0.9 of a year. Rather it indicates that *approximately* $0.9 \times \$500,000 = \$450,000$ will be paid as a final payment at the end of the twenty-first year. The method of calculating the correct amount of the last payment is outlined in the next paragraph.

AMOUNT OF THE FINAL PAYMENT IN A SERIES

As illustrated in Example 4.4, when n is the derived quantity in a problem involving repayment, the last payment in the series usually will be different than the rest of the series. To find the odd final payment use the following three-step procedure.

1. Given the present worth, P_1, of the loan, and the amount of the periodic repayment, A_1, find the present worth, P_2, of the series except the final payment.
2. Subtract P_2 from P_1 to find P_3.
3. Find the future value, F_1, of P_3 at the time the last payment is due. The amount due as the odd last payment is F_1.

To find the amount of the last payment due on the civic auditorium of Example 4.4 at EOY 21, the three steps are applied as follows:

Given: The amount of the loan is $P_1 = \$5{,}000{,}000$

The amount of the periodic repayment is $A_1 = \$500{,}000$

Then,

1. Find the present worth P_2 of the first 20 payments. This yields the amount that could be financed by those 20 payments alone.

$$P_2 = \$500{,}000 \, (P/A, 8\%, 20) = \$4{,}909{,}050$$
$$9.8181$$

2. Subtract P_2 from the original loan of $P_1 = \$5{,}000{,}000$ and find P_3. This yields the amount borrowed, P_3, that must be paid off by the final payment F_1 at the end of 21 years.

$$P_3 = \$5{,}000{,}000 - \$4{,}909{,}050 = \$90{,}950$$

3. Find $F_1 = \$90{,}950(F/P, 8\%, 21) = \$457{,}824$, the amount of final payment at the end of the twenty-first year.

IF PROBLEM TIME ZERO DIFFERS FROM EQUATION TIME ZERO

Problem situations sometimes occur that involve a different time base than that specified for the P/A and A/P factors. (For example, see Figure 4.2.) The general approach to a solution of this type of problem is to subdivide the problem into simple components, solve each component individually, and then find the sum or product of the parts. The following two examples will illustrate the method. In Example 4.5, time zero for the problem occurs several periods *before* the start of the series of A payments, while in Example 4.6, problem time zero occurs *after* the A payments start.

Example 4.5 *ETZ ≠ PTZ (ETZ later than PTZ)*

Your firm wants to borrow $50,000 now to buy land the company will need in the future. The proposed terms of the loan are

1. No cash payments on the loan (neither interest nor principal) until the end of year 5, although interest accrues.
2. The loan will be paid off in eight equal annual installments, with the first installment due at the end of year 5.
3. Interest is earned on the outstanding balance at 9 percent compounded annually on the loan including, of course, the period during which no payments are made.

You are requested to find the size of the annual repayments, A, required from year 5 through 12. A cash flow diagram of the problem appears in Figure 4.5.

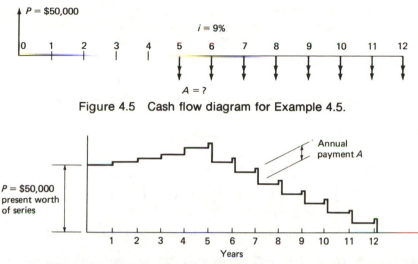

Figure 4.5 Cash flow diagram for Example 4.5.

Figure 4.6 Balance-in-the-account diagram for Example 4.5.

SOLUTION

This type problem may be solved by dividing it into two parts. The first few years are periods during which the interest accumulates as in an F/P problem. The second series of years involves an A/P series during which the loan is paid off by a series of equal annual payments, A. A graph of the loan balance appears in Figure 4.6.

At this point in their studies students sometimes encounter difficulty in finding which year in the problem corresponds to time zero for the equation (ETZ), and a consequent difficulty in determining the correct value of n. Such difficulties should be overcome (by further review, practice, and consultation) before proceeding further since the determination of proper n values is essential to the solution of many problems throughout the remainder of the text. By referring to the cash flow diagram of the loan balance and remembering that equation time zero is one time period before the first payment, the ETZ for the A/P equation can be found as the end of year 4. After this point is found, the problem can be solved in two steps as follows:

1. Find F_1, the amount owed at the end of year 4. The cash flow diagram is shown in Figure 4.7.

$$F_1 = P(F/P, i, n) = \$50,000 \underset{1.4116}{(F/P, 9\%, 4)} = \$70,580$$

2. Find A, the annual payment required to pay out the loan in eight years. The value of F_1 found in step (1) is considered as the P value of the loan to be paid off. The cash flow diagram is shown in Figure 4.8.

$$A = P(A/P, i, n) = \$70,580 \underset{0.1807}{(A/P, 9\%, 8)} = \$12,754 = \underline{\underline{\$12,800}}$$

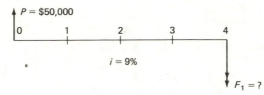

Figure 4.7   Cash flow diagram for solution to Example 4.5.

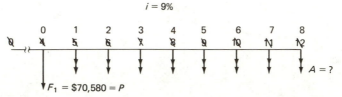

Figure 4.8   Cash flow diagram for solution to Example 4.5.

Example 4.6   (*ETZ≠PTZ, ETZ earlier than PTZ, find the unpaid balance.*)

A year ago your firm bought a tractor and financed it with a $20,000 loan, repayable in 36 monthly installments with 1 percent interest per month charged on the unpaid balance. Now the firm has some extra cash and would like to pay off the loan. You are requested to determine (1) the balance due on the loan, and (2) the total amount of interest that would be paid as part of the last 24 payments. The twelfth monthly installment has just been paid.

SOLUTION 1
The cash flow diagram for this problem is shown in Figure 4.9. The graph of the loan balance appears in Figure 4.10. Referring to the cash flow diagram and the graph of the loan balance, the problem can be subdivided into two parts, the series before problem time zero and the series after problem time zero. We need to find the payoff amount which is the same as the present worth of the series after problem time zero. Since problem time zero occurs at the end of the twelfth month, the value of n for the last series is $36-12=24$. With this information, the problem may be solved in two steps as follows:

Step 1: Find the amount, A, of the monthly payments.

$$A=P(A/P,i,n)$$

$$A=20{,}000(A/P,1\%,36)$$

$$A=20{,}000\times0.0332=\$664 \text{ monthly payment}$$

Step 2: Find the present worth, P, of the 24 remaining monthly payments after the twelfth payment has been made. The cash flow diagram is shown in Figure 4.11.

$$P=A(P/A,i,n)$$

$$P=664(P/A,1\%,24)=\underline{\$14{,}106}$$

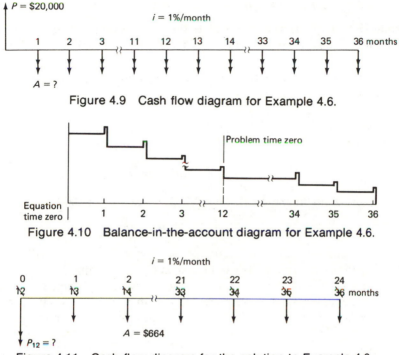

Figure 4.9   Cash flow diagram for Example 4.6.

Figure 4.10   Balance-in-the-account diagram for Example 4.6.

Figure 4.11   Cash flow diagram for the solution to Example 4.6.

SOLUTION 2

Find the amount of interest only, apart from principal, on the last 24 payments. Notice that the amount of cash required to pay off the loan ($14,106) is considerably smaller than the total cash flow if the loan were to be paid using monthly installments ($24 \times \$664 = \$15,936$). The difference is the amount of interest that would be paid ($\$15,936 - \$14,106 = \$1,830$) as part of the last 24 payments. (Remember the numerical answer is only significant to the number of digits used in the factors from Appendix A.)

CAPITALIZED COSTS FOR INFINITE LIVES ($P/A \rightarrow 1/i$ as $n \rightarrow \infty$)

Occasionally problem situations occur involving practically infinite values of n. For instance a government department of transportation (DOT) may need land for right of way (R/W). The land will require maintenance (mowing, etc.) for as long as the DOT owns it and the DOT intends to own it for a very long time, or perpetually for all practical purposes. If the annual cost of maintenance is valued at A, then the present worth of that annual cost charged at the end of every year forever, from Equation 4.1, is

$$P = A\left[\frac{(1+i)^{\infty}-1}{i(1+i)^{\infty}}\right] = A\left[\left(\frac{(1+i)^{\infty}}{i(1+i)^{\infty}} - \frac{1}{i(1+i)^{\infty}}\right)^{0}\right]$$

$$P = A\left(\frac{1}{i}\right) = \frac{a}{i}, \quad \text{or} \quad A = Pi \qquad \text{for } n \rightarrow \infty \tag{4.3}$$

Example 4.7 (*Given* $P_1, A_1, i, n \rightarrow \infty$, *find* P_2, A_2)

A state DOT requests you to calculate the present value of acquiring and maintaining a new R/W which costs $300,000 to purchase and $7,200 per year to maintain. They intend to hold the R/W in perpetuity and ask for equivalent costs both in terms of (1) present worth (*P*), and (2) annual worth (*A*). Use $i = 6$ percent.

SOLUTION 1

The equivalent total present cost in terms of present worth, *P*, is

$$\text{cost of acquisition, } P_1 = \$300,000$$

$$\text{cost of maintenance, } P_2 = A/i = \$7,200/0.06 = \$120,000$$

$$\text{total equivalent present worth, } P_t = \$420,000$$

(P_t is sometimes referred to as capitalized value or capitalized cost.)

Comment: The $120,000 equivalent present worth of maintenance may be visualized as the amount of money needed in a fund drawing 6 percent interest so that the interest alone is sufficient to pay all of the annual costs.

SOLUTION 2

The equivalent annual cost is

$$\text{cost of acquisition, } A_1 = Pi = \$300,000 \times 0.06 = \$18,000/\text{yr}$$

$$\text{cost of maintenance, } A_2 = \underline{7,200}$$

$$\$25,200/\text{yr}$$

Comment: The $18,000 per year equivalent annual cost of acquisition may be viewed as the cost of annual interest if the $300,000 is borrowed and not repaid. In other words, the $300,000 could be borrowed, exchanged for the land, and then repaid whenever the land is resold at the same purchase price. The only cost of ownership would be the interest cost on the borrowed money.

MORE THAN ONE COMPOUNDING PERIOD OCCURS BETWEEN PAYMENTS IN A SERIES

Sometimes a project will require a series of expenditures at intervals of time greater than the compounding period, such as a dam that needs dredging of accumulated silt every five years. The series may be delayed in starting. The capitalized value may be found by one of several methods, two of which are shown in the following example.

Example 4.8 *(Find P, given i, n→∞, a series of payments with the period between payments, some multiple of the interest compounding period, and a delayed start)*

A dam will be completed three years from now, and water impounded. Five years after that at EOY 8 and every subsequent five years, the accumulated silt will be dredged out at a cost of $500,000 for each dredging. This series will continue for a long time to come ($n \to \infty$). Find the present worth if $i = 10$ percent.

SOLUTION

Method 1, sinking fund. Find the annual series amount A required for deposit into an account that will build up to an F of $500,000 every five years. Then find the present value of this A series with $n = \infty$. The *ETZ* for this series is EOY 3, so a further P/F transfer to EOY 0 is required (see Figure 4.12).

$$A = F(A/F, 10\%, 5) = 500,000 \times 0.1638 = \$81,900/\text{yr}$$

$$P = A(P/A, 10\%, \infty) = \$81,900 \times 10.0 = \$819,000 @ \text{ EOY 3}$$

$$P = F(P/F, 10\%, 3) = \$819,000 \times 0.7513 = \underline{\underline{\$615,300}}$$

Conclusion. If a lump sum (capitalized value) of $615,300 is deposited into an account bearing interest compounded annually at 10 percent, then $500,000 may be withdrawn at EOY 8, and every five years thereafter forever.

Method 2, effective interest. Find the effective interest for a five-year compounding period that is equivalent to 10 percent compounded per year and solve the problem as a regular periodic series.

$$i_e = (1+i)^m - 1 = 1.1^5 - 1 = 0.6105$$

$$P = A(P/A, i_e, n/m) = \$500,000(P/A, 61.05\%, \infty)$$

$$P = \$500,000/0.6105 = \$819,000 @ \text{ EOY 3}$$

$$P = F(P/F, 10\%, 3) = \$819,000 \times 0.7513 = \underline{\underline{\$615,300}}$$

Conclusion. The capitalized value is $615,300, the same as for Method 1.

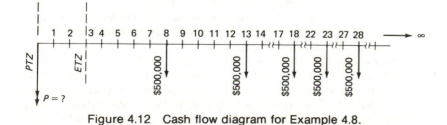

Figure 4.12   Cash flow diagram for Example 4.8.

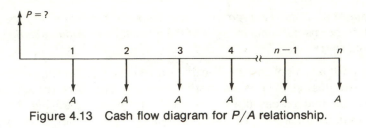

Figure 4.13   Cash flow diagram for P/A relationship.

SUMMARY

This chapter developed equivalence factors relating a present value (P) to a uniform series of values (A). The factors and their functional forms are

$$P=A\left[\frac{(1+i)^{n}-1}{i(1+i)^{n}}\right]=A(P/A,i,n)$$

$$A=P\left[\frac{i(1+i)^{n}}{(1+i)^{n}-1}\right]=P(A/P,i,n)$$

The cash flow diagram for these relationships is shown in Figure 4.13.

Also introduced was the concept of infinite life $(n=\infty)$, in which the capitalized cost P was shown to be

$$P=\frac{A}{i}, \qquad \text{for} \quad n=\infty$$

PROBLEMS FOR CHAPTER 4 $(P/A, A/P)$

Note: Assume all payments are made at the end of the period unless otherwise noted in the problem.

Problem Group A

Given three or four variables, find A or P.

A.1 A contracting firm needs a new addition to its equipment maintenance shop. The firm estimates that the maximum repayment it can afford to make is. $8,000 per year. How much can they afford to borrow for the shop if
  a. Loan money is available at 15 percent for 15 years?
  b. Loan money is available at 20 percent for 15 years?

A.2 You purchase a $\frac{3}{4}$ yd^{3} hydraulic backhoe for $20,000. The terms are 10 percent down and 1 percent per month on the unpaid balance for 60 months.
  a. How much are the monthly payments?
  b. What annual effective interest rate is being charged?

A.3 A city estimates that for each new citizen who moves into town it must eventually add about $1,000 worth of new improvements (water, sewer pipes and treatment facilities, vehicle lane capacity, R/W, schools, recreation, library, enlarged government buildings, etc.). In order to assess this cost directly to the new resident it is suggested that a "hook-on" charge be made for new residential utility services to newly constructed residences (both single and multifamily). Ordinarily, this would then be included in the mortgage. Assuming 3.2 residents per single family dwelling, how much would annual mortgage payments be increased to cover the $3,200 added charge, if $i=9$ percent and $n=25$ years?

A.4 Your firm has an opportunity to buy a five-year lease on a rock quarry (for concrete aggregate) that is expected to yield a net profit of $58,000 per year (end of year) for the next five years. How much can your firm pay for this lease (lump sum payment now) in order to make 14 percent on the remaining worth of the lease each year? (How large an investment could be paid off with payments of $58,000 per year with 14 percent interest on the unpaid balance?)

Problem Group B

Find i.

B.1 A contractor invests $1,500,000 in a scraper fleet with an expected life of ten years with zero salvage value. He expects to move an average of 100,000 yd^3 of earth per month at $1.00/yd^3, of which $0.80 is the cost of operating the fleet, leaving $0.20/yd^3 as return on the investment in the machines. Using $n=120$ months,

   a. What is the actual monthly rate of interest earned by this equipment?
   b. What is the annual nominal interest earned?
   c. What is the annual effective interest earned?

B.2 Your firm owns a large earthmoving machine that may be renovated to increase its production output by an extra 10 yd^3/hr, with no increase in operating costs. The renovation will cost $100,000. The earthmoving machine is expected to last another eight years, with zero salvage value at the end of that time. Earthmoving for this machine is currently being contracted at $1/yd^3. The company is making a 15 percent return on their invested capital. You are asked to recommend whether or not this is a good investment for the firm. (Does the return on this investment exceed 15 percent?) Assume the equipment works 2,000 hours per year.

Problem Group C

Find n. (May also request F.)

C.1 A proposed toll parking structure is estimated to cost $1,000,000. Loan funds are available at 6 percent. Net income (after other expenses) available to pay principal and interest is anticipated at $100,000 per year.

   a. How long will it take to pay off the loan?
   b. Find the amount of the final payment.

C.2 A left-turn lane may be installed at an intersection at a cost of $125,000. The turning lane is expected to save motorists an average of $0.01 (stopping, delays, and accident costs) for each vehicle traversing the intersection. Traffic is estimated as 10,000 vehicles per day every day of the year. Assuming $i=7$ percent, how many years will it take for the savings to amortize (pay out) the cost of the improvement? (Same situation as if the $125,000 were being repaid by the annual savings to motorists.)

C.3 A developer client of yours purchased some property for $75,000. The terms were $20,000 down and a note for $55,000, payable at the rate of $650 per month at a nominal annual interest rate of 12 percent on the unpaid balance.
   a. How long will it take to pay off the loan?
   b. What annual effective interest rate is being charged?

Problem Group D

$ETZ \neq PTZ$ (equation time zero does not equal problem time zero).

D.1 Assume you need a dragline. You can purchase one for $50,000 fully financed at 8 percent for ten years.
   a. Find the annual end-of-year payments.
   b. Find the hourly charge that should be included to pay off the dragline, assuming 2,000 hours work per year.
   c. Assume you would like to trade this one in and purchase another new dragline five years from now. You expect a 5 percent increase in price each year. What would the new dragline cost?
   d. What is the unpaid balance on the current dragline after five years?
   e. Assume the existing machine can be sold in five years for $40,000, and a new machine can be purchased for the amount calculated in part (c). If the existing loan is paid off in one lump sum and a new loan is taken out for the balance (cost new $-\$40,000$) at $i=8$ percent and ten more years, what are the payments on the new loan?

D.2 Five years ago your firm bought a concrete batch plant. To finance it they borrowed $100,000 on a 20-year mortgage bearing interest at 1 percent per month on the unpaid balance. Repayment is scheduled in 240 equal monthly installments. The firm now has money available and wants to pay off the entire balance due. You are asked to calculate the balance due. The sixtieth monthly payment has just been made.

D.3 Your hometown bank offers to loan you money to finance your last two years in college. They will loan you $4,000 today, and another $4,000 one year from today. They will accept repayment in ten equal annual installments with 15 percent interest on all unpaid balances. The first annual repayment installment is due three years from today (at the end of your first full year of work). What is the amount of each installment payment?

D.4 A transportation authority asks you to check on the feasibility of financing for a toll bridge. The bridge will cost $2,000,000 at problem time zero. The authority can borrow this amount to be repaid by tolls. It will take two years to construct, and will be open for traffic at the end of the second year. Tolls will be accumulated throughout the third year and will be available for the initial annual repayment at the end of the third year. If the authority is expecting an average of 10,000 cars per day every day of the year, how much must each car be charged in order to repay the borrowed funds in 20 equal annual installments (first installment due at end of year 3), with 8 percent compound interest on the unpaid balance?

D.5 Your client is trying to arrange financing for the construction of a new $1,200,000 office building. The building will take 12 months to construct. In order to make monthly payments to the contractor, your client needs to borrow $100,000 per month at the end of each of the 12 months during which construction is underway. He agrees to repay the entire amount owed in 240 monthly payments, backed by a 20-year mortgage. The lender charges a nominal 12 percent interest (1%/month) on the balance owed (at *all* times, including the construction period). After the building is completed, the tenants will move in during the thirteenth month, and start paying rent at the beginning of the fourteenth month. Your client wants the first repayment to the lender to coincide with the first rental payments from the tenants, so he will make his first repayment to the lender at the beginning of the fourteenth month.

   a. Draw a line diagram illustrating the cash flow for this problem.

   b. How much does your client owe the lender at the end of the twelfth month after making the last payment to the contractor?

   c. How much are the monthly repayments on the client's loan?

D.6 An enterprising college graduate wishes to set up a retirement fund which will pay $5,000 per year for ten years commencing 20 years from today (10 payments). What annual deposit must be made starting today and continuing for the next 19 years (20 deposits), if $i=9$ percent?

Problem Group E

Perpetual or very long lives. Find P, A, or i.

E.1 A grateful college graduate wishes to set up a $1,000 a year perpetual scholarship in her name at her alma mater. If she can receive 8 percent on this investment, how much money will be required to set up the scholarship?

E.2 A dam costing $17,000,000 is designed for a 100-year life. Annual maintenance expenses on the dam are expected to be $2,000. If $i=7$ percent:

   a. What is the present value of the annual maintenance expenses over the 100-year period?

   b. What is the present value of the annual maintenance expenses if the dam lasts forever?

   c. What is the equivalent annual value of the purchase price of the dam over the 100-year period?

   d. What is the equivalent annual value of the purchase price of the dam if the dam lasts forever?

E.3 A major real estate investor is considering the purchase of a land site near downtown Houston for $10,000,000. He plans to construct a 50-story building which is estimated to cost $80,000,000. He assumes the building will last 50 years with zero salvage value. For $i=15$ percent:

   a. What annual cost will be incurred for the building and land assuming the land will not change in value during the 50-year period?

   b. How far off would the estimate of the annualized cost be, if: (1) the salvage value of the land were zero; (2) the salvage value of the land increased five-fold (worth $50,000,000 in 50 years)?

Chapter 5
Arithmetic Gradient, *G*: The Constant Increment to a Series of Periodic Payments

KEY EXPRESSIONS IN THIS CHAPTER

G = Arithmetic gradient increase in funds flow at the end of each period for n periods.

P & I = Principle and interest comprising the periodic payments for an amortization schedule.

PAYMENT SERIES WITH A CONSTANT INCREMENT

Engineers frequently encounter situations involving periodic payments that increase or decrease by constant increments from period to period. For example, annual maintenance costs on a proposed engineering project are estimated at

YEAR	MAINTENANCE
1	$ 0
2	1,000
3	2,000
etc.	etc.

The growth increment *G* in this example is $G=\$1,000$ per year and is termed a "gradient." Since the increment is a *constant amount* at $1,000, the progression is described as arithmetic (compared to geometric if the growth increases by the same *percent* each year, as in Chapter 6).

This situation occurs often enough to warrant the use of special equivalence factors relating the arithmetic gradient to other cash flows. In this chapter *G* will be related to the future value, *F*, the uniform series value, *A*, and the present value, *P*.

FUTURE VALUES OF GRADIENT AMOUNTS (*F*/*G*)

Notice that the maintenance table above starts with a value of $0 at year 1 and then increases by a *G* of $1,000 per year. The cash flow for such a series is shown in Figure 5.1.

The gradient amount is designated as *G* in the generalized problem and can be either positive or negative, but always increases in absolute value with increasing time (the values grow larger from left to right on the cash flow diagram), as shown in Figure 5.2.

Generalized Problem Relating *G* to Future Value *F* (*Find F, given G, i, n*)

An investor decides to invest on an arithmetic gradient schedule as shown below:

YEAR	AMOUNT INVESTED
1	$0G$
2	$1G$
3	$2G$
⋮	⋮
n	$(n-1)G$

If the investor earns an interest rate of *i* on the investment, what is the future value of this investment at year *n*?

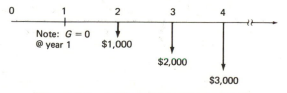

Figure 5.1   Arithmetic gradient series.

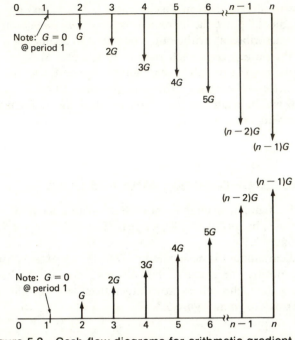

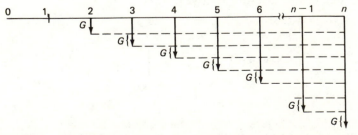

Figure 5.2   Cash flow diagrams for arithmetic gradient (G).

SOLUTION
One legitimate (but time consuming) approach is to consider this problem
to be the sum of an overlapping group of uniform series of annual
payments, each of value $A = G$, but for decreasing time periods beginning
with $n-1$, as shown in Figure 5.3. Thus,

$$F = G(F/A, i, n-1) + G(F/A, i, n-2) + \cdots + G(F/A, i, n-n+1)$$

From Equation 3.2

$$(F/A, i, n) = \frac{(1+i)^n - 1}{i}$$

$$F = G \left[\frac{(1+i)^{n-1} - 1}{i} + \frac{(1+i)^{n-2} - 1}{i} + \cdots + \frac{(1+i) - 1}{i} \right]$$

(5.1)

Figure 5.3   Overlapping group of uniform series annual payments.

Simplifying,

$$F=\frac{G}{i}\left[(1+i)^{n-1}+(1+i)^{n-2}+\cdots+(1+i)-(n-1)\right] \tag{5.2}$$

Equation 3.1 is

$$F=A\left[(1+i)^{n-1}+(1+i)^{n-2}+\cdots+(1+i)^{2}+(1+i)+1\right] \tag{3.1}$$

Subtracting 3.1 from 5.2 yields

$$\frac{Fi}{G}-\frac{F}{A}=-n$$

or

$$F=\frac{G}{i}\left(\frac{F}{A}-n\right) \tag{5.3}$$

But $F/A=[(1+i)^{n}-1]/i$ (from Equation 3.3). Therefore,

$$F=\frac{G}{i}\left[\frac{(1+i)^{n}-1}{i}-n\right] \tag{5.4}$$

or, in notation form

$$F=G(F/G,i,n) \tag{5.4A}$$

Values for F/G are listed in Appendix A for commonly occurring values of i and n.

The use of this equivalence factor is demonstrated in the following examples.

Example 5.1   *(Find F, given G, i, n)*

A student scheduled to graduate next year is planning on saving for a major purchase, and decides to invest on the following schedule. The investment will earn 6 percent compounded annually. How much will he have saved after five years?

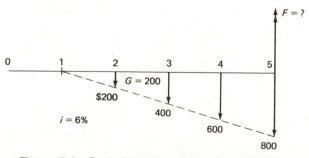

Figure 5.4   Cash flow diagram for Example 5.1.

SOLUTION
The cash flow diagram is shown in Figure 5.4, and the problem is solved as
follows:

$$F_2 = G(F/G, i, n) = \frac{200}{0.06}\left[\frac{(1+0.06)^5 - 1}{0.06} - 5\right] = \underline{\underline{\$2,124}}$$

Graph of Increase in *F* as *n* Increases

Figure 5.5 illustrates the timing of the cash flows of G, as well as the
relationships between G, F, i, and n, as illustrated by Example 5.1. Note
that the first deposit of G ($200 in this example) occurs at the *end* of the
second period. Therefore the first interest is not earned until the end of
the third period. Gradient deposits are made at the end of every year except
the first year, so there are a total of $(n-1)$ deposits in the G series (as
contrasted to n deposits in all of the A series). Deposits are increased by the
amount G ($200 in this case) for each period (year for this example).
Therefore, the amount of the last deposit is $G(n-1)$, or in this example
$200(5-1) = \$800$. An ability to determine the correct value of n is essential
in solving many problems commonly encountered by the professional
engineer. Since there are always $(n-1)$ gradient deposits in the G series, the
value of n may be determined as the number of deposits plus one.

Graph of *F/G* and *G/F*

A graph of the equation

$$F/G = \frac{1}{i}\left[\frac{(1+i)^n - 1}{i} - n\right]$$

and its inverse is shown in Figure 5.6.

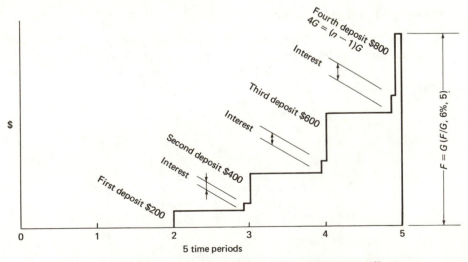

Figure 5.5   Timing of the cash flows of $F = \$200(F/G, 6\%, 5)$.

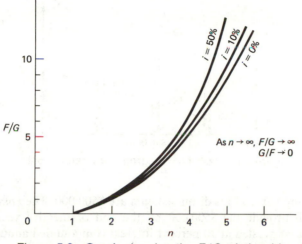

Figure 5.6　Graph showing the F/G relationship.

Example 5.2　(*Find F, given A, G, i, n with a decreasing gradient, G*)

Your client is a contractor who needs to save $500,000 to replace his worn out equipment upon completion of a five-year contract for a mining company. In his bid on the contract he includes an extra $100,000 the first year to be deposited at EOY 1 in a fund earning 10 percent compounded annually. Due to declining productivity, as the equipment grows older, he expects to include only $90,000 the second year, $80,000 the third year, and so on. He asks you to determine whether or not he will accumulate $500,000 in the fund by the end of the fifth year.

SOLUTION
The problem is solved by separating it into two parts (the A series, and the G series), solving for the respective future value (F) and summing. The cash flow diagram for this problem is shown in Figure 5.7.

Notice the use of signs to indicate a gradient G in the opposite direction from the uniform series A. The $100,000 annual series payments are shown as downward arrows and negative, in the calculations, while the $10,000 gradient G series is shown as upward arrows and positive.

$A = -\$100,000/\text{yr},$　$i = 10\%,$　$n = 5$ for both A and G
$G = +\$10,000/\text{yr}$

The future value of an annual disbursement of $100,000 per year is

$$F_1 = A(F/A, i, n) = -\$100,000(F/A, 10\%, 5) = -\$610,510$$
$$6.1051$$

Then the future value of the positive gradient of $+\$10,000$ per year is

$$F_2 = G(F/G, i, n) = +\$10,000 \times 11.051 = +\$110,510$$
$$F = F_1 + F_2 = -\$610,510 + \$110,510 = -\$500,000$$

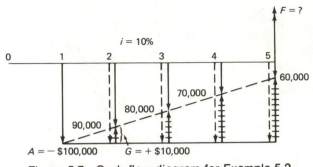

$$A = -\$100,000 \qquad G = +\$10,000$$

Figure 5.7   Cash flow diagram for Example 5.2.

The total future value of his disbursements are $500,000. The answer is yes, this plan will provide the $500,000 by the end of the five years if each disbursement is invested at 10 percent interest compounded annually.

Example 5.3   (*Decreasing gradient; find G and A in terms of dollars per cubic yard; given F, i, n, with G and A in terms of cubic yards per hour*)

Your contractor client estimates she will need a total of $80,000 ten years from today in order to replace a dragline she now owns. She expects to move about 100,000 yd³ of earth this year with the dragline but, due to age and increasing "down time," she expects this amount to decrease by about 5,000 yd³/yr each year over the ten-year period. Her other operating and maintenance (O & M) costs run about $0.50/yd³. She asks you to determine the additional charge per cubic yard required to accumulate the $80,000 in ten years if she can invest all end-of-year deposits and earn 10 percent interest compounded on the amount invested.

SOLUTION
First, identify the variables:

$$F = \$80,000$$

$$n = 10 \text{ yr}$$

$$A = (100,000 \text{ yd}^3/\text{yr})(\$/\text{yd}^3)$$

$$G = (-5,000 \text{ yd}^3/\text{yr})(\$/\text{yd}^3)$$

$$i = 10\%$$

The "other O&M costs" of $0.50/yd³ is extraneous information not needed to solve the problem. The inclusion of such information here simulates the real world where myriads of useless information is often available to confuse the problem.

One basic problem here is translating cubic yards per year into dollars per year. This may be done by letting y represent the extra charge in terms of dollars per cubic yard. Then find how dollars per cubic yard times cubic yards per hour are needed to accumulate the $80,000.

$$A = (100,000 \text{ yd}^3)(y), \text{ the annual base income}$$

$$G = (-5,000 \text{ yd}^3)(y), \text{ the declining gradient}$$

$$F = A(F/A, i, n) - G(F/G, i, n)$$

$$\$80,000 = (100,000 y)(F/A, 10\%, 10) - 5,000 y(F/G, 10\%, 10)$$
$$\qquad\qquad\qquad\qquad 15.937 \qquad\qquad\qquad\qquad 59.374$$

$$80,000 = 1,593,700 y - 296,870 y$$

$$y = \frac{80,000}{1,296,830} = \$0.0617 = \underline{\underline{\$0.062/\text{yd}^3}}$$

By charging an extra $\$0.062/\text{yd}^3$ and investing the resulting funds at the end of each year at 10 percent compounded, the contractor will accumulate $80,000 at the end of ten years to purchase a new dragline.

Examination of Examples 5.1, 5.2, and 5.3 reveal that a uniform series A was involved with the gradient G. This is often the case in real life problems as an investment, or disbursement, is initially made and then decreased (or increased) a constant amount in succeeding years.

UNIFORM SERIES VALUES OF GRADIENT AMOUNTS (*A/G*)

Many times the problem involves finding an equivalent uniform series A that is equivalent to a gradient series G. The cash flow diagram for this problem is given in Figure 5.8.

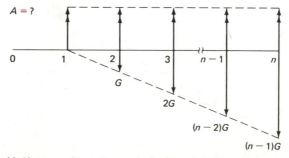

Figure 5.8　Uniform series values equivalent to arithmetic gradient series.

Referring to Equation 5.3,

$$F = \frac{G}{i}\left(\frac{F}{A} - n\right)$$

and dividing both sides by G, then

$$\frac{F}{G} = \frac{1}{i}\left(\frac{F}{A} - n\right) \tag{5.5}$$

Multiplying both sides by A/F yields

$$\frac{A}{F} \times \frac{F}{G} = \frac{1}{i}\left[\frac{A}{F} \times \frac{F}{A} - \frac{A}{F}(n)\right]$$

$$\frac{A}{G} = \frac{1}{i}\left[1 - \frac{A}{F}(n)\right] \tag{5.6}$$

Substituting the equation for A/F (Equation 3.4) yields

$$A = G\left[\frac{1}{i} - \frac{n}{(1+i)^n - 1}\right] \tag{5.7}$$

In functional notation this relationship becomes

$$A = G(A/G, i, n) \tag{5.7A}$$

Values for $(A/G, i, n)$ are tabulated in Appendix A for commonly occurring combinations of i and n.

Graph of A/G

A graph of the equation

$$A/G = \frac{1}{i} - \frac{n}{(1+i)^n - 1}$$

is shown in Figure 5.9. Knowing how A/G values respond to changes in i

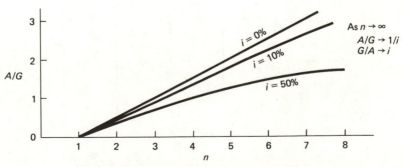

Figure 5.9 Graph of A/G relationship.

and *n* will prove helpful in evaluating the effect on problem solutions of possible variations in *i* and *n*.

Account-Balance Graph of *A* Equivalent to *G*

Figure 5.10 illustrates an *A* series that is equivalent to a *G* series. The graph shows a balance in the account over a five-year period resulting from annual deposits of the *A* series and withdrawals of the *G* series. The *A* payments are considered as a cash flow into the account balance (graph moves upward), and the *G* payments are cash flow out (down on the graph). Interest is paid at the end of each period on whatever balance has been in the account for the duration of that period. Note that the first *A* payment occurs at the end of the first period, whereas the first *G* payment does not occur until the end of the second period. At the end of the final period three payments occur, (1) an interest payment, (2) a payment of *A*, and (3) a *G* payment amounting to $(n-1)G$. Since the *A* series is exactly equivalent to the *G* series the graph ends with a zero balance in the account at the end of the final period.

The examples that follow illustrate the *A/G* relationship. Example 5.4 shows a negative annual income, *A*, coupled with a positive gradient, *G*. The method of solution is to subdivide the problem into its components and solve as usual.

Example 5.4 (*Given A, G, i, n, find equivalent A*)

A bridge authority collects $0.25 per vehicle and deposits the revenues once a month in an interest bearing account. Currently 600,000 vehicles a month cross the bridge and the number is expected to increase at the rate of 3,500 vehicles per month. What equivalent uniform monthly revenue would be generated over the next five years if the interest in the account were (1) 6 percent nominal interest compounded monthly (0.5%/month), and (2) 6 percent compounded annually?

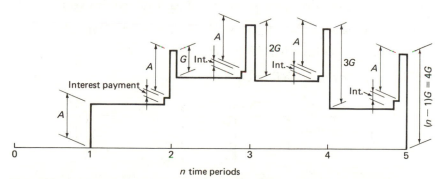

Figure 5.10 Graph of balance in the account for an *A* series equivalent to a *G* series.

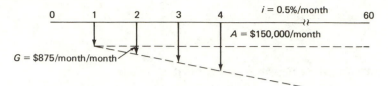

Figure 5.11   Cash flow diagram for Example 5.4, Solution 1.

SOLUTION 1

The monthly costs for the problem in terms of A and G are determined as follows:

$A = 600,000$ vehicles/month $\times \$0.25$/vehicle $= \$150,000$/month

$G = 3,500$ vehicles/month $\times \$0.25$/vehicle $= \$875$/month

$n = 5$ yr $\times 12$ months/yr $= 60$ months

The cash flow diagram is shown in Figure 5.11.

$$A_1 = \$150,000/\text{month} = \$150,000$$

$$A_2 = 875(A/G, 0.5\%, 60) = \underline{\$\ 24,500}$$
$$28.0064$$

$$A = A_1 + A_2 = \underline{\$174,500}$$

SOLUTION 2

Since interest in part (2) is earned only once each year the payments can be grouped together at the end of each year, negating any benefit from depositing at monthly periods (see Chapter 4 for a discussion of this situation). Thus, the base annual income deposited at EOY 1,

$$A_1 = \big[600,000 \text{ vehicles/month} \times 12 + 3,500 \text{ vehicles/month/}$$
$$\text{month} \times (1+2+\cdots+11)\$0.25/\text{vehicle}\big] = \$1,857,750/\text{yr}$$

The gradient income deposited at the end of each subsequent year,

$$G = 3,500 \text{ vehicles/month/month } (1+2+3+\cdots+12)\text{yr}$$
$$\times \$0.25/\text{vehicle} = 68,250/\text{yr/yr}$$

See Figure 5.12 for the cash flow diagram.

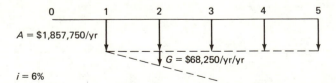
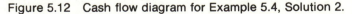

Figure 5.12   Cash flow diagram for Example 5.4, Solution 2.

Solving for the *annual A* yields

$$A_1 = \qquad\qquad \$1,857,750/\text{yr}$$

$$A_2 = \$68,250(A/G,6\%,5) = \underline{\quad 128,656}$$
$$\qquad\qquad\qquad 1.8836$$

annual $A = \qquad\qquad \$1,986,300$

Since interest is earned only at the end of each *annual* period, the monthly equivalent is simply one-twelfth of the annual equivalent. Thus

monthly $A = \$1,986,300/12 = \underline{\underline{\$165,500/\text{month}}}$

(Compare to $A = \$174,500/\text{month}$ in Solution 1.)

This answer is significantly different from Solution 1, even though the interest rate does not appear to be too different. (For Solution 1 the annual $i_e = (1.05)^{12} - 1 = 6.17\%$ versus 6% in Solution 2.) The reason for this difference is the simplifying assumption that receipts that occur between the points in time at which compounding occurs (intermediate funds flow) earn *no* interest until the *next* complete compounding period ends. Thus, the precise terms of interest compounding periods and earnings on intermediate funds flow become very important in actual practice.

Example 5.5 *(Find G, given P, A, i, n)*

An apartment project is estimated to cost $40,000 per apartment. Financing is available at 11 percent for 25 years on the full $40,000 cost. Due to competition in the area, rents must begin at $300 per month but may be raised some every year. Operating costs, taxes, and insurance amount to $100 per month, leaving $200 per month to meet principal and interest on the mortgage payments at the start. What *annual* increase in rents is necessary to pay off the 11 percent, 25-year mortgage, assuming the lender is willing to take sliding scale payments on your terms? To simplify the calculations, use an annual basis instead of monthly.

SOLUTION
The annual mortgage payments to amortize the $40,000 cost are found as

$$A_1 = \$40,000(A/P, 11\%, 25) = \$4,748/\text{yr}$$
$$\qquad\qquad 0.1187$$

The initial rental income available to pay the mortgage is

$$A_2 = \$200 \times 12/\text{month} = 2,400/\text{yr}$$

The equivalent annual payment deficit to be made up by periodic rent increase is

$$A_1 - A_2 = \text{deficit} = \$2,348/\text{yr}$$

The required annual gradient increase in rent is found as

$$G = \$2,348 (G/A, 11\%, 25) = \underline{\underline{\$330.50}}$$

(Equivalent to raising the rent approximately $\$330.50/12 = \$27.54$/month each year from a base of $\$200$/month the first year.)

PRESENT WORTH VALUES OF GRADIENT AMOUNTS (P/G)

This P/G series provides for solution of problems involving a payment (cost or income) that changes (increases or decreases) by $\$G$ per period for n periods. Interest is earned on any balance that remains in the loan or deposit. The equation is derived by finding the present sum required to finance this G series for n periods (given G, i, n, find P).

Derivation of the P/G Equivalence Factor

It was previously found in Equation 5.3 that

$$\frac{F}{G} = \frac{1}{i}\left(\frac{F}{A} - n\right)$$

Both sides may be multiplied by P/F as follows:

$$\frac{F}{G} \times \frac{P}{F} = \frac{1}{i}\left[\frac{F}{A} \times \frac{P}{F} - n\left(\frac{P}{F}\right)\right]$$

The F's cancel out to provide the general equation relating P, G, i, and n.

$$\frac{P}{G} = \frac{1}{i}\left[\frac{P}{A} - n\left(\frac{P}{F}\right)\right] \quad \text{or} \quad P/G = \frac{1}{i}\left[\frac{(1+i)^n - 1}{i(1+i)^n} - \frac{n}{(1+i)^n}\right] \quad (5.8)$$

or, in notation form

$$P = G(P/G, i, n) \tag{5.8A}$$

Values for P/G are listed in Appendix A.

Graph of P/G

A graph of the equation

$$P/G = \frac{1}{i}\left[\frac{(1+i)^n - 1}{i(1+i)^n} - \frac{n}{(1+i)^n}\right]$$

is shown in Figure 5.13. Figure 5.14 illustrates a graph of the balance of investment P with $(n-1)$ repayments periodically increasing by an amount G. The next example (5.6) illustrates an application of the P/G equation in finding a present value, P, given a gradient, G.

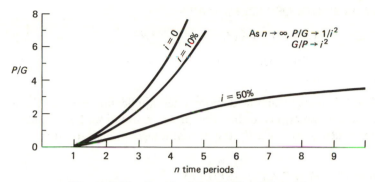

Figure 5.13   Graph of the *P/G* relationship.

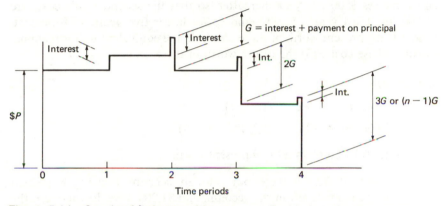

Figure 5.14   Graph of balance of investment *P* with (*n*−1) repayments periodically increasing by amount *G*.

Example 5.6   (*Find P, given G, i, n*)

The repair costs on a water works pump are expected to be zero the first year and increase $100 per year over the five-year life of the pump. A fund bearing 6 percent on the remaining balance is set up in advance to pay for these costs. How much should be in the fund? Costs are billed to the fund at the end of each year.

SOLUTION BY EQUATION

$$P = \frac{G}{i} \left[\frac{(1+i)^n - 1}{i(1+i)^n} - \frac{n}{(1+i)^n} \right]$$

$$P = \frac{100}{0.06} \left[\frac{(1+0.06)^5 - 1}{0.06(1+0.06)^5} - \frac{5}{(1+0.06)^5} \right]$$

$$P = 100 \times 7.935 = \underline{\underline{\$793.50}}$$

SOLUTION BY USE OF TABLES IN APPENDIX A

$$P = G(P/G, 6\%, 5) = 100(7.9345) = \$793.40$$

The slight difference in the answers is due to rounding off of the factors.

The following example demonstrates applications of all three arithmetic gradient equations to the same problem situation. Each application is then explained to help clarify the mental image of each solution.

Example 5.7 (*Find P, A, F, given G, i, n*)

To operate, maintain, and repair a motor grader costs $1,000 the first year and increases $500 per year thereafter, so that the second year's costs are $1,500, the third year's, $2,000, and so forth, for five years. $i = 8$ percent. Find: (1) the present worth of these costs; (2) the equivalent uniform annual worth of these costs; (3) the future worth of these costs.

SOLUTION 1

$$P = A(P/A, i, n) + G(P/G, i, n)$$
$$P = 1,000(P/A, 8\%, 5) + 500(P/G, 8\%, 5)$$

$$P = \$3,993 + \$3,686 = \$7,679 \text{ present worth}$$

Comment: If $7,679 were put into an account drawing 8 percent interest on the balance left in the account, and $1,000 were drawn out at the end of the first year to pay for operating and maintenance, $1,500 the second year, and so forth, the account would be drawn down to zero at the end of the fifth year.

SOLUTION 2

$$A = A + G(A/G, i, n)$$
$$A = \$1,000 + 500(1.8465)$$

$$A = \$1,000 + 923 = \$1,923/\text{yr annual worth}$$

Comment: If $1,923 were paid into an account drawing 8 percent interest every year, $1,000 could be drawn out the first year, $1,500 the second year, and so forth, and the fifth year the account balance would be zero.

SOLUTION 3

$$F = A(F/A, i, n) + G(F/G, i, n)$$
$$F = \$1,000(5.8666) + 500(10.832)$$

$$F = \$5,867 + 5,416 = \$11,283 = \$11,280 \text{ future worth}$$

Comment. If $1,000 were borrowed at 8 percent at the end of the first year to pay for operating and maintenance expenses, $1,500 borrowed for the second year, and so forth, at the end of the fifth year, the amount owed would be $11,280.

NONCONTINUOUS SERIES

The arithmetic gradient series begins with an income or cost at the end of the second period after *ETZ* as previously noted. If the income or cost begins at some other point in time, an additional calculation must usually be made to find the answer in terms of present worth or annual worth, as illustrated by Example 5.8.

Example 5.8 (*ETZ≠PTZ, find P, given G, i, n*)

A county engineer is thinking of lining a certain drainage ditch in order to save on maintenance costs. She anticipates the ditch lining will save the following amounts over its 11-year life. With $i = 12$ percent, find the present worth of the savings.

EOY	INCOME
1 through 4	$ 0
5	1,000
6	2,000
7	3,000
8	4,000
9	5,000
10	6,000
11	7,000

Assume there is no salvage value and that after the eleventh year the ditch lining will be worthless.

SOLUTION
To help clarify the problem a cash flow diagram is sketched as shown in Figure 5.15.

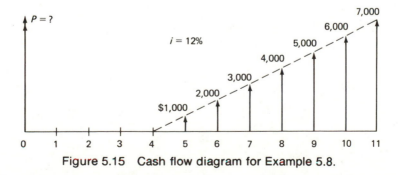

Figure 5.15   Cash flow diagram for Example 5.8.

Note: The arithmetic gradient series always has its zero point at two periods before the first payment occurs (see derivation of equation for proof). In this problem the first payment occurs at the end of year 5, so equation time zero is the end of year 3. To find n for the gradient series, subtract the first 3 years from the total time of 11 years and find $n=8$. Another way of finding the same n is to remember that $n=$ the number of gradient payments $+1$. Here the number of payments is 7, so $n=7+1=8$. Then let P_3 represent the present worth of the gradient series at equation time zero (end of year 3) at the beginning of the gradient series. Now P_3 is also the future value, F_3, of the investment after three years. Therefore,

$$P_3=1,000(P/G,12\%,8)$$

$$F_3=P_3$$

$$P=F_3(P/F,12\%,3)=1,000(P/G,12\%,8)(P/F,12\%,3)$$

$$P=(1,000)(14.471)(0.7118)=\underline{\underline{\$10,300}}$$

PERPETUAL LIFE WITH GRADIENT INCREASES

Some perpetual life problems involve a gradient increase in cost. For instance, assume the maintenance costs on each mile of right-of-way increase by $100 per year over a perpetual life. To find the present worth of this arithmetic gradient, simply solve Equation 5.8, letting n approach infinity.

$$P/G=\frac{1}{i}\left[\frac{(1+i)^n-1}{i(1+i)^n}-\frac{n}{(1+i)^n}\right]$$

Consider the elements of the equation one at a time. As n approaches infinity both the numerator and the denominator of both parts of the equation become very large. To analyze, use L'Hospital's rule, as n approaches infinity, $(1+i)^n$ becomes very large and also approaches infinity.

As $n\to\infty$

$$\frac{n}{(1+i)^n}=\frac{d(n)/dn}{d(1+i)^n/dn}=\frac{1}{n(1+i)^{n-1}}=0$$

Therefore

$$\frac{n}{(1+i)^n}\to0 \quad\text{as}\quad n\to\infty$$

and

$$\frac{(1+i)^n-1}{(1+i)^n}\to1 \quad\text{as}\quad n\to\infty$$

so that

$$P/G\to\frac{1}{i}\left(\frac{1}{i}-0\right) \quad\text{as}\quad n\to\infty,$$

or, for $n=\infty$,

$$P/G=\frac{1}{i^2} \tag{5.9}$$

Applying this solution to the foregoing right-of-way problem (assuming $i=6\%$), the present worth of the perpetually increasing maintenance cost of $100 per year is found as

$$P=\frac{G}{i^2}=\frac{\$100}{0.06^2}=\$27,778$$

In other words, if a trust fund of $27,778 earns 6 percent interest, then withdrawals of $100 may be commenced at EOY 2 and increased by $100 per year each year forever.

Perpetual Life for Other Time-Value Equations

As $n\to\infty$, the other time-value equations are evaluated in a manner similar to the evaluations in the preceding paragraph. The results are as follows:

$A/F=0$

$A/P=i$

$A/G=1/i$

$P/F=0$

$P/A=1/i$

$P/G=1/i^2$

$F/P=\infty$

$F/A=\infty$

$F/G=\infty$

SUMMARY

In this chapter the equivalence factors equating the arithmetic gradient G to future worth F, annual worth A, and present worth P, were developed. The equations for relating G to A, F and P are

$$A=G\left[\frac{1}{i}-\frac{n}{(1+i)^n-1}\right]=G(A/G,i,n)$$

122,080

$$F=\frac{G}{i}\left[\frac{(1+i)^n-1}{i}-n\right]=G(F/G,i,n)$$

$$P=\frac{G}{i}\left[\frac{(1+i)^n-1}{i(1+i)^n}-\frac{n}{(1+i)^n}\right]=G(P/G,i,n)$$

As $n\to\infty$ then $A/G\to1/i$, and $P/G\to1/i^2$.

PROBLEMS FOR CHAPTER 5, ARITHMETIC GRADIENT

Note: Assume all costs and income are incurred at the end of the year designated, unless otherwise specified in the problem.

Problem Group A

Given three variables, find the fourth.

A.1 Five years from now, how much will there be in an account that receives the deposits listed below, and bears interest at a rate of 6 percent compounded annually on the balance on deposit?

EOY	DEPOSIT
1	$ 0
2	300
3	600
4	900
5	1,200

A.2 What is the equal annual equivalent to the following series of costs, if $i=7$ percent?

EOY	COST
1	$ 0
2	1,000
3	2,000
4	3,000

A.3 What is the present worth of the following series of income payments, if $i= 9$ percent?

EOY	INCOME
1	$ 0
2	2,500
3	5,000
4	7,500
5	10,000
6	12,500

Problem Group B

Find i (or whether i is greater than or less than given).

B.1 A five-year franchise for a certain patented prefabricated steel building is available in your area. Since the system is new, sales are expected to be slow at first and pick up in future years. The estimated profit at the end of each year is shown in the table. An investor friend of yours says that he will buy the franchise if he can make a 20 percent return on this speculative enterprise.

Can he make 20 percent on his invested capital if the selling price for the franchise is $50,000, with no salvage value at the end of five years?

EOY	NET PROFIT
1	$ 0
2	10,000
3	20,000
4	30,000
5	40,000

B.2 A piece of property is available for $120,000. The property can be subdivided and lots sold. The net return after expenses is projected as follows:

EOY	NET INCOME
1	$ 0
2	10,000
3	20,000
4	30,000
5	40,000
6	50,000
7	60,000
8	70,000
9	80,000

According to the schedule, by the end of the ninth year, all the lots will be sold and the investment will have no further income or value. Find the equivalent interest rate of return on the $120,000 investment if all goes according to the schedule.

Problem Group C

Find *n*.

C.1 A fund of $100,000 was set up, earning 8 percent on the balance, to pay for maintenance of a facility at the rate of $14,903 per year over a ten-year period. The first year's maintenance did cost $14,903 per year, but maintenance the second year rose to $15,903 per year. Due to inflation, it is now estimated the costs will rise by $1,000 per year each year. If the costs continue to increase by $1,000 per year, how long will the fund last?

Problem Group D

These may require two or more equations to find *A*, *P*, comparisons, rates.

D.1 The operating cost for a proposed waste water treatment plant is estimated as $500,000 per year, increasing by $100,000 each year after the first. Assume $i = 6$ percent. Find:
a. Equivalent annual cost of all operating costs, for a life of 25 years.
b. Present worth of all operating costs, for a life of 25 years.

c. Equivalent annual cost of all operating costs, for a perpetual life.

d. Present worth of all operating costs, for a perpetual life.

D.2 The plant in Problem D.1 is estimated to cost $10,000,000 to build. After 25 years it will have a $2,000,000 salvage value. Assume the same $n=25$ and $i=6$ percent, and that the money to construct is borrowed. Find:

a. The total equivalent annual cost of (i) P&I payments, plus (ii) annual operating costs, plus (iii) salvage value.

b. This plant will serve 25,000 housing units. What annual charges per year per unit is required if the sewer rate remains constant for 25 years?

c. What annual rate per housing unit is required to cover only the first year's operating costs plus P&I payment on the capital cost (deduct annual equivalent to salvage value for P&I payment)?

d. What annual increment G is necessary per housing unit to cover the increasing operating costs?

e. If it is necessary because of competition to begin with an annual rate of $36 per unit per year, what annual increment, G, is necessary to reach the breakeven point (on operating costs plus P&I, plus salvage) over the period of 25 years?

D.3 A friend of yours, just turned 23, wants to become a millionaire by the time she is 63. She plans to make regular deposits into an investment account that she hopes will continue to bear interest compounded at 10 percent. Whatever her initial investment deposit, she hopes to increase the amount deposited by $200 per year throughout the 40-year period. She asks you to calculate how much her initial investment should be on her twenty-fourth birthday.

D.4 An engineering firm wishes to establish a trust fund to provide one engineering scholarship every year for a deserving student. The amount of the first scholarship to be awarded at the end of the first year is $1,000. However, due to the historical trend of generally rising costs, it seems wise to provide for an increase in the award of $100 per year. Therefore the award will be $1,100 at the end of the second year, $1,200 at the end of the third year, and so on. The trust fund bears interest at 7 percent on the unspent balance. The firm asks you to find how much would be needed to set up a trust fund for each of the following options.

a. To provide for annual awards for the next 20 years.

b. To provide annual awards forever ($n \rightarrow \infty$).

c. Assuming the amount needed in part (a) is available to set up the trust fund, and that this amount could be invested at 9 instead of 7 percent, how much could the initial award be raised to, assuming the gradient is still $100 and $n=20$?

D.5 An intersection now has a traffic count of 10,000 cars daily, expected to increase by 1,000 cars per day each year. A new traffic signal system for the intersection is estimated to cost $50,000 and will save each car passing through an average of $0.01 worth of fuel, wear, and time. With $i=6$ percent and a ten-year life, compare the present worth of the savings versus the cost of the proposed system. (For simplicity, assume the traffic count increases and savings to motorists occur at EOY each year.)

D.6 Your client, a toll turnpike authority, is considering the feasibility of a 100-mile extension to the present turnpike at a cost of $2,000,000 per mile. The traffic is estimated to average 25,000 cars per day (365 days/yr) for the first year. Each year thereafter the traffic is expected to increase by an additional 2,500 cars per day for the full 20-year life of the project. (Use $n=20$.) The money to build the road can be borrowed on road bonds at 7 percent. Sixty percent of the toll receipts will be needed to pay operating and maintenance costs and the remaining 40 percent will be available to retire the bonds. The authority asks you to determine how much toll must be charged each car in terms of dollars per mile in order to pay O&M costs as well as retire the bonds.

D.7 Your firm is in need of a new dragline. Two competing brands are under consideration. All other factors have been checked and balance out, except the estimated O&M costs, which are listed below. You are asked to determine which of the following two schedules of costs is the lesser cost ($i=9\%$).

EOY	O&M COSTS BRAND A	O&M COSTS BRAND B
1	$40,000	$62,500
2	45,000	62,500
3	50,000	62,500
4	55,000	62,500
etc.	etc.	etc.
10	85,000	62,500

D.8 A manufacturing firm selling a new type of form tie is expected to ose $4,000 the first year, lose $2,000 the second year, and just break even the third year. Every year thereafter the firm is expected to earn $2,000 more than the previous year for the life of the manufacturing equipment (ten years). For $i=15$ percent, find the present value of the total earnings over the ten-year period.

D.9 The cost for a handheld calculator is $139 this year. Due to technological advances, the costs are expected to decrease by $14 each year for the next five years. An engineering firm estimates it will purchase eight calculators each year for the next five years. If $i=15$ percent, what is the present value of the total purchase of calculators? (Assume each group of calculators are purchased at the end of each year.)

D.10 A contractor is considering the purchase of a new concrete pumping machine. He estimates the machine will save $5,000 the first year over the costs of placing concrete by his present method. He estimates savings should increase by $500 per year every year thereafter. The estimated life of the pump is nine years with negligible salvage value at that time. The contractor aims at a 15 percent return on all invested capital. What is the most he can afford to pay for the concrete pump and still earn his 15 percent?

Problem Group E

$ETZ \neq PTZ$.

E.1 A graduating class of 100 engineers decide they want to present a gift of $500,000 to the university at their twenty-fifth class reunion. They set up a trust fund that earns 8 percent per year compounded. They plan to start with a small donation of G dollars to the fund on the first anniversary of their graduation and increase the amount by regular increments every year $(2G, 3G, \ldots, nG)$. If they put nothing into the fund at the time of their graduation, but continue increasing their donations by a regular amount, G, each year, find how much G amounts to in terms of the donation each classmate should give on his or her first anniversary.

Chapter 6
Geometric Gradient, The Constant Percentage Increment

KEY EXPRESSIONS IN THIS CHAPTER

C = First end-of-period payment of an incremental increasing series, base payment.

r = Rate of growth (or decline) of the series. [For example, a growth rate of 15% is shown as $r=0.15$, a decline rate (decay rate or negative growth) of 3% is shown as $r= -0.03$.]

w = Derived quantity as shown below.

i, n, P, F, and A all have the same meanings as in previous chapters.

A PAYMENT SERIES CHANGING BY A CONSTANT PERCENTAGE

A geometric gradient series is a series of end-of-year payments with each payment increasing (or decreasing) by a fixed percentage.

Change has become a normal feature of many cost analysis problems encountered by practicing engineers. Costs change and incomes change along with a great array of changes in other variables. The two basic tools available for evaluating the effects of these changes are the two sets of

gradient equations, (a) the arithmetic gradient discussed in Chapter 5, and
(b) the geometric gradient developed in this chapter. The arithmetic gradient
in Chapter 5 dealt with constant gradient amounts, G, such as an annual
increase in the cost of maintenance of $1,000 per year. The gradient G does
not include any annual base amount A, which is accounted for separately.
This chapter will deal with geometric increments, where the percentage
increase is constant but the numerical amount of increase varies, such as an
annual increase in the cost of maintenance of 10 percent per year, or
increase in traffic count of 8 percent per year. The base amount C will be
included in the geometric gradient instead of excluded as in the arithmetic
gradient. As an example of a geometric gradient, if maintenance costs for a
particular project are $10,000 per year the first year and increase 10 percent
per year, the costs for each year are:

END OF YEAR	MAINTENANCE COST INCREASING 10% EACH YEAR	MAINTENANCE COST EACH YEAR IN TERMS OF BASE YEAR COST C (C IN THIS CASE IS $10,000)	MAINTENANCE COST EACH YEAR IN TERMS OF RATE OF GROWTH r (r IN THIS CASE IS 0.10)	
1	10,000	$C \times 1$	C	$= C(1+r)^0$
2	11,000	$C \times 1.1$	$C + Cr$	$= C(1+r)^1$
3	12,100	$C \times 1.21$	$C(1+r) \times (1+r) = C(1+r)^2$	
4	13,310	$C \times 1.331$		$= C(1+r)^3$
etc.	etc.	etc.	etc.	etc.
n	$10^4 \times (1.1)^{n-1}$	$C \times (1.1)^{n-1}$		$= C(1+r)^{n-1}$

Graphic Representation

Figure 6.1 shows a graph of the balance in the account showing the present
worth of a series of geometrically increasing annual payments that start at
first year level C and increase by r each period for n periods. The payment
for each subsequent period is $(1+r)$ times the payment for the previous
period. Notice that the first payment occurs at the end of the first period, as
in the A series, rather than at the end of the second period as in the G series.

Derivation of Equations for the Present Worth

Following is the derivation of an equation to determine the present worth P
of the series of costs, with an initial cost C, and increasing at a constant rate
r (or decreasing at a constant rate, $-r$) for n periods with interest at i on
the balance remaining in the fund.

The cash flow diagram for this situation is given in Figure 6.2. In the
extreme case where $n=1$, then only one payment C is made at the end of
the first period. In this case C is a single future sum and its present worth

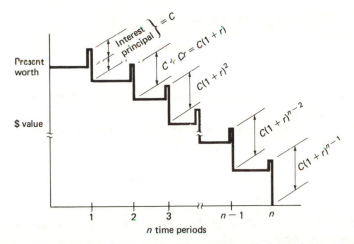

Figure 6.1   A graph of the balance-in-the-account with initial balance, P, and a series of geometrically increasing payments, beginning with an initial payment, C.

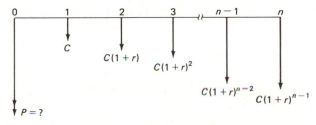

Figure 6.2   Cash flow diagram for geometric gradient.

may be found by

$$P = \frac{F}{(1+i)^n} = \frac{C}{1+i}$$

If $n=2$, the present worth is the sum of the two individual future sums

$$P = \frac{F_1}{(1+i)^{n-1}} + \frac{F_2}{(1+i)^n} = \frac{F_1}{(1+i)^1} + \frac{F_2}{(1+i)^2}$$

The two successive payments, F_1 and F_2, can be replaced respectively by C and $C(1+r)$. Then the same present worth, P, is now expressed in terms of the initial payment C and the growth rate r.

$$P = \frac{C}{1+i} + \frac{C(1+r)}{(1+i)^2}$$

With this example of $n=2$ in mind, the series logically may be expanded to accommodate larger values of n, as follows.

$$P=C\left[\frac{1}{1+i}+\frac{1+r}{(1+i)^2}+\cdots+\frac{(1+r)^{n-1}}{(1+i)^n}\right]$$

To condense the equation into a more manageable form, the following procedures are employed.

Multiply the right hand side by $(1+r)/(1+r)$ and obtain

$$P=\frac{C}{1+r}\left[\frac{1+r}{1+i}+\frac{(1+r)^2}{(1+i)^2}+\cdots+\frac{(1+r)^n}{(1+i)^n}\right] \tag{6.1}$$

• **Case 1.** $r>i$. When $r>i$, let $1+w=(1+r)/(1+i)$ and substitute into Equation 6.1 to obtain

$$P=\frac{C}{1+r}\left[(1+w)+(1+w)^2+\cdots+(1+w)^n\right]$$

The term $(1+w)$ may be extracted from the brackets, as

$$P=\frac{C}{1+r}(1+w)\left[1+(1+w)+\cdots+(1+w)^{n-1}\right]$$

Then, multiply both sides by $(1+i)/w$, subtract the new equation from the old, and multiply both sides by w/i to find the final standard form relating P, C, i, and n.

$$P=\frac{C}{1+i}\left[\frac{(1+w)^n-1}{w}\right] \quad \text{for} \quad r>i, \quad \text{and} \quad w=\frac{1+r}{1+i}-1 \tag{6.2}$$

As a further convenience, Chapter 3 showed that

$$F/A=\frac{(1+i)^n-1}{i}$$

and that the right-hand side may be designated by $(F/A, i, n)$.

Therefore, using w in place of i to indicate a derived equivalent,

$$P=\frac{C}{1+i}(F/A, w, n) \quad \text{for} \quad r>i \tag{6.2A}$$

where

$$w=\frac{1+r}{1+i}-1, \quad \text{for} \quad r>i$$

Case 2. $r<i$. When $r<i$, let

$$\frac{1}{1+w}=\frac{1+r}{1+i}$$

Then substituting $1/(1+w)$ for $(1+r)/(1+i)$ in Equation 6.1 yields

$$P = \frac{C}{1+r}\left[\frac{1}{1+w} + \frac{1}{(1+w)^2} + \cdots + \frac{1}{(1+w)^n}\right]$$

Since

$$\left[\frac{1}{1+w} + \frac{1}{(1+w)^2} + \cdots + \frac{1}{(1+w)^n}\right] = \frac{(1+w)^n - 1}{w(1+w)^n}$$

Then

$$P = \frac{C}{1+r}\left[\frac{(1+w)^n - 1}{w(1+w)^n}\right] \quad \text{when} \quad r<i \quad \text{and} \quad \frac{1+r}{1+i} = \frac{1}{1+w} \qquad (6.3)$$

Chapter 4 showed that

$$P/A = \frac{(1+i)^n - 1}{i(1+i)^n}$$

The right-hand side can be designated by $(P/A, i, n)$. Therefore, substituting w in place of i yields

$$P = \frac{C}{1+r}(P/A, w, n) \quad \text{for} \quad r<i \qquad (6.3A)$$

where

$$w = \frac{1+i}{1+r} - 1, \quad \text{for} \quad r<i$$

· *Case 3.* $r=i$. If $r=i$, a simplified equation results. Examining Equation 6.1, if $r=i$, then

$$P = \frac{C}{1+r}(1^1 + 1^2 + \cdots + 1^{n-1} + 1^n)$$

or

$$P = \frac{Cn}{1+r} \quad \text{for} \quad r=i \qquad (6.4)$$

Applications of Present Worth of a Geometric Gradient

The simplest problem situation occurs where $r=i$, as in the following example.

Example 6.1

Assume your client is selling some land to a developer who is short of cash and offers to pay in three annual installments with each installment 10 percent greater than the previous one. He offers $60,000 at the end of year

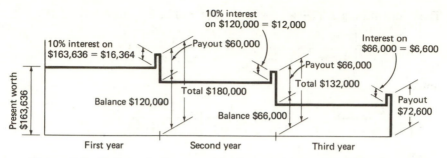

Figure 6.3   Balance-in-the-account diagram illustrating Example 6.1.

1, $66,000 at the end of year 2, and $72,600 at the end of year 3. Your client wants to know what this is equivalent to in terms of cash now (present worth). He borrows and lends at 10 percent.

SOLUTION
Setting up the problem,

$$r=i=10\%$$
$$C=\$60,000$$
$$n=3$$

Since $r=i$, use Equation 6.4

$$P=\frac{Cn}{1+r}=\frac{\$60,000\times 3}{1.10}=\underline{\underline{\$163,636}}$$

The present worth of the three payments is $163,636. The present sum of $163,636 cash could be deposited now at 10 percent and the series of payments made as shown. Therefore the present worth (cash equivalent) of the three payments shown is $163,636. The equivalence can be represented graphically as shown in Figure 6.3.

FINDING THE FUTURE SUM OR EQUIVALENT ANNUAL PAYMENTS OF A GEOMETRIC GRADIENT SERIES

The future or annual worth of the geometric gradient can be determined by first finding the present worth, and then applying the F/P or A/P factor using i (not r). The following is a continuation of Example 6.1, which illustrates the method.

Example 6.1a

Assume your client agrees to the proposed sale outlined in Example 6.1 but has another question. He does not need the cash right now but will at the end of three years. If he deposits each of the three payments as they are received and gets interest at 10 percent on the balance, how much will be in the account at the end of the third year?

SOLUTION

The future sum can be obtained directly from the present worth by the following relationship:

$$F = P(F/P, i, n)$$

$$F = \$163,636 \underset{1.331}{(F/P, 10\%, 3)} = \$217,800$$

There would be $217,800 in the account at the end of three years.

The relationships may be portrayed graphically in two ways:

1. Assume the equivalent cash is on deposit now, and will receive interest at the rate of 10 percent per year, as shown in Figure 6.4.
2. Assume the payments will be deposited as made and earn interest at 10 percent from the date of deposit, as shown in Figure 6.5.

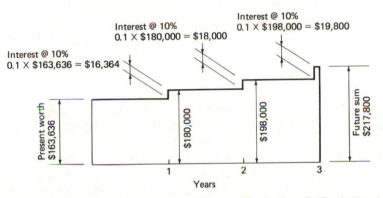

Figure 6.4   Balance-in-the-account diagram illustrating P/F relationship.

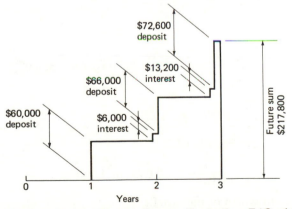

Figure 6.5   Balance-in-the-account diagram illustrating F/C relationship of Example 6.1a.

RATE OF GROWTH IS MORE OR LESS THAN INTEREST RATE

The next two examples illustrate solutions to problems involving rates of growth that are not the same as the interest rate.

Example 6.2   (*Find P, given C, r, i, n*)

A school crossing guard currently costs the school district $6,000 per year which includes salary and fringe benefits. The costs for the guard are expected to increase at the rate of 5 percent per year. If interest is 6 percent, find the present worth of the cost of the guard over a ten-year period.

SOLUTION
Since $r < i$, use Equation 6.3.

$$w = \frac{1+i}{1+r} - 1 = \frac{1.06}{1.05} - 1 = 0.009524 = 0.9524\%$$

$$P = \frac{C}{1+r}\left[\frac{(1+w)^n - 1}{w(1+w)^n}\right] = \frac{6,000}{1.05}\left[\frac{1.009524^{10} - 1}{0.00952 \times 1.00952^{10}}\right] = \$54,262$$

Example 6.3   (*Find P, given C, r, i, n*)

A traffic intersection improvement is under study. The improvement is estimated to cost $100,000. The benefit to the motoring public is estimated at an average saving of 0.5¢ per vehicle if the improvement is made. The current traffic count averages 8,000 cars per day for 365 days per year and is expected to increase 10 percent per year over the ten-year anticipated life of the installation. Money to finance the improvement comes from road bond funds bearing an interest rate of 5.77 percent. What is the present worth of the benefits to the public?

SOLUTION
Identify the variables.

   benefit, C = 8,000 vehicles/day × 365 days/yr × $0.005/vehicle

      = $14,600/yr

   $r = 10\%$

   $i = 5.77\%$

Since $r > i$, use Equation 6.2.

$$w = \frac{1+r}{1+i} - 1 = \frac{1.10}{1.0577} - 1 = 0.0400 \text{ or } 4.000\%$$

then the present worth of the benefit is

$$P = \frac{C}{1+i}\left[\frac{(1+w)^n - 1}{w}\right] = \frac{14,600}{1.06}\left(\frac{1.0400^{10} - 1}{0.0400}\right) = \$165,300$$

DECLINING GRADIENT ($r<0$)

Sometimes a decline in net income will occur, due to declining productivity, increasing competition, or costs increasing more rapidly than income. The following example illustrates a situation with $r<0$.

Example 6.4 (*Find C, given P, F, i, r*)

Due to increasing age and down time the productivity of a contractor's excavator is expected to decline with each passing year. Using the data below, calculate the price in terms of dollars per cubic yard that the contractor must charge to cover the cost of buying and selling the excavator.

$$\text{cost new} = \$50,000$$
$$\text{resale price @ EOY } 6 = 20,000$$
$$\text{contractor borrows @ } i = 10\%$$
$$\text{production declines @} r = -6\%$$
$$\text{production for year } 1 = 200,000 \text{ yd}^3/\text{yr}$$

Assume all funds are credited at EOY.

SOLUTION

$$r = -0.06$$
$$i = +0.10$$

Since $r<i$, then use Equation 6.3. Therefore

$$w = \frac{1+i}{1+r} - 1 = \frac{1.10}{0.94} - 1 = 0.1702$$

Then

$$P = \frac{C}{1+r}\left[\frac{(1+w)^n - 1}{w(1+w)^n}\right] = \frac{C}{0.94}\left[\frac{1.1702^6 - 1}{0.1702 \times 1.1702^6}\right] = 3.8164C$$

The value P represents the cost new, less the present worth of the resale value, or,

$$P = -\$50,000 + \$20,000 \, (P/F, 10\%, 6) = -\$38,710$$
$$0.5645$$

Solving for C yields

$$P = 3.8164C$$
$$-\$38,710 = 3.8164C$$
$$C = \$10,143/\text{yr} \quad \text{for the first year}$$

Since 200,000 yd^3 are excavated the first year, the price to cover the cost of buying and selling the excavator is calculated as

$$\$10,143/200,000 \text{ yd}^3 = \underline{\$0.051/\text{yd}^3}$$

The cash flow diagram for these items is shown in Figure 6.6.

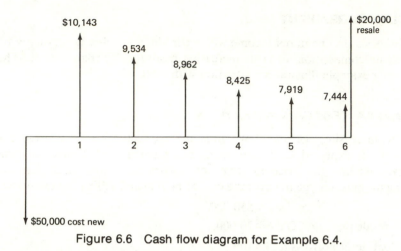

Figure 6.6 Cash flow diagram for Example 6.4.

Geometric Gradient Passing Through Zero

Ordinarily the geometric gradient never passes through zero since any number increased or decreased by an exponential factor never can quite reach zero.

An exception can occur when a constant is added or subtracted to the gradient. The following example illustrates this type of problem.

Example 6.5 (*Geometric gradient passing through zero. Find P, r, C, given the cash flow series, n, i*)

Assume the income for a certain project increases as a geometric gradient, but the costs remain 'constant. As a result the net income (shown below as income minus cost) could begin as a negative value dominated by high costs, but subsequently rise through zero into the positive range as income improves. Find the present worth of the net income, with $i = 10$ percent.

YEAR	NET INCOME IN $1,000's (INCOME—COSTS)
1	− $50
2	− 35
3	− 17.75
4	+ 2.09
5	+ 24.90
⋮	⋮
25	+ 2712.52

SOLUTION

Since it is known that this growth is geometric (or a reasonable approximation thereto) but the base amount of cost, income, and growth rate are unknown, the constants r and C may be found as follows.

1. List the net income for the first three years.
2. Find the change in net income for the years 1 to 2, and from 2 to 3 and list these under Δ net as shown.
3. Find the change in the change and list under $\Delta\Delta$ net as shown.
4. The growth rate $r = \Delta\Delta/\Delta$.
5. The first value, C in the geometric gradient series is found as $C = \Delta/r$. In this case C is the first year's income.

YEAR	NET INCOME	Δ NET	$\Delta\Delta$ NET
1	$-\$50$		
		15	
2	-35		2.25
		17.25	
3	-17.75		

Then $r = \Delta\Delta/\Delta = 2.25/15 = 0.15$ and $C = \Delta/r = 15/0.15 = 100$.
From this information the income and cost columns can be reconstructed as

YEAR	INCOME ($1,000's)	COST ($1,000's)	NET INCOME ($1,000's)
1	$100	$-\$150$	$-$ $50
2	115	-150	$-$ 35
3	132.25	-150	$-$ 17.75
4	152.09	-150	$+$ 2.09
5	174.90	-150	$+$ 24.90
$\vdots$	$\vdots$	$\vdots$	$\vdots$
25	$100 \times 1.15^{n-1}$	-150	$+2712.52$

The next step is to find the present worth of the net income with $i = 10$ percent. This is done by adding the present worth of the income, P_1, to the present worth of the cost, P_2. The sum of $P_1 + P_2$ is the present worth of the net income for the project.

Since $r > i$ then use Equation 6.2.

$$w = \frac{1+r}{1+i} - 1 = \frac{1.15}{1.10} - 1 = 0.04545$$

and

$$P_1 = \frac{C}{1+i}\left[\frac{(1+w)^n - 1}{w}\right] = \frac{100}{1.10}\left(\frac{1.04545^{25} - 1}{0.04545}\right) = +\$4,076$$

$$P_2 = -150\,(P/A, 10, 25) = - \underset{9.077}{1,362}$$

$$+\$2,714$$

At EOY 0 the income from the project is worth $P_1 + P_2 = \underline{\$2,714,000}$.

SUMMARY

In this chapter the use of the geometric gradient is described. The geometric gradient is a series of gradient payments that increase (or decrease) in sequence by a fixed percentage from the first end-of-period payment C.

The equations for geometric gradient are:

1. When $r > i$, then

$$w = \frac{1+r}{1+i} - 1 \qquad \text{as } n \to \infty$$

$$P = \frac{C}{1+i}\left[\frac{(1+w)^n - 1}{w}\right] = \frac{C}{1+i}(F/A, w, n) \qquad P/C \to \infty$$

2. When $r < i$, then

$$w = \frac{1+i}{1+r} - 1$$

$$P = \frac{C}{1+r}\left[\frac{(1+w)^n - 1}{w(1+w)^n}\right] = \frac{C}{1+r}(P/A, w, n) \quad P/C \to \frac{1}{(1+r)w}$$

3. When $r = i$, then

$$P = \frac{Cn}{(1+r)} \qquad P/C \to \infty$$

The future worth or annual worth of the geometric gradient can easily be determined by first finding the present worth through the use of the above equations and then applying the appropriate F/P or A/P factor using i (not r).

PROBLEMS FOR CHAPTER 6, GEOMETRIC GRADIENT

Note: Assume all payments are made at the end of the year unless otherwise noted in the problem.

Problem Group A

Given C, i, r, n, find P using basic equation.

A.1 Find the present worth of the geometric gradient series shown in the table below, using the geometric gradient equations with $i = 6$ percent.

EOY	MAINTENANCE COSTS INCREASING 10%/YR
1	$10,000
2	11,000
3	12,100
4	13,310
5	14,641

A.2 A contractor is considering purchase of a new concrete pumping machine. She estimates the machine will save $5,000 the first year over costs of placing concrete by the present method. She estimates savings should increase by 10 percent per year, every year thereafter. The estimated life of the pump is nine years with negligible salvage value at that time. The contractor aims for a 15 percent return on all invested capital. What is the most she can afford to pay for the concrete pump and still earn her 15 percent?

A.3 A municipal power plant is expected to experience an increase in net profit of 10 percent per year for the next 20 years. The base net profit this year is $1,000,000. Interest is 6 percent on borrowed funds.
   a. What is the present worth of this income?
   b. If current long-term debt now stands at $10,000,000, how much additional borrowing could be secured by the estimated present worth of the net profit if lenders will lend on 100 percent of that value?

Problem Group B

Find P, A, or F involving two or more steps.

B.1 Annual maintenance costs on a certain road are $2,000 per mile this year and are expected to increase 4 percent per year. $i = 7$ percent.
   a. What is the present worth of the maintenance costs for the next five years (including the cost incurred at the end of the fifth year) assuming that all costs are billed at the end of the year?
   b. What is the annual equivalent (equivalent equal annual costs) of these increasing costs?

B.2 A student scheduled to graduate at age 23 is making plans to retire by age 48. He plans to begin by investing $600 per year ($50/month) the first year, an additional $660 the second year, and increase the annual amount added to the investment by 10 percent every year to $726 the third year, and so on. He hopes by careful management to average 10 percent return compounded on reasonable investments.
   a. How much would he accumulate by age 48 (25 years)?
   b. If he could raise his average interest rate of return on the investment to 12 percent, what would the amount be by age 48?
   c. How much has the annual amount required to be invested per year grown in 25 years?

B.3 Your client, the Metro Solid Waste Authority, asks you to design a sanitary land fill with capacity to receive all the solid waste from the area for the next ten years. Collections are averaging 300 yd^3 of land fill per week and the amount is increasing on the average of 5 percent per year. You calculate that you need about 1 acre of disposal site for each 10,000 yd^3 of solid waste received. What area in acres will you recommend to the authority as being adequate for the next ten years, just for the disposal site? (This is a special case where $i = 0$.)

Problem Group C

Find C, involving two or more steps.

C.1 A rate study is underway to determine water rates to pay for a new well field and treatment plant financed by $2,000,000 in revenue bonds at 6 percent. It is assumed customers would rather pay less now and more later since incomes in the community are expected to rise on the average. Assume that rates can be raised 4 percent per year and the pay-out period is 20 years.
   a. How much should the first payment on the bonds be at the end of the first year?
   b. How much will the last payment be at the end of the twentieth year?

C.2 An engineering student decides to accumulate $500,000 by her sixty-fifth birthday. She expects to start by investing a certain amount, C, on her twenty-third birthday and increasing the payment by 10 percent each year. She feels she can safely invest her funds at 12 percent compounded.
   a. How much should her initial investment, C, be?
   b. n for this problem is _____ years.

C.3 A young engineer on his twenty-fifth birthday decides to accumulate the equivalent of $100,000 by his sixty-fifth birthday, but is concerned about the effects of inflation on the purchasing power of $100,000.
   a. If he assumes that inflation will increase the price of the goods and services he normally buys (thus reducing the purchasing power of the dollar) by 6 percent compounded annually, how much money will he have to accumulate in order to have the same purchasing power 40 years from now that $100,000 has at the present time?
   b. If he starts his savings program with a deposit now on his twenty-fifth birthday, and continues making equal annual deposits on each birthday including his sixty-fifth, how much should he deposit annually in order to accumulate the amount found in (a) above if his savings draw $9\frac{1}{4}$ percent interest compounded annually?
   c. The young engineer reasons that his salary should be increasing as time goes on and it makes more sense for him to start depositing a small amount in the earlier years, with increasing deposits in later years. He estimates that his salary will increase by 8 percent per year, and decides to increase the amount of his deposits by that percentage each year also. How much should he deposit the first year in order to reach the sum found in (a) above?
   d. Same as (c) except, how much is the amount of the last deposit made on his sixty-fifth birthday?

Problem Group D

Involves comparisons and unit costs.

D.1 An overpass and traffic interchange is proposed for a certain intersection. The traffic this year will average 10,000 vehicles per day and is expected to increase at the rate of 10 percent per year. This growth rate is expected to continue for the next 20 years. The interchange is estimated to cost $2,000,000. Maintenance costs are estimated to be about the same with the interchange as without it. The savings in time, fuel, wear and tear, accidental property damage, and personal injury is expected to average about $0.02 per vehicle. Assume all savings accrue at EOY. The project can be financed with 7 percent road bond money. Is the project justifiable on the basis of the information given?

D.2 A fleet of earthmoving equipment may be purchased for $2,000,000 cash. In an average year the fleet is expected to move 5,000,000 yd^3 of earth. The O&M costs are currently running about $0.60/$yd^3$ but are expected to increase about 8 percent per year for the next five years. Earthmoving of this type is currently being successfully bid at about $0.90/$yd^3$ but is expected to increase at the rate of about 5 percent per year. Overhead (supervision, clerical, home and field office, etc.) costs are running about $500,000 per year and are expected to remain fairly constant at that level. It is estimated that at the end of five years the fleet can be sold for $600,000. A prospective buyer says he will buy the fleet if it will return at least 15 percent on his invested capital. Should he buy the fleet? Show all the calculations.

D.3 Solar hot water heaters are available for $600 installed. No maintenance is expected for the first 15 years. After 15 years they are expected to have a value of $100. A developer asks your advice on whether to install the solar heaters in all the new houses in a development in place of conventional gas hot water heaters costing $100 installed. Fuel costs are expected to average $8 per month at the start, with an increase of 5 percent per year over the 15-year period. Maintenance costs on the gas heaters are expected to be negligible, and the resale value after 15 years will be $10. Mortgage money is used to finance either installation, at 8 percent. Assume end-of-year payments.
   a. Which is the most economical alternative for the home buyer, if the house with the solar heater costs an extra $500?
   b. In your opinion, do you think customers will pay an extra $500 for a house with the solar heater? Why?

D.4 A dealer is trying to lease you a new dozer, rented and operated for $20,000 per year on a five-year contract. Your present dozer will bring $20,000 on the used equipment market. Operating and maintenance costs are $15,000 a year and are increasing 10 percent per year. You can borrow money from your bank at 8 percent. Estimated life of the dozer is five more years with a resale value at that time of $5,000. Compare the equivalent annual cost of owning and operating your present dozer with the dealer's offer of $20,000 per year.

D.5 Your firm asks you to calculate how much should be charged per ton of rock processed by the rock crusher in order to have sufficient funds on hand to replace the rock crusher in eight years. At the end of every year the extra amount charged during the year will be placed in an account bearing interest at 7 percent. Due to generally increasing prices, it is expected that the extra charge can be increased by 5 percent per year. The rock crusher is expected to process 6,000 tons of rock per year. The estimated replacement cost eight years from now is $46,000. The present rock crusher should have a salvage value of $5,500 at that time.

D.6 A contractor has a contract to construct a 12,000-ft-long tunnel in 30 months. He is trying to decide whether to do the job with his own forces or subcontract the job. He asks you to calculate the equivalent *monthly cost* of each alternative. His $i = 1$ percent per month. Production under both alternatives will be 400 ft/month.

Alternative A: Buy a tunneling machine and work with contractor's own forces.
Cost of tunneling machine: $500,000 paid now.
Salvage value of machine @ EOM 30: $100,000.
Cost of labor and materials: $80/ft for the first ten months, increasing by
0.5%/month at EOM each month thereafter ($80.40/ft at EOM 11, $80.802/ft
at EOM 12, etc.).

Alternative B: Subcontract the work.
Cost $200/ft of advance of the tunnel heading, for the full 12,000 ft.

Part II
COMPARING ALTERNATIVE PROPOSALS

Chapter 7
Present Worth Method of Comparing Alternatives

KEY EXPRESSION IN THIS CHAPTER

NPW = Net present worth, the equivalent lump sum value now after summing the present worths of all incomes, costs, and savings.

REVIEW AND PREVIEW

The previous chapters dealt with the concept of equivalence involving the time value of money. Equations, involving an interest rate i per period and a number of periods n were developed to relate the various forms of cash flow (P, F, A, G, C). Those relating to present worth are summarized as follows:

$$(P/F, i, n) = \frac{1}{(1+i)^n}$$

$$(P/A, i, n) = \left[\frac{(1+i)^n - 1}{i(1+i)^n} \right]$$

$$(P/G, i, n) = \frac{1}{i} \left[\frac{(1+i)^n - 1}{i(1+i)^n} - \frac{n}{(1+i)^n} \right]$$

$$(P/C, i, n) = \frac{1}{1+i}\left[\frac{(1+w)^n - 1}{w}\right] \quad \text{for} \quad r > i \quad \text{and} \quad w = \frac{1+r}{1+i} - 1$$

$$(P/C, i, n) = \frac{1}{1+r}\left[\frac{(1+w)^n - 1}{w(1+w)^n}\right] \quad \text{for} \quad r < i \quad \text{and} \quad w = \frac{1+i}{1+r} - 1$$

$$(P/C, i, n) = \frac{n}{1+r} = \frac{n}{1+i} \quad \text{for} \quad r = i$$

Values are tabulated in Appendix A for commonly used combinations of i and n.

A clear understanding of the use of these equations is essential for comprehension of the concepts in the next section. Once such an understanding is achieved, the remainder of the text explains how to compare engineering cost and saving and income alternatives to obtain optimum economy. As discussed in Chapter 1, many engineering problems lead to a variety of alternative solutions and involve selecting the optimum alternative from among a number of different options.

COMPARING ALTERNATIVES

In selecting among alternatives there are two types of data input—those which *can* be expressed in economic terms (often called "reducibles" or "tangibles") and those which *cannot* be expressed in economic terms (often called "irreducibles" or "intangibles"). For the most part this text deals with those considerations that can be expressed in economic terms (dollars, interest, time), but intangibles are important and will be discussed where appropriate.

Mutually Exclusive Alternatives

In this chapter alternatives from which selections are made are all mutually exclusive, that is, the selection of one alternative *excludes* the selection of any other alternative. An example would be a choice between building a lift station to carry water over a hill or making a deep trench to carry the water through the hill. Future chapters will consider mutually independent alternatives as well as interdependent alternatives.

The usual goal of engineering economy studies is to minimize costs or maximize profits or savings. These studies usually fall into either of the following two problem situations:

SITUATION	CRITERION
Alternatives involve cost only	Select alternative with lowest equivalent costs
Alternatives involve both costs and benefits (income or savings)	Select alternative which maximizes the net equivalent benefit (the do-nothing alternative should always be considered if available)

Methods of Comparing Alternatives

In comparing alternatives several methods are available including:

Present worth analysis (Chapter 7)
Annual worth analysis (Chapter 8)
Future worth analysis (Chapter 9)
Rate of return analysis (Chapters 10 and 11)
Benefit cost analysis (Chapter 14)

Each of these methods has advantages and limitations. These are discussed, with examples, to assist the student in determining the preferred method in analyzing any particular problem.

The Present Worth Method of Comparing Alternatives

In this approach alternatives are compared on the basis of equivalent present worth. To make such a comparison the interest rate i *must* be known or assumed, as the resulting choice may depend upon the particular value of interest rate chosen to evaluate the problem.

The length of life for each alternative also should be carefully considered. There are three comparative life span situations encountered.

1. Each alternative has the same useful life span.
2. Alternatives with different life spans are being considered.
3. The lives are very long ($n \to \infty$).

Each of these situations will be considered in turn.

COMPARISON OF ALTERNATIVES WITH EQUAL LIFE SPANS

Example 7.1 (*Alternatives with equal life spans; find the NPW of each and compare, given P, A, i, n*)

An engineer is in need of an automobile for business purposes, and finds that a suitable one can be obtained by either of two methods: (a) lease a car for $150 per month for two years, paid monthly at the end of each month, or (b) purchase the same type car for $5,000 now and sell it in two years for $3,000.

In either case, the engineer pays all operating, maintenance, and insurance costs. He can borrow funds for either alternative from the local bank at 9 percent interest. Which is the least costly alternative, assuming interest is compounded monthly?

SOLUTION
Since the life spans of both alternatives are equal, the alternatives can be compared directly by finding the net present worth of each and selecting the one with the least cost.

a. Lease. The NPW of the lease alternative is the present worth of the 24 payments of $150 per month.

$$P_1 = -\$150(P/A, 9/12\%, 24) = -\$150\frac{(1+0.0075)^{24}-1}{\underset{21.889}{0.0075(1+0.0075)^{24}}} = -\$3,283$$

b. Purchase. The NPW of the purchase alternative is the present worth of the purchase price minus the present worth of the income from the resale at the end of two years.

$$P_2 = \hspace{4cm} -\$5,000 \text{ purchase price}$$

$$P_3 = +\$3,000\,(P/F, 9/12\%, 24) = +2,507 \text{ resale}$$
$$0.84168 \hspace{3cm} \underline{\hspace{2cm}}$$

NPW of this alternative
$$P_2 + P_3 = \hspace{4cm} -\$2,493 \text{ net}$$

Comparing the present worth, it is apparent that the cost of the purchase alternative is much less expensive than the cost of the lease alternative. In the above example, alternatives involving payments of different amounts made at different points in time are compared by reducing each series or lump sum payment to an equivalent present worth and then simply comparing the equivalent NPW of each alternative. (Of course each equivalency derived is valid only for the one interest rate used. If the interest rate changes, the present worth changes also.) It is evident in this example that the monthly lease rate is higher than the purchase alternative. To find how much higher per month, it is only necessary to find the monthly lease rate that is exactly equivalent to the purchase alternative and compare. This monthly lease rate is found by setting the present worth of the purchase alternative equal to the monthly lease rate, A, times the P/A ratio found for $(P/A, 9/12\%, 24)$. Then the monthly lease rate, A, may be determined as follows:

present worth of purchase alternative, $P = \$2,493$

$$A = P(A/P, 9/12\%, 24) = 2,493 \times 0.045685 = \underline{\$113.89/\text{month}},$$

monthly lease rate equivalent to purchase alternative

Thus, if the monthly lease rate were $113.89 per month, either alternative would be equally costly. A break-even point is reached. The lump sum cash flow out of $5,000 now together with the cash flow of the $3,000 at the end of two years is equivalent to a periodic series cash flow out of $113.89 per month for 24 months. The equivalent cash flow diagrams are illustrated in Figure 7.1.

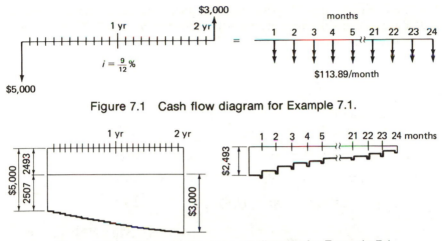

Figure 7.1   Cash flow diagram for Example 7.1.

Figure 7.2   Balance-in-the-account diagram for Example 7.1.

Another way of visualizing the comparison is to picture the account balance versus time for each case as shown in Figure 7.2. In case of the purchase alternative, the account is drawn down by $-\$5,000$ immediately, and then brought back up to the $-\$2,493$ level by the $+\$3,000$ income from sale of the car at the end of two years. Thus, the present worth of the engineer's cost of purchase is $-\$2,493$. For the lease, the 24 payments of $113.89 per month is the amount that will pay off a deficit of $2,493. Thus, the present worth of the cost to the engineer in this case is also $2,493 and the two alternatives are equivalent (at $i=9\%$ only).

Alternatives with Equal Life Spans, Second Example

Gradients frequently occur in professional practice. For instance, as labor for strenuous physical tasks becomes more expensive and difficult to obtain, mechanization becomes a logical alternative in some cases. The following example is typical of such situations.

Example 7.2 (*Compare the NPW of two alternatives with equal lives*)

Assume you are asked to compare current garbage pickup costs versus the costs of a new mechanical arm extending from the garbage truck which picks up garbage cans at curbside and empties them into the truck.
The estimated costs are

new mechanical arm $=\$22,000$
salvage value in 10 yr $=\$2,000$
O & M first year $=\$500/\text{yr}$
increase in O & M per year $=\$100/\text{yr}/\text{yr}$

The mechanical arm will replace one person on the crew who now draws $4,000/yr with anticipated increases of $300/yr/yr. Find NPW of each alternative and compare for $i = 6$ percent.

SOLUTION

mechanical

arm, $P_1 =$ $\quad\quad\quad\quad\quad\quad\quad\quad\quad$ $-\$22,000$

salvage, $P_2 = +2,000(P/F, 6\%, 10) = +$ $\quad$ $1,120$

O&M, $P_3 = -500(P/A, 6\%, 10)$ $\quad = -$ $\quad$ $3,680$

gradient, $P_4 = -100(P/G, 6\%, 10)$ $\quad = -$ $\quad$ $2,960$

$\quad\quad\quad\quad\quad\quad\quad\quad\quad\quad\quad\quad$ $-\$27,520$ $\quad$ *NPW* of
$\quad\quad\quad\quad\quad\quad\quad\quad\quad\quad\quad\quad\quad\quad\quad\quad\quad\quad$ mechanical arm

manual loading,

person, $P_5 = -4,000(P/A, 6\%, 10) = -\$29,440$

gradient, $P_6 = +300(P/G, 6\%, 10)$ $\quad = -$ $\quad$ $8,800$

$\quad\quad\quad\quad\quad\quad\quad\quad\quad\quad\quad\quad$ $-\$38,320$ $\quad$ *NPW* of
$\quad\quad\quad\quad\quad\quad\quad\quad\quad\quad\quad\quad\quad\quad\quad\quad\quad\quad$ manual loading

Using these costs and projections, mechanical loading appears more economical. In actual practice a check of intangible benefits and sensitivity analysis of all important variables should be made to see if all other aspects have been considered.

Sensitivity Check

A sensitivity check involves ranging of all the input variables over their most probable range of values and observing the effect on the resulting answer. In this case, for instance, a sensitivity check reveals the solution is quite sensitive to change in estimated life of the mechanical arm. Changing the life from ten years to six years reverses the outcome.

Example 7.2A

Compare the *NPW* of each alternative with $i = 6$ percent and n changed to six years.

SOLUTION

mechanical
arm, $P_1 -$ $-\$22,000$

salvage, $P_2 = +2,000(P/F,6\%,6) = +$ $1,410$

O&M, $P_3 = -500(P/A,6\%,6)$ $= -$ $2,460$

gradient $P_4 = -100(P/G,6\%,6)$ $= -$ $\underline{1,150}$

$-\$24,200$ *NPW* of
mechanical arm

manual loading,
person, $P_5 = -4,000(P/A,6\%,6) = -\$19,670$

gradient, $P_6 = -300(P/G,6\%,6)$ $= -$ $\underline{3,440}$

$-\$23,110$ *NPW* of
manual loading

If a life of only six years (instead of ten years) were likely then manual loading would be more economical.

COMPARISON OF ALTERNATIVES WITH DIFFERENT LIFE SPANS

Frequently, two or more alternatives occur that do not have identical life expectancies. These cannot be compared directly by means of present worth. For instance, alternative A with an *NPW* of $20,000 and a life of ten years is probably less costly than alternative B with an *NPW* of $15,000 and a life of three years. Even though alternative A has a higher *NPW*, the much longer life of A more than compensates for the higher cost (using a normal range of interest rates). Thus the life expectancy is an important factor and must be taken into account. To account for the differences in life expectancy, two methods are available.

1. Assume future replacements for the shorter lived alternative to obtain equal life spans. In the case of a ten-year A and five-year B, assume B will be replaced in five years with a similar installation. This compares one ten-year A with two sequential five-year B alternatives. If A were ten years and B were four years, then two ten-year A's (20 years) could be compared to five sequences of B at four years each (also 20 years).
2. The second method involves shortening the comparison period to the expected life span of the shorter alternative and attributing a salvage value to the longer lived alternative. For instance, in com-

paring the ten-year A to the five-year B, the life of A would be shortened to five years and some estimated salvage value placed on A at the end of five years. This salvage value of course is treated as a lump sum income at the end of five years and credited to A in the comparison.

Example 7.3 (Unequal life spans, compare NPW)

A county engineer has a choice of paving with either type A pavement or type B pavement. Type A pavement has a life expectancy of ten years, after which part of the material can be salvaged and reused. Type B pavement only lasts five years but is much less expensive. Which is the better alternative? Two sequential five-year installations of type B are compared to one ten-year life of type A.

	PAVEMENT COST PER MILE	
	TYPE A	TYPE B
Cost new	$20,000	$5,000
Annual maintenance	1,000	2,000
Estimated life	10 yr	5 yr
Salvage value at end of life	2,500	0
i	6%	6%

SOLUTION
Find NPW of each.

TYPE A	PRESENT WORTH
Cost new	$-$ \$20,000
Annual maintenance $P = -1,000 \times (P/A, 6\%, 10) =$ 7.360	$-$ 7,360
Less salvage $P = 2,500 \times (P/F, 6\%, 10) =$ 0.5584	$+$ 1,396
	$-$\$25,964 NPW type A

TYPE B	PRESENT WORTH
Cost new first application	$-$\$ 5,000
Second application* $P = -5,000 \times (P/F, 6\%, 5) =$ 0.7473	$-$ 3,736
Annual maintenance $P = -2,000 \times (P/A, 6\%, 10) =$ 7.360	$-$ 14,720
	$-$\$23,456 NPW type B

*Assume that the second application costs the same as the first application.

Comparing the two alternatives, type B is the less expensive. Notice the assumption was made that the second application costs the same as the first. This is usually not the case due to inflation. If inflation raises prices by 3 percent compounded every year, the price of the second application could be computed as

$$F=\$5,000(F/P,3\%,5)=\$5,000\times1.159=\$5,795$$

The present worth of the second application then becomes

$$P=\$5,795(P/F,6\%,5)=\$5,795\times0.7473=\$4,330$$

and the total NPW for type B becomes $24,050, still the lowest of the two options.

If the annual maintenance costs also rise for the second five-year period, for example to $2,500 per year, the present worth may be found as

$$P=\$2,500(P/A,6\%,5)(P/F,6\%,5)+\$2,000(P/A,6\%,5)$$

$$P=\$2,500\times4.212\times0.7473+\$2,000\times4.212=\$16,293$$

Explanation: First find the present worth at the beginning of the second five-year period of the annual expenditure, then treat this value as a future lump sum at the end of the first five years and reduce it to present worth at the beginning of the first five-year period. First, the $2,500 annual expenditure is worth $2,500\times4.212=\$10,530$ at the beginning of the second five-year period (end of year 5) and $10,530\times0.7473=-\$7,869$ at the beginning of the first five-year period (beginning of year 1). Then, the $2,000 annual expenditure for the first five years is worth $2,000\times4.212=\$8,424$ at the beginning of the first five years (beginning of year 1). The total NPW of both expenditures is $7,869+\$8,424=\$16,293$.

The Importance of Replacement Costs When Comparing Unequal Lives

When comparing different life spans, it is well to recognize certain features inherent in the comparison system. For instance, if item A has one-half the life and one-half the cost of B, then the NPW of A will be less (for all $i>0$). Why? Because no interest will be paid on the purchase of the second A until the first life of A is ended. By contrast, interest on the funds representing the second half of B's life is paid over the entire life span of B.

CAPITALIZED COST OF PERPETUAL LIVES, $n \rightarrow \infty$

This term has two popular meanings. Accountants use capitalized cost to describe expenditures that may be depreciated over more than one year, as contrasted to expenditures that may be written off entirely in the year they were made. The second meaning for capitalized cost is found in the traditional literature of engineering economy. Here capitalized cost refers to the lump sum amount required to purchase and maintain a project in perpetuity (present worth for $n=\infty$). Capitalized cost problems involve one or more of the following categories of payments:

1. An infinite series of annual (periodic) costs.
2. An infinite gradient series of annual (periodic) costs.
3. First cost.
4. Cost of replacements required for perpetual service or periodic major maintenance (interval between payments>interest compounding period).
5. Periodic gradients to replacement costs or periodic maintenance.

1. Where only an infinite series of annual (periodic) costs are involved, the capitalized cost is simply the lump sum amount required in an interest bearing account now which will produce interest payments sufficient to meet all annual costs. Thus, annual cost, A, equals the capitalized cost times the interest rate, i, or, capitalized cost$=A/i$. An example follows.

Example 7.4 *(Find the capitalized cost, $n \to \infty$)*

The annual cost of maintaining a certain right-of-way is $10,000. The right-of-way is expected to be required for an unknown but long period of time. Thus, for all practical purposes n is infinite. Assume $i=8$ percent. Find the capitalized cost.

SOLUTION
Mental picture: If funds could be invested at 8 percent, how much should be invested so that the interest alone would pay for the right-of-way maintenance?

$$\text{capitalized cost} = \frac{\$10,000}{0.08} = \$125,000$$

2. Where an arithmetic gradient is involved, simply apply $P=G/i^2$. For instance, in Example 7.4, if the maintenance cost increases by $1,000/yr/yr, the capitalized cost increases to $P=\$1,000/0.08^2=\$156,250$. If a geometric gradient occurs, then apply the P/C equation. Inspection reveals that valid solutions occur only when $r<i$.

3. Where a first cost is involved in addition to the capitalized cost of annual and gradient payments, simply add the sums to obtain the total capitalized cost of the project. For instance, in Example 7.4, if the first cost (cost of acquisition) of the right-of-way were $47,000, then the total capitalized cost of the project would be $125,000 (for maintenance) plus $156,250 (for gradient maintenance) plus $47,000 for first cost$=\$328,250$.

4. Where a facility is needed for perpetual service, but needs replacement from time to time, additional annual interest income must be available

to accumulate to the replacement cost at the appropriate time. For instance, in the above example, assume drainage structures in the right-of-way have an estimated life of 20 years and cost $120,000 to construct now, and $100,000 to replace at the end of every 20-year period. The cost of $120,000 to construct now is simply added to the capitalized cost as in (2) above. The $100,000 cost to replace at the end of every 20 years is accumulated by annual interest payments, A, deposited into a sinking fund (savings account bearing compound interest i). The amount required for annual deposit is calculated as $A=\$100,000(A/F,8\%,20)=\$2,190$. The amount of capitalized cost required to generate this annual income at 8 percent is

$$P=\frac{A}{i}=\frac{\$2,190}{0.08}=\$27,400$$

5. When the replacement costs increase by either an arithmetic or geometric gradient, the equivalent capitalized cost may be calculated using the effective interest rate, $i_e=(1+i)^m-1$. For instance, assume the cost of the replacement drainage structure previously described in (4) doubles every 20 years ($r_{20}=100\%$). Then the effective interest rate for compounding once every 20 years (instead of every year at 8%) is found as $i_e=1.08^{20}-1=3.661$ or 366.1 percent. Since $r<i$, then $P=C(P/A,w,n)/(1+r)$, and $w=(1+i)/(1+r)-1=4.661/2-1=1.330$, and $P=\$100,000/(1.330\times2)=\$37,580$.

Thus, the total capitalized cost of the project for parts (1) through (5) is

capitalized cost of: (1) annual maintenance	$=\$125,000$
(2) arithmetic gradient maintenance	$=\ 156,250$
(3) first cost R/W	$=\ 47,000$
(4) first cost drainage	$=\ 120,000$
(5) geometric gradient 20-yr replacement of drainage	$=\ \underline{37,580}$
total capitalized cost	$=\$485,830$

This is the total lump sum amount required to pay all the costs of the project for an infinite period of time, provided that costs in the future rise at the estimated rate and the interest rate on the capitalized funds remains at 8 percent.

Unfortunately, the term capitalized cost when applied to engineering economy sometimes gives rise to confusion since it is more frequently encountered in the accounting sense. The engineering use of the term could be profitably laid aside and neglected by engineering students were it not for its frequent appearance on licensing exams.

PRESENT WORTH WHEN A PERIODIC SERIES
OF PAYMENTS STARTS AT THE PRESENT, PAST,
OR FUTURE

In order to make the best possible use of their available resources, many people spend a great deal of time analyzing, managing, operating, buying, selling, or trading property. This property may consist of machines, vehicles, structures, or real estate, whatever happens to be of greatest use or value to the interested party at the time. Property ownership and operation usually involves cash flows of income and/or cost, both periodic series and lump sums occurring at various points in time. The basic approach is to separate the problem into simple components, solve each component, transform each component into an equivalent present worth, and then sum the results to determine the net present worth of the entire investment. In applying this method, single lump sum payments seldom pose any difficulty since they can be reduced readily to present worth at PTZ by the simple P/F relationship. Periodic series, however, sometimes occur in more complex surroundings, as do gradient series. However, if the student can master the reduction of the periodic series, the gradient series usually can be solved in a similar manner. Therefore, the discussion that follows deals with the reduction to present worth only of periodic series.

The nomenclature introduced in Chapter 3 to discuss these noncoincident time problems is reviewed and expanded as follows:

PTZ = Problem time zero, the present time according to the statement of the problem narrative.

ETZ = Equation time zero, with reference to a particular payment series, the time at which $n = 0$. Where the beginning or end of a series does not conveniently correspond with PTZ, a separate ETZ may be used to designate the beginning point in time for the particular series under consideration. For a periodic series of equal payments, the ETZ is always either one period before the first payment occurs (P/A), or at the time of the last payment (F/A).

Problems involving the present worth of a periodic series can be divided conveniently into three groups, as listed below:

Group I: Periodic series of payments begins at the present time, $PTZ = ETZ$.

Group II: Periodic series of payments began in the past, the ETZ is earlier than the PTZ.

Group III: Periodic series of payments will begin sometime in the future, PTZ earlier than ETZ.

The cash flow diagram for each situation is shown in Figures 7.3, 7.4, and 7.5.

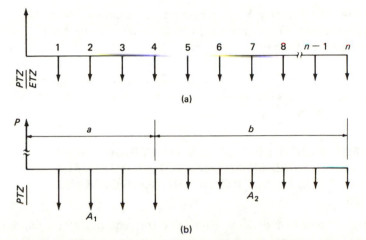

(a)

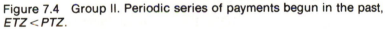

(b)

Figure 7.3   Group I. Periodic series of payments begins now, at present time, $PTZ = ETZ$. (a) Periodic series coincides with economic life. $P = A(P/A, i, n)$. (b) Periodic series is divided into two or more parts, a and b. $P_a = A_1(P/A, i, a)$. $P_b = A_2(P/A, i, b)(P/F, i, a)$.

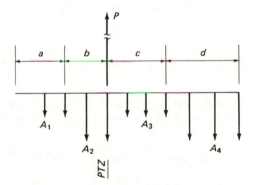

Figure 7.4   Group II. Periodic series of payments begun in the past, $ETZ < PTZ$.

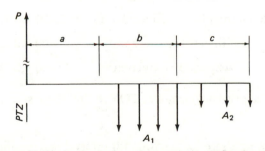

Figure 7.5   Group III. Periodic series of payments will begin at some future date $PTZ < ETZ$.

Note: The letters a, b, c, d represent some number of time periods, as shown in Figures 7.3–7.5. For the periodic series, the number of payments is always equal to the number of time periods, and the last payment always occurs at the end of the last period. All payments in a given series are of equal amounts, designated A. Thus ETZ always occurs either one period before the first payment in the series for P/A or at the time of the last payment for F/A.

PRESENT WORTH OF A SERIES OF PERIODIC AND ARITHMETIC GRADIENT PAYMENTS AT INTERVALS EQUAL TO SOME MULTIPLE OF THE COMPOUNDING PERIOD

At times a series of periodic payments are required at regular but long intervals. For instance, heavy construction equipment may require periodic overhaul every six months while the compounding period is monthly, or certain pavement may require resealing at five-year intervals while the compounding period is annual. In addition, the payments may increase each time as costs continue to rise. To account for these long interval gradient payments, the equivalent interest rate is used, as illustrated in the following example.

Example 7.5 (*Periodic overhaul every six months with gradient, compounded monthly*)

Find the cost of excavation in terms of dollars per cubic yard over the 60-month life of the excavator whose data are listed below. Use $i = 1\%/$month.

$$\text{excavator cost new} = -\$200{,}000$$
$$\text{resale value at EOM } 60 = + \quad 50{,}000$$
$$O \& M = - \quad 4{,}000/\text{month}$$
$$\text{gradient in } O \& M = - \quad 50/\text{month}/\text{month}$$
$$\text{periodic overhaul every 6 months} = - \quad 12{,}000/\text{each}$$
$$\text{gradient in periodic overhaul} = - \quad 800/\text{each}/6 \text{ months}$$
$$\text{expected production} = \quad 20{,}000 \text{ yd}^3/\text{month}$$

The first periodic overhaul occurs at EOM 6 and the last at EOM 60.

SOLUTION
The effective interest rate for a six-month interval is determined first as

$$i_e = (1.01)^6 - 1 = 6.15\%$$

Then the present worth of each cost item is calculated and summed as follows.

$$\text{cost new, } P_1 = \qquad\qquad\qquad = -\$200{,}000$$

$$\text{resale, } P_2 = +50{,}000(\,P/F,1\%,60) \quad = + \quad 27{,}523$$
$$0.5505$$

$$\text{O\&M, } P_3 = -4{,}000(\,P/A,1\%,60) \quad = - \quad 179{,}820$$
$$44.955$$

$$\Delta\,\text{O\&M, } P_4 = -50(\,P/G,1\%,60) \qquad = - \quad 59{,}640$$
$$1192.81$$

$$\text{periodic overhaul, } P_5 = -12{,}000(\,P/A,6.15\%,10) = - \quad 87{,}720$$
$$7.31$$

$$\Delta\,\text{periodic overhaul, } P_6 = -800(\,P/G,6.15\%,10) \quad = - \quad 23{,}456$$
$$29.32$$

$$NPW = P_T = -\$523{,}114$$

$$\text{monthly equivalent, } A = \;523{,}114(\,A/P,1\%,60) \;=\; \$\;11{,}634/\text{month}$$
$$0.02224$$

$$\text{unit cost} = \frac{\$11{,}634/\text{month}}{20{,}000 \text{ yd}^3/\text{month}} \quad = \quad \$0.582/\text{yd}^3$$

Present Worth of Geometric Gradient Payments at Intervals of a Multiple of the Compounding Period

A series of payments made at some multiple of the compounding period may rise geometrically in many instances. For instance, if a generator or pump is overhauled once every year for 15 years, the costs may increase, say 10 percent each year, while the interest may compound monthly. This type of situation can be accommodated by employing the effective interest rate for the interval between payments, as the following example illustrates.

Example 7.6 (*Geometric gradient in periodic maintenance*)

A public water utility department overhauls one of their pumps once a year, and finds their overhaul costs are increasing by 10 percent every year as indicated on the cash flow diagram shown. They expect to overhaul the pump every year for 15 years. Find the present worth of these costs assuming $i = 1$ percent per month. See Figure 7.6.

SOLUTION
Find the effective interest rate as

$$i_e = 1.01^{12} - 1 = 0.127$$

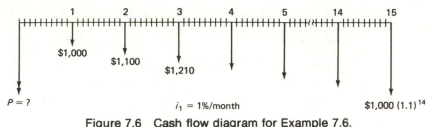

Figure 7.6 Cash flow diagram for Example 7.6.

Then the present worth is calculated using the appropriate P/C equation. Since $r=10$ percent, and $i_e=12.7$ percent, then $r<i$ and

$$w=\frac{1.127}{1.10}-1=0.0244$$

Then

$$P=\frac{C}{1+r}\left[\frac{(1+w)^n-1}{w(1+w)^n}\right]=\frac{1,000}{1.1}\left[\frac{1.024^{15}}{0.0244(1.024)^{15}}\right]$$

$$P=\underline{\underline{\$11,300}}$$

FINANCING A LOAN, LIFE OF LOAN ≠ LIFE OF PROPERTY

For loans secured by property, the banks or other lending institutions usually insist that the pay-out period for the loan be shorter than the expected life of the property. Naturally the lenders are concerned with the security of their loans. If the collateral for the loan has been fully used up (depreciated) before the loan is paid out, the risk of default is increased since the loan no longer can be foreclosed against property of equal or greater value than the balance of the loan. For instance, on new cars with an average life expectancy of 12 years, lenders typically allow a maximum 4-year payment period. For used cars, a 3-year maximum is common. On buildings with a life expectancy of 40 to 50 years and more, mortgage lenders usually permit 25- to 30-year repayment schedules. Frequently, however, the borrower does not hold the property for the full life of the loan. The owner will often decide to sell the property and buy something else before the loan is paid out. In the case of houses, the average homeowner moves about every five years. He or she sells the former residence and the new buyer may take over payments on the old loan contract or make a new loan (refinance) so that the balance of the old contract can be paid off. Thus, it is important to be able to calculate both the amount that the loan has been paid down, as well as the balance due at any point in time on the loan contract.

Example 7.7 (*Find the current balance owed on a partially amortized loan.*)

For a typical example, suppose that 18 months ago your firm borrowed $5,000 at 9 percent with equal monthly repayments for 48 months. Now, after the eighteenth payment has been made, the boss asks you to calculate (a) how much of the loan has been paid off, and (b) how much is the balance due if paid in cash now.

SOLUTION

This example involves a periodic series of payments, where the equation time zero (*ETZ*) (18 months ago) began before the problem time zero (*PTZ*) which occurs now after the eighteenth payment. It is the same type of situation as shown in Group II previously illustrated in this chapter. The first step is to find the amount of the monthly payment, A ($i = 9/12 = 0.75\%$).

$$A = P(A/P, i, n) = 5{,}000(A/P, 0.75\%, 48) = \$124.50/\text{month}$$
$$0.0249$$

Then the balance due can be calculated as the present worth of the remaining 30 payments,

$$P_1 = A(P/A, i, n) = 124.50(P/A, 0.75\%, 30) = \$3{,}331.50 = \underline{\$3{,}332}$$
$$26.7751$$

Then the amount of the loan already paid off is simply the original amount less the balance remaining

$$P_2 = P - P_1 = \$5{,}000 - \$3{,}332 = \underline{\$1{,}668}$$

Life of Loan ≠ Life of Property, Plus Present Worth Used to Obtain Periodic Rates (Dollars per Month, Hour, etc.) or Unit Costs (Dollars per Cubic Yard, etc.)

Rental rates or unit costs often need to be calculated for construction equipment or other property used in a productive enterprise. For instance, a dragline may be rented out by the month and a rental fee must be determined that will cover all costs plus a margin of profit. Or a contractor may need to determine break-even costs on a unit price basis ($/yd^3) for excavating equipment. One method of obtaining these answers involves the use of present worth. First, each component cost item is reduced to an equivalent present worth. Second, all the components are summed to obtain an equivalent lump sum single payment net present worth. Third, a periodic equivalent (monthly, weekly, daily, hourly) rate is determined by use of A/P factors. Then if a unit cost is desired, this periodic rate (e.g., $/hr) can be divided by the production rate (e.g., yd^3/hr) to obtain costs per unit of production (e.g., $/yd^3). The following situation serves as a simplified example of the method involved.

Example 7.8 (*Life of loan ≠ life of property, find unit costs*)

A contractor has just bought a dragline for $50,000. He expects to keep the dragline for *seven years* and then sell it. His installment payment contract on the machine runs for *five years*. He paid $10,000 down and will pay the balance of $40,000 in 60 monthly installments with 1% per month interest on the unpaid balance. He asks you to compute how much he must charge per cubic yard of earth moved in order to cover the following costs:

> P&I (principal and interest) payments
> O&M costs $2,000/month
> Resale value at the end of seven years $15,000
> $i=1.5\%$ month on invested capital
> Production is estimated to average 10,000 yd³ of material per month

SOLUTION

Find the present worth of all costs, then find equivalent monthly cost and divide by monthly production to obtain dollar per cubic yard charges. The monthly payments for the dragline in 60 months are

$$A_1 = \$40,000(A/P, 1\%, 60) = \$890/\text{month}$$
$$0.02224$$

The present worth of all costs and resale value with $i=1.5$ percent per month is

down payment, $P_1 =$ $\qquad$ $-\$10,000$

monthly payments, $P_2 = -\$890(P/A, 1.5\%, 60)$ $\quad = -\$35,050$
$$39.380$$

O&M, $P_3 = -\$2,000(P/A, 1.5\%, 84) = -\$95,140$
$$47.570$$

resale, $P_4 = \$15,000(P/F, 1.5\%, 84)$ $\quad = +\$\ 4,290$
$$0.2863$$
$$\overline{\qquad\qquad \$135,900}$$

To provide a simplified mental picture of these payments and income, imagine the contractor as having the sum of $135,900 invested in a savings account that pays 1.5 percent per month on whatever is left in the account. From time to time he withdraws whatever amount is required for each of the costs or payments listed above. At the end of the seven years he has paid all his bills, but is overdrawn by $15,000. The resale of the dragline provides the $15,000 income needed to bring the account back up to zero.

For an alternate mental picture, imagine the same $135,900 invested at 1.5 percent per month. An equal amount may be withdrawn monthly as shown below to bring the account to zero at the end of 84 months. This is the same amount that the contractor has to charge each month to cover the expenses above.

$$\text{monthly equivalent } A_2 = \$135,900 \underset{0.02102}{(A/P, 1.5\%, 84)} = \$2,857/\text{month}$$

The charge per cubic yard is obtained as

$$\frac{\$2,857/\text{month}}{10,000 \text{ yd}^3/\text{month}} = \$0.286/\text{yd}^3$$

Conclusion. If the contractor charges $\$0.286/\text{yd}^3$ for excavation by this machine, he will cover his costs listed above plus 1.5 percent per month on all amounts (costs) invested in the machine for the months during which they are invested.

TRADE-INS

When an older machine is replaced by a new one, the older machine may be disposed of by selling outright on the competitive market, or by trading it in on the new machine. Salespeople sometimes promote the sale of a new machine by offering an inflated trade-in allowance for the older machine. The salesperson's loss from this part of the deal must then be recouped by extra charges of other types, such as financing charges, credit investigation, dealer's preparation fee, special add-on features, and so on. A variety of methods for acquisition of new machines is available and the wary customer must carefully weigh each alternative to determine which is better. Three common methods of acquiring the use of machines are

1. Cash purchase.
2. Trade-in, and/or finance with equal periodic repayments.
3. Lease.

In evaluating the alternatives, two basic concepts should be kept in mind.

1. Account for all of the cash flow or equivalent trade-in value into each alternative.
2. Trade-ins are worth the best cash offer on the open competitive market, not the fictitious trade-in allowance offered by the salesperson.

The following example illustrates a comparison between the three common methods of acquiring the use of new equipment.

Example 7.9 (*Illustrates use of trade-in value and use of actual cash flow resulting from dealer's calculations*)

Your firm uses a small fleet of cars and pickups. As a matter of company policy all vehicles are replaced upon reaching three years of age. This year your firm needs to replace ten three-year-old vehicles whose current cash market value is $15,000. Specifications for the new vehicles have been distributed and three dealers have submitted proposals which you are asked to evaluate. It is estimated that three years from now the new fleet will have a cash market value of $17,500. Your firm can borrow money to finance the acquisition of this fleet at 9 percent nominal annual interest compounded monthly.

Proposal A: Lease the ten vehicles for 36 months at $1,400 per month, including insurance. All other operating and maintenance (O&M) costs are paid by the customer (your firm).

Proposal B: Purchase for cash. The sum of the list prices for all ten vehicles is $60,000. For a fleet purchase, this dealer is offering a 15 percent discount. In addition, insurance costs $75 per month.

Proposal C: Trade in the existing fleet of ten three-year-old vehicles. The dealer will allow $18,000 for the fleet and give financing for 33 months at 7 percent, calculated as follows:

list price	$60,000
less trade-in allowance	18,000
balance owed	$42,000
dealer prep charges	2,300
required insurance at $90/month × 33 months (paid in advance)	2,970
balance to be financed	$47,270

$$7\% \text{ interest} \times \frac{33 \text{ months}}{12 \text{ months/yr}} = 19.25\%$$

interest charges = 19.25% × $47,270	9,099
total amount owed	$56,369

total monthly payments for 33 months $56,369/33 = $1,708/month

SOLUTION

The three proposals are compared by finding the present worth of all of the cash flow or equivalent trade-in into each alternative over the 36-month anticipated life of the ten vehicles.

Proposal A: Lease.

$$P = -\$1,400(P/A,9/12\%,36) \qquad = -\$44,025$$

Proposal B: Purchase for cash.

cash price, $P = -0.85 \times \$60,000 \qquad\qquad = -\$51,000$

insurance, $P = -\$75(P/A,9/12\%,36) \qquad = - \quad 2,359$

resale value at end of 3-yr life
$$P = +\$17,500(P/F\,\%,36) \qquad = + \quad 13,373$$

total present worth of purchase for cash $\qquad = -\$39,986$

Proposal C: Trade in.

cash market value of trade-in
  (*not* the salesman's figure) $\qquad\qquad = -\$15,000$

monthly payments for 33 months
$$P = -\$1,708(P/A,9/12\%,33) \qquad = - \quad 49,776$$

Proposal C includes only 33 months insurance so
  the cost of the last three months insurance at \$75/
  month paid in advance is found as
$$P = -\$75(P/A,9/12\%,3)(P/F,9/12\%,32) = - \quad 175$$

resale value at end of 3-yr life,
$$P = +\$17,500(P/F,9/12\%,36) \qquad = + \quad 13,373$$

total present worth of trade-in and finance $\qquad = -\$51,578$

Comparison of the three alternatives indicates that purchase for cash is the most economical in this instance.

Value Attributed to Trade-In

When considering a trade-in, remember that the value traded in is just as much an expenditure of assets as any cash that would have to replace the trade-in. For instance, assume a new car can be purchased for either \$5,500 cash, or \$3,000 cash plus your present car worth \$2,500. The total expenditure for the new car in either case is \$5,500, not \$3,000. If the present car is truly worth \$2,500 it represents \$2,500 worth of assets that can be sold on the open market (e.g., through the classified ads in the newspapers) and converted into \$2,500 cash. Therefore, trading in the present car is equivalent to trading in (or spending) \$2,500 cash. By the same reasoning, in any trade-in or replacement-decision problem, the asset traded in or replaced constitutes an expenditure spent to acquire the replacement alternative.

For example, a contractor has an old bulldozer worth $20,000. A dealer offers to take the old dozer in trade plus $35,000 for a new dozer. Regardless of the dealer's figures, the contractor is being offered a new dozer for $55,000. Another dealer may tell the contractor he will allow $30,000 for the trade-in and then mark the new dozer up to $65,000 in order to compensate. But if the old dozer is really worth $20,000 on the open market then the contractor is still spending $55,000 worth of his assets in the form of $20,000 worth of bulldozer plus $35,000 worth of cash.

Where a replacement results in the release of an existing facility for conversion to cash, a similar outlook prevails. A facility, such as a bridge or a building, usually represents money invested in a certain location. If the existing facility has any cash salvage value this cash is usually available only if the new replacement facility is selected. Therefore, if the existing facility remains in place, the salvage value remains invested in the existing facility and is *not* available and is charged as one of the costs of keeping the existing facility. This salvage value is *not* an income credited to the selection of the new replacement facility.

Example 7.10

An existing bridge with a present salvage value of $50,000 may be renovated at a cost of $500,000 and left in service, or the existing bridge may be sold for salvage value and replaced with a new bridge costing $525,000. If both alternatives provide equal service, which is the least costly?

SOLUTION

The existing bridge could be converted into $50,000 cash if salvaged, so leaving it in service will cost that $50,000 cash plus another $500,000 for renovation for a total of $550,000. The new bridge costs only $525,000 and is therefore less costly. The $50,000 cash made available by selecting the new bridge alternative is not new earned income, but simply a conversion of assets already owned, from $50,000 worth of bridge to $50,000 worth of cash.

The basic approach to trade-in problems is to treat the existing asset as equivalent cash. If a car worth $1,500 is traded in, the trade-in is equivalent to spending $1,500 cash. If an existing bridge currently owned by the DOT could be sold for $15,000 salvage, the DOT is *foregoing* the cash in order to leave the bridge in service. Therefore, if the DOT selects the alternative of leaving the existing bridge in service, that alternative costs them $15,000. Further, in the event that the existing bridge costs $10,000 to demolish (it has a negative $10,000 salvage value), then leaving the bridge in service actually saves $10,000 during the period of time over which the bridge is retained in service. In effect, the expenditure is postponed until the bridge finally is demolished. As a consequence, interest charges are saved for the extra years of service, but escalating costs due to inflation may negate this gain.

SUNK COSTS

Usually there is a difference between cost and current market value, since cost is the price paid for an item, whereas current market value is the price the item would sell for on the current market. Such items as construction equipment, tools, automobiles, clothes, and so forth, typically decrease in value after purchase, so the cost is usually higher than the current market value. Other items such as real estate frequently appreciate in value so that the cost is often below the current market value. Once the cost has been paid and the transaction is completed the cost becomes past history and is usually irretrievable. Any drop from cost to current market value is lost and cannot be regained and is referred to as "sunk cost." For instance, suppose a new pickup truck is purchased for $12,000. Shortly thereafter it is wrecked without insurance and the junk yard offers $100 for the wrecked remains. The difference of $11,900 ($12,000 cost less $100 salvage value) is termed "sunk cost." If the pickup were financed so that the payments continue on despite the absence of the pickup itself, the $11,900 is still sunk cost. By the same reasoning, if a bridge were completed just yesterday at a cost of $1,000,000 and if today it has a current market (salvage) value of $100,000, then the current market value for the bridge is $100,000 and the $900,000 drop is the sunk cost. If the bridge is financed by bonds that are not yet paid off, the bridge is still worth just today's salvage value. As a result, when comparing the *NPW* of the existing bridge with other alternatives, the capital cost of the existing bridge (or other similar facilities) is very low, giving it a decided cost advantage (at least for the capital cost portion of the *NPW*).

Replacement Cost

The value of the existing bridge is *not* the replacement value since the replacement bridge is actually another alternative. Using the replacement value as the actual present worth of the existing bridge erroneously favors the alternative as illustrated in the following example, simplified to clarify the point.

Example 7.11 (*Comparing alternative bridges using the current market value and the replacement cost of the existing bridge*)

The state DOT is considering the removal of an existing bridge A, and replacing it with either proposed bridges B or C. The traffic carrying capacity, maintenance costs, and salvage values 25 years from now are the same for all three bridges. Only the capital costs are different, and are listed below. Determine which is the lower cost alternative of the three.

	EXISTING BRIDGE A	PROPOSED BRIDGE B	PROPOSED BRIDGE C
Present market (salvage) value (in $1,000's)	$ 100		
Replacement value (in $1,000's)	1,500		
Cost new (in $1,000's)		$1,200	$1,000

SOLUTION

If the existing bridge A is valued at the replacement value, $1,500,000, then proposed bridge C is the lower cost selection at $1,000,000. However, the only actual cost involved in retaining existing bridge A is the loss of $100,000 of present market (salvage) value. So retaining A really only costs $100,000, not $1,500,000. Therefore the lowest cost alternative is to retain existing bridge A at a true cost of $100,000 rather than construct proposed bridge C at a cost of $1,000,000.

Examples of Sunk Costs, Current Market Value, and Periodic Gradient

Example 7.12 (*Existing bridge with current salvage value, periodic maintenance and gradient, loan life $\neq$ project life*)

Due to an unexpected growth in traffic counts, an existing DOT-owned bridge must either be widened or replaced with a new bridge.

Existing bridge: The existing bridge was built just 12 years ago at a cost of $1,000,000, financed by money borrowed at $4\frac{1}{2}$ percent which is being repaid in 30 equal annual installments. The current maintenance costs are $25,000 per year (charged at the end of this year) with an anticipated escalation rate of $2,000 per year each year. If removed now the existing bridge would yield a salvage value of $150,000. The replacement value of the existing bridge is estimated at $2,100,000 if it were constructed today.

Alternative A   Proposed widening: If the existing bridge is widened, a federal government loan is available to pay the full construction cost of widening—estimated at $1,500,000. The loan must be repaid in ten equal end-of-year payments, with 3 percent interest on the unpaid balance, with the first payment due at EOY 1. None of the loan money may be used for operating or maintenance or any other costs except construction. Other costs of the widened bridge are as shown below. All of these costs are paid from DOT funds with $i = 10$ percent.

O&M (first payment charged @ EOY 1, last @ EOY 40)
  (these O&M costs completely replace   = − $ 50,000/yr
  the current maintenance costs for the
  existing bridge)

gradient O&M   =   5,000/yr/yr

periodic heavy maintenance every 4 yr   = −   100,000/4 yr
  first periodic maintenance @ EOY 4, last
  @ EOY 36

gradient in periodic maintenance   = −   40,000/each/4 yr

salvage value at EOY 40   = +   20,000

Alternative B   New replacement bridge

design and construction cost of new bridge = − 2,300,000

O&M costs   = −   60,000
  equivalent annual costs, including
  gradients and periodic maintenance

salvage value at EOY 40   = +   100,000

Assume both alternatives have life spans of 40 more years from now, and compare the NPW's of each at $i = 10$ percent.

SOLUTION
The $1,000,000 cost of the existing bridge 12 years ago, together with the financing and repayments, is sunk cost and together with replacement cost and some other data, has no effect on the problem. This type of data is included here to simulate the myriad of useless background information that frequently surrounds projects of this type in actual practice. The NPW of each alternative is found as follows (all figures are in 1,000's).

Alternative A   Proposed widening of existing bridge

current cash value, $P_1 = -\$150$ $= -\$ \ 150.0$

present worth of
  government loan repayments
  $P_2 = -\$1,500 \, (A/P, 3\%, 10) \, (P/A, 10\%, 10) = - \ 1,080.2$
              0.1172   6.1446

O&M $P_3 = -\$50 \, (P/A, 10\%, 40)$ $= - \ \ 489.0$
           9.779

ΔO&M $P_4 = -\$5 \, (P/G, 10\%, 40)$ $= - \ \ 444.8$
            88.952

periodic maintenance every 4 yr

$$P_5 = -\$100\,(A/F,10\%)\,(P/A,10\%,36) \qquad = - \quad 208.5$$
$$0.2155 \qquad\quad 9.6765$$

Δ Periodic maintenance $i_e = (1+i)^n - 1 = 1.1^4 - 1 = \qquad\quad 0.464$

$$P_6 = \frac{G}{i}\left[\frac{(1+i)^n - 1}{i(1+i)^n} - \frac{n}{(1+i)^n}\right]$$

$$= \frac{-40}{0.464}\left[\frac{1.464^9 - 1}{0.464 \times 1.464^9} - \frac{9}{1.464^9}\right] \qquad = - \quad 154.7$$

salvage $P_7 = 20(P/F,10\%,40)$ $\qquad\qquad = + \quad\underline{0.4}$

NPW of alternative A P_t $\qquad\qquad = - \$2,526.8$

Alternative B   New replacement bridge

cost new P_1 $\qquad\qquad\qquad = - \$2,300.0$

O&M $P_2 = -\$60\,(P/A,10\%,40)$ $\qquad\qquad = - \quad 586.7$
9.7791

salvage $P_3 = \$100\,(P/F,10\%,40)$ $\qquad\quad = + \quad\underline{2.2}$
0.0221

NPW of alternative B, P_t $\qquad\qquad -\ \$2,884.5$

compare to NPW of alternative A @ $\qquad -\ \$2,526.8$

Alternative A is least costly.

SUBSIDY BY MEANS OF LOW INTEREST RATES

The federal government sometimes aids needy groups (such as those in disaster areas or low-income groups) by making low-interest loans available. The low interest constitutes a subsidy (a grant of money) because the present value of the amount loaned by the government is greater than the present value of the amount paid back by the needy group. The following example illustrates the method of determining the value of the subsidy.

Example 7.13   (*Find NPW of interest rate subsidy*)

Farmers in a certain drought-stricken area are entitled to government loans at 3 percent with repayments required in equal annual installments over a 20-year period. The first installment is due one year from now. If the government is borrowing at 7 percent and $10,000,000 is loaned at 3 percent under this program, what is the present worth of the subsidy?

SOLUTION

The amount of the subsidy is determined by a simple three-step procedure.

Step 1. Find the annual amount, A, that the government will receive back for repayment of the low-interest loan.

$$A = \$10,000,000(A/P, 3\%, 20) = \$672,160/\text{yr}$$
$$0.067216$$

Step 2. Find the present worth to the government of this series of 20 annual repayments. The i used in this step is the i at which the government borrows money to replace the funds used in this program. Thus i in this equation is 7 percent.

$$P = \$672,160/\text{yr} \ (P/A, 7\%, 20) = \$7,120,900$$
$$10.5941$$

Step 3. To find the cost of the program (the amount of the subsidy), subtract the present value of the repayment income from the present value of the total loaned under the program.

P of total amount loaned $= -\$10,000,000$

P of all repayments $= + \ \ \ 7,120,900$

NPW (cost) of the program
  (subsidy) $= -\$ \ 2,879,100$

Comment: Other additional features are sometimes added to such programs. For instance, suppose the repayments were deferred for five years with no additional interest charges. The approach to solving for the value of the subsidy is basically the same. The present worth of the 20 repayments begun after a five-year delay with no extra charge is calculated as

$$P = \$672,160/\text{yr} \ (P/A, 7\%, 20)(P/F, 7\%, 5) = \$5,077,200$$
$$10.5941 \qquad 0.71299$$

The amount of the subsidy is found as before:

P loaned $= -\$10,000,000$

P repaid $= + \ \ \ 5,077,200$

P subsidy $= -\$ \ 4,922,800$

The amount of the subsidy increased in this case due to the extension of time.

A variety of similar provisions can be added to the terms of such programs depending on the aims and desires of the legislators. Whenever possible, these provisions should be translated into present worth costs, so that these grants of tax-payer dollars can be weighed against the benefits envisioned.

SUMMARY

In this chapter alternative engineering projects or equipment are analyzed by comparing their net present worth.

The general criterion for engineering cost economy studies is to maximize profits or savings, or minimize costs. One method of comparison is by calculation of equivalent net present worth. Each alternative must be compared over equal time periods. Three situations occur regarding lives:

1. Useful lives of each alternative are the same. In this situation each alternative's net present worth is calculated and compared directly.
2. Useful lives of each alternative are *not* the same. In this situation the most widely used approach is to assume equal replacement costs and benefits, then select the lowest common multiple of the lives involved to compare each alternative over the same time span of multiple lives.
3. The useful lives of alternatives are very long (perhaps infinite). In this situation the net present worth is often termed "capitalized cost."

Present worth techniques are widely utilized in many engineering and business considerations but present worth techniques are only one of several methods employed to compare alternatives. Other techniques are presented in subsequent chapters.

PROBLEMS FOR CHAPTER 7, PRESENT WORTH ANALYSIS

Problem Group A

Given three or more of the variables P, A, F, i, n, find the present worth of each proposal. Compare alternative proposals.

A.1 Parking meters are being considered for a certain district. Two types are available with the following estimated costs:

	TYPE A	TYPE B
Estimated life	5 years	5 years
Cost new	$100/each	$200/each
Annual O&M	$300/yr	$250/yr
Resale value at end of life	0	$ 75

Using present worth analysis, determine which type of parking meter should be selected for an $i = 12$ percent.

A.2 Same as Problem A.1 except type B has an estimated life of ten years.

A.3 A small utility company is for sale. The estimated net income is $40,000 per year for the next 40 years after which its franchise expires and the system is virtually worthless. If an interest rate of 15 percent is agreed upon as adequate, what should the selling price of the utility be at the present time (or what is the present worth)?

A.4 A contractor needs to choose between two makes of motor grader. Money can be borrowed at 18 percent. Which is the most economical?

	BRAND A	BRAND B
Cost new	$20,000	$15,000
Repair costs	2,000/yr	3,000/yr
Resale value in 10 yr	10,000	5,000

A.5 Same as Problem A.4 except brand B will last five years (same resale value).

A.6 A tree spade is proposed for a public works department to replace hand labor. Estimated costs for the next five years are

	TREE SPADE	HAND LABOR
Cost new	$20,000	—
Annual O & M	10,000/yr	$15,000/yr
Resale in 5 yr	5,000	—

Using present worth analysis, which alternative should be chosen for $i=7$ percent?

A.7 A contractor needs a new power shovel which she can acquire by either of the following plans:

a. Pay $50,000 cash. To do this she would have to cash in some bonds currently paying 6 percent.

b. Nothing down and $8,500 per year for nine years.

What is the more advantageous plan for the contractor? Compare by present worth analysis with $i=6$ percent.

A.8 Three different types of outside covering are available for a concrete frame and block building. If $i=7$ percent, and the building life is 20 years, which is the lowest cost covering? Use present worth analysis.

	CAST STONE FACING	STUCCO	PAINT
Cost new	− $15,000	− $5,000	− $1,000
Add to resale value of building in 20 yr	+ 5,000	+ 1,000	0
Add to rental income for 20 yr	+ 1,200/yr	+ 600/yr	0
Maintenance cost (including repainting)	− 100/yr	− 200/yr	− 300/yr

A.9 Same as Problem A.8 except $i=15$ percent.

A.10 A representative selection of roof beam spans for a store are available as follows. If $i=12$ percent and the building life is 30 years, which is lowest cost?

	25-FT SPANS	50-FT SPANS	100-FT SPANS
Cost new	− $2.50/ft^2	− $2.74/ft^2	− $3.25/ft^2
Add to resale value of building in 30 yr	0	+ 0.15/ft^2	+ 0.50/ft^2
Add to rental income for 30 yr	0	+ 0.10/ft^2/yr	+ 0.20/ft^2/yr

A.11 Find which of the two types of floor covering has the lowest NPW if $i=15$ percent.

	TYPE A	TYPE B
Estimated life	8 yr	4 yr
Maintenance	\$3,000/yr	\$2,000/yr
Cost new	\$150,000	\$100,000

A.12 A municipal parking lot to be owned and operated by the city government is proposed for downtown. Costs and income are estimated to be

$$\text{Land cost} = \$100,000$$
$$\text{improvement costs} = \$15,000 \quad \text{(cost of construction)}$$
$$\text{maintenance} = \$50/\text{space}/\text{yr}$$
$$\text{parking lot capacity} = 40 \text{ cars}$$
$$\text{revenue} = \$0.10/\text{hr}/\text{space}$$
$$\text{occupancy} = \times 9 \text{ hr}/\text{day} \times 250 \text{ days}/\text{yr}$$
$$n = 20$$
$$i = 6\%$$
$$\text{resale value} = \$100,000 \text{ at end of life}$$

Using present worth analysis determine whether or not this lot should be built. Since the city will be taking over land now privately held, there will be a loss of tax revenue (30 mills) from property assessed at $\$100,000 \times 0.03 = \$3,000$ per year. This property has held a constant value for some time, and for purposes of estimating lost income assume the assessment and millage rate would remain constant over the 20-year project life.

A.13 Same as Problem A.12 except that $n=40$ years.

A.14 Same as Problem A.12 except that $n=30$ years and $i=8$ percent.

A.15 You have an opportunity to construct a small neighborhood shopping center for an estimated cost of \$100,000. The property would consist of three small stores. The estimated income is \$12,000 per year. Average taxes, insurance, and so forth, you estimate to be about \$2,700. You estimate the building can be sold in ten years for \$50,000 (its depreciated value). You need to earn at least 10 percent on your investment. Should you construct this building?

A.16 Comparing alternatives for commercial building design, one alternative calls for extensive landscaping, costing \$100,000 but resulting in an increased rental income of \$10,000 per year over the 20-year life of the building. Assume zero salvage value. If $i=6$ percent, is the landscaping a good investment? Compare using present worth analysis.

A.17 Same as Problem A.16 except that $i=12$ percent.

A.18 As a contractor you can purchase a front end loader under any of the following options. Which is least cost in terms of present worth ($i=12\%/\text{yr}$, compounded monthly)?
a. \$10,000 cash.
b. \$200 per month for 60 months, no down payment, first payment due EOM 1.
c. \$3,000 cash and \$133 per month for 60 months, first payment due EOM 1.

A.19 A building owner wants you to build a \$1,000,000 building and gives you a choice of methods of payment. You can build it on a cost plus 5 percent contract whereby he will finance the job and pay you $50,000 profit when the building is completed in 12 months. Or you can take your profit in ownership shares expected to yield $5,000 per year for 8 years after which the building will be sold with your share worth an estimated $60,000. If money is worth 12 percent to you, which is the most profitable in terms of present worth?

A.20 A sanitary sewer may be run either through a deep cut or a lift station with costs estimated as follows:

	DEEP CUT	LIFT STATION
Extra cost of CI Pipe	$50,000	
Extra cost of lift station		$20,000
Annual operating expenses	$ 500/yr	$ 2,000/yr
Expected life	40 yr	20 yr
i	6%	6%

Which alternative should be selected using present worth analysis?

A.21 Two types of roofing materials are available to cover a building scheduled for demolition in 20 years. The building has no salvage value.

	TYPE A	TYPE B
Estimated life	10 yr	20 yr
Maintenance	$3000/yr	$2000/yr
Cost new	$50,000	$100,000

With the interest rate on borrowed money at $i=17$ percent, which alternative should be selected? Use present worth analysis.

A.22 Flooring. Assume $i=10$ percent. Find lowest present worth.

	TYPE A	TYPE B
Estimated life	12 yr	20 yr
Maintenance	$0.10/ft^2/yr	$0.08/ft^2/yr
Cost new	$0.30/ft^2	$0.50/ft^2

A.23 Paint. Assume $i=12$ percent. Find the lowest present worth.

	TYPE A	TYPE B
Estimated life	3 yr	5 yr
Cost materials (paint)	$0.03/ft^2	$0.045/ft^2
Cost labor and equipment	$0.06/ft^2	$0.06/ft^2

A.24 You have an opportunity to invest in a hamburger establishment to be constructed adjacent to the campus. The investment would be $10,000, for which you should receive annual receipts of $1,650 for ten years. If you can earn 8 percent interest from a certificate of deposit, should you invest in this venture? Assume zero salvage value.

A.25 An equipment rental firm has $30,000 to invest and is considering the addition of a backhoe to its rental inventory. If the firm uses a 15 percent return on investment which (if either) of the following alternatives should be selected?

Explain.

	ALTERNATIVE A	ALTERNATIVE B
First cost	− $20,000	− $30,000
Salvage value, 5 yr	+ 8,000	+ 10,000
Annual maintenance	− 5,000	− 6,000
Annual rental income	+ 9,000	+ 14,000

Solve by present worth analysis.

Problem Group B

This group involves gradients.

B.1 A dragline is being considered for a long-term canal drainage project. Estimate the cost of ownership in terms of present worth ($i=15\%$).

$$\text{cost of dragline}=\$40,000$$
$$\text{resale value in 5 yr} = 20,000$$
$$\text{maintenance, repairs, etc.} = 10,000/\text{yr increasing } \$1,000/\text{yr}$$

B.2 Find the present worth of a dozer with $i=12$ percent.

$$\text{original cost}=\$20,000$$
$$\text{annual O\&M increasing at a rate of } \$2,000/\text{yr}= 10,000/\text{yr/first yr}$$
$$\text{resale value in 5 yr}= 5,000$$

B.3 Same as Problem B.1 except cost of maintenance and repairs increase by 10 percent per year compounded.

B.4 An existing dragline may be replaced with a new model. Over the next five years costs are estimated as follows ($i=10\%$):

	EXISTING DRAGLINE	NEW DRAGLINE
Present market value	$10,000	$20,000
Estimated market value, 5 yr from now	5,000	10,000
Annual maintenance and repairs	4,000	2,000
Increase in O&M costs each year	5%	5%
Annual costs of insurance, taxes, storage as a percentage of present market value	4%	4%

Which alternative should be selected?

B.5 A bus company and franchise with ten years to run is up for sale. Net income the first year is estimated at $50,000 per year. Each year thereafter the net is expected to drop $5,000 per year.

YEAR	NET INCOME
1	$50,000
2	45,000
3	40,000
etc.	etc.

Resale value at the end of ten years is estimated at $20,000. You are looking for at least a 10 percent return on your investment. What is your maximum bid for this company?

B.6 Your client asks you to check on the feasibility of constructing an apartment house with data as follows:

$$current\ income = \$100,000/yr$$
$$gradient\ increase\ in\ income = 10,000/yr/yr$$
$$current\ costs = 60,000/yr$$
$$gradient\ increase\ in\ costs = 5,000/yr/yr$$

Whatever it costs now, you estimate the value will *increase* at a compound rate of 2 percent per year until resale in ten years. What can she afford to pay for the apartments now if she requires a 20 percent return on investment?

B.7 A young high school graduate just turned 18 is contemplating future career alternatives based on the probable salaries to be earned. Which of the alternatives yields the higher present worth (assume $i = 10\%$, expenses occur at BOY, while income occurs at EOY)?

a. Go directly to work and continue until age 65, starting annual salary $10,000 per year increasing 5 percent per year.

b. Invest two years and $6,000 per year in a junior college education. Work the remaining years until age 65. Starting annual salary of $14,000 per year increasing 6 percent per year.

c. Invest four years and $6,000 per year in a four-year college education. Work remaining years until age 65, starting annual salary of $20,000 per year increasing 8 percent per year.

d. Invest five years and $6,000 per year for a graduate degree and earn a starting salary of $22,000 per year increasing 10 percent per year.

Chapter 8
Annual Payments Method of Comparing Alternatives

(OR, EVALUATING ALTERNATIVES BY COMPARING EQUIVALENT SERIES OF EQUAL PERIODIC PAYMENTS)

KEY EXPRESSION IN THIS CHAPTER

NAW = Net annual worth, the equivalent annual (or other periodic) worth after summing the equivalent annual worth of all costs, incomes and savings.

COMPARING ANNUAL PAYMENTS

In any organization, whether government or private enterprise, when funds become available, there are often a number of competing suggestions for spending the money. Each suggestion usually has its own dedicated sponsors and attractive features. Numerical comparisons of all competing proposals need to be made so the better ones can be selected. One common method of comparing competing alternative proposals is to reduce all cash flows involved in each proposal to an equivalent series of periodic payments. For example, alternative A with cost equivalent to $10,000 per year can be compared to alternative B with a cost of $13,000 per year. Alternative A is the least costly, and may be the better selection provided that productivity, safety, appearance, and other factors are at least equal to the other alternative. The period selected for this example is yearly, but any suitable interval that fits the problem will do (quarterly, monthly, weekly, daily, biennially, etc.).

The equivalent series of periodic payments method may be used for even the most sophisticated analysis. However, since the results of this method are intuitively understandable, it is particularly useful when making presentations to the public or to other decision-making groups with limited background in time-value-of-money concepts. While present worth (PW) might be a difficult concept to present without lengthy preparation, a simple comparison of annual payments is often grasped by even the most casual audience with relative ease.

The major advantage of the annual worth (AW) method of comparing alternatives on the basis of periodic payments is that the complication of unequal lives between alternatives is automatically taken into account without any extra computations. This automatic feature is predicated on the assumption of *equal replacement costs* (see Chapter 7). With this assumption, the AW technique will yield the same decision as the PW technique.

Basic Approach

The recommended approach to solving even the most complex problems involves a three-step approach: (a) subdivide into simple components, (b) solve the components, and (c) fit together the component solutions into a solution for the original complex problem. Applying this basic procedure to annual worth problems involves the recognition of the standard types of cash flow typically encountered. These cash flows occur in three basic types:

1. Single-payment lump sum
2. Periodic series of equal payments
3. Periodic series of nonequal payments
    a. Arithmetic gradient
    b. Geometric gradient
    c. Irregular amounts

In addition to the types of cash flows, the relative points in time at which the cash flows occur must be considered. For instance, when converting a single payment to an equivalent annual series, the single payment may occur before or after the beginning or end of the annual worth series. The periodic series of one component of the problem may not coincide in beginning or ending date with the periodic series needed for solution of the entire problem. Each of these possibilities is discussed in the following paragraphs.

CONVERTING THE THREE TYPES OF CASH FLOW INTO AN EQUIVALENT PERIODIC SERIES

1. *A single-payment lump sum converted into an equivalent periodic series.*
    a. Lump sum payment occurs at the beginning of the series, so that problem time zero (PTZ)=equation time zero (ETZ).

Example 8.1 (*Given P, i, n, find AW*)

A road pavement costing $40,000 per mile is expected to last 20 years with no salvage value. If $i = 6$ percent what is the equivalent annual cost per mile?

SOLUTION

$$A = P(A/P, i, n) = \$40,000(A/P, 6\%, 20) = \underline{\underline{\$3,488/\text{yr}}}$$
$$0.0872$$

     b. Time base variations. If the lump sum payment occurs at any time other than the beginning or end of the periodic series, ($PTZ \neq ETZ$), the lump sum amount is (i) converted to an equivalent lump sum payment at the beginning or end of the desired periodic series by means of the P/F or F/P factors. (ii) The new lump sum amount is converted into an equivalent periodic series as in (a) above by using the A/P factor or the A/F factor. Of course, the two steps may be combined, as illustrated in the equations below.

Typical problem solution equation for cash flow shown in Figure 8.1(a) is

$$A = P(F/P, i, n_1)(A/P, i, n_2 - n_1)$$

Typical problem solution equation for cash flow shown in Figure 8.1(b) is

$$A = F(P/F, i, n_1)(A/P, i, n_2)$$

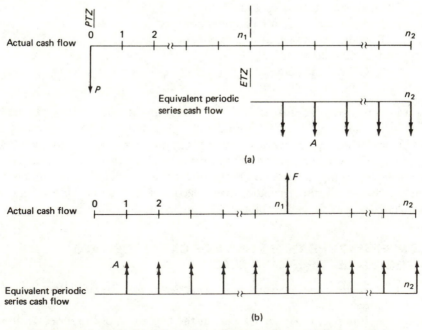

Figure 8.1(a) and (b) Cash flow diagrams from single payments.

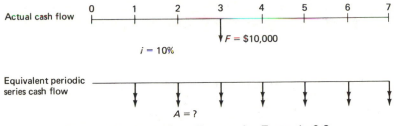

Figure 8.2 Cash flow diagram for Example 8.2.

Example 8.2 (*Given P, i, n, ETZ≠PTZ, find AW*)

A bulldozer is scheduled for a major overhaul costing $10,000 at the end of
the third year of its seven-year economic life. What is the equivalent cost
over the seven-year period for an $i=10$ percent?

SOLUTION
The cash flow diagram for this problem is illustrated in Figure 8.2. The
example problem can be solved by converting the lump sum to an equivalent lump sum at the beginning of the periodic series, and then converting
again to a periodic series, as follows:

$$A=F(P/F, i, n_1)(A/P, i, n_2)$$

$$A=\$10,000\ (P/F, 10\%, 3)\ (A/P, 10\%, 7)$$
$$\qquad\qquad\quad 0.7513 \qquad\qquad 0.2054$$

$$=\$1,543/\text{yr equivalent annual cost}$$

2. *One periodic series converted into another equivalent periodic series.*
   Sometimes one component of a problem involves a series of equal
   payments over a time period that does *not* coincide with the time
   period for the entire problem. Either the beginning or end (or both
   beginning and end) of the series occurs at a time period different
   from that of the main problem. When this occurs the following
   solution sequence is suggested:
   a. The series may be transformed into an equivalent lump sum;
   b. The lump sum is converted to either the beginning or end of the
      problem time;
   c. The new lump sum is converted into an equivalent series of
      periodic payments, with n equal to the n needed to solve the
      whole problem.
   These steps may be combined into one as illustrated by the following equation:

Typical problem solution equation for cash flows shown in Figure 8.3 is

$$A_2=A_1(P/A, i, n_2-n_1)(P/F, i, n_1)(A/P, i, n_3)$$

Note where the periods start and end to make the factors applicable.

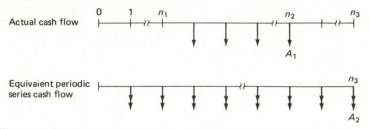

Figure 8.3   Cash flow diagrams—two equivalent periodic series.

Example 8.3   (*Given A, i, n, ETZ≠PTZ, find AW*)

A marine dredge has an expected life of 12 years. The estimated maintenance costs for years 3 through 6 are estimated at $15,000 per year payable
in this case at the *beginning* of the year. What is the equivalent annual cost
over the 12-year life of the dredge ($i=7\%$)? The cash flow diagram for this
problem is shown in Figure 8.4.

SOLUTION
The problem may be solved by converting the series to an equivalent lump
sum at the beginning (or end) of the series, then converting the lump sum to
another equivalent lump sum at the beginning (or end) of the desired
equivalent periodic series, and then converting that last lump sum into the
desired equivalent periodic series with n equal to the proper n for the whole
problem. This procedure is illustrated by the following calculations.

$$A_2 = A_1(P/A, i, n_1)(P/F, i, n_2)(A/P, i, n_3)$$

$$A_1 = \$15,000 \, (P/A, 7\%, 4) \, (P/F, 7\%, 1) \, (A/P, 7\%, 12)$$
$$ 03.3872 0.9346 0.1259$$

$$A_1 = \$5,979 \text{ annual equivalent over 12 yr period}$$

3. *Periodic series of nonequal payments* (*gradient or irregular series*)
   converted into an equivalent periodic series. A gradient series, either
   arithmetic or geometric, may be converted into an equivalent periodic series in a manner similar to paragraph 2 above. The major
   difference is that the first payment in the arithmetic gradient series

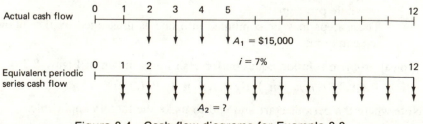

Figure 8.4   Cash flow diagrams for Example 8.3.

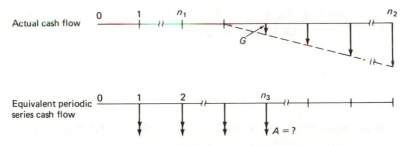

Figure 8.5   Cash flow diagrams—equivalent series.

occurs at the end of the second period (instead of at the end of the first period as in the geometric gradient and equal payment series). If the series involves regular periods but irregular payment amounts (or regular payments amounts but irregular times), then the payments usually must be treated as individual lump sum payments as illustrated in paragraph 1 above.

Typical problem solution equation (arithmetic gradient) for the cash flow shown in Figure 8.5 is

$$A = G(P/G, i, n_2 - n_1)(P/F, i, n_1)(A/P, i, n_3)$$

Example 8.4   (*Given G, i, n, ETZ $\neq$ PTZ, find AW*)

An airport runway is expected to incur no maintenance costs for the first five years of its life. In year 6 maintenance should cost $1,000, in year 7 maintenance should cost $2,000, and each year thereafter until resurfacing the runway is expected to increase in maintenance costs by $1,000 per year. If resurfacing is expected after 15 years of service, what equivalent uniform annual maintenance cost is incurred if $i = 7$ percent?

SOLUTION
The cash flow diagrams for this problem are shown in Figure 8.6. The equations for the solution of the cash flows shown are as follows:

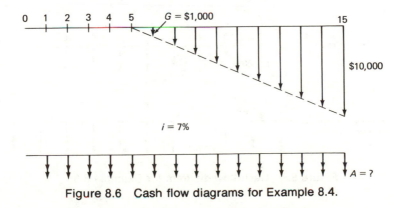

Figure 8.6   Cash flow diagrams for Example 8.4.

$$A = 1,000 \underset{68.3371}{(F/G,7\%,11)} \underset{0.03979}{(A/F,7\%,15)}$$

$$= \$2,719/\text{yr equivalent annual cost}$$

Example 8.5 (*Given geometric gradient C, r, i, n, ETZ≠PTZ, find AW*)

Operating and maintenance costs for one part of a traffic signal system are expected to be $10,000 at the end of the third year and to increase 10 percent per year thereafter for five additional years. The entire system has an expected life of 20 years. With $i = 10$ percent, find the equivalent annual cost of this part of the system for the entire 20-year period.

SOLUTION
The cash flow diagram for this example is shown in Figure 8.7. The example problem illustrated in Figure 8.7 may be solved by converting the geometric gradient series into an equivalent lump sum at the end of year 2, then converting the lump sum into another equivalent sum at the beginning of year 1, and then converting the last lump sum into an equivalent annual series over the 20-year life of the installation.

$$\text{since} \quad r=i, \quad \text{then} \quad P = \frac{Cn}{1+r}$$

$$\text{then} \quad A = \frac{Cn_1}{1+r}(P/F, i, n_2)(A/P, i, n_3)$$

$$A = \frac{10,000 \times 6}{1.10} \underset{0.8264}{(P/F, 10\%, 2)} \underset{0.1175}{(A/P, 10\%, 20)}$$

$$A = \$5,296/\text{yr annual equivalent over the 20-yr life}$$

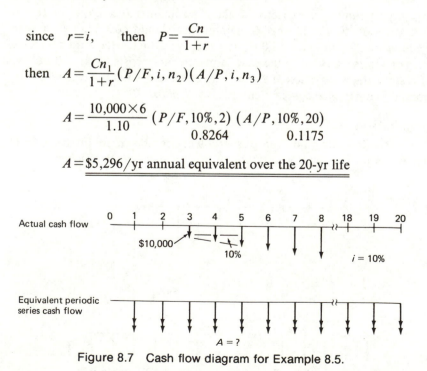

Figure 8.7   Cash flow diagram for Example 8.5.

PERPETUAL LIVES

Some investments, such as land, typically do not depreciate, but maintain their value perpetually. The annual cost of such an investment may be pictured as the annual amount required to pay interest on a loan used to acquire the investment. If at some future time the investment is sold for the same amount as the purchase price, then the loan is paid off, and the only costs have been the interest payments. (The present worth of all of these perpetual annual costs is sometimes referred to as capitalized value.)

Example 8.6   (*Given P, i, n=∞, find AW*)

Some land is needed for right-of-way that will be used for a very long time ($n \rightarrow \infty$). Find the equivalent annual cost of the land valued at \$100,000 if $i = 6$ percent.

SOLUTION

$$A_1 = iP = 0.06 \times \$100,000 = \underline{\underline{\$6,000}}$$

Mental picture: The only cost involved is the annual payment of 6 percent interest on the \$100,000 loan. The \$100,000 is termed capitalized cost by some engineering economists. This is unfortunate because the term capitalized cost is better known in accounting circles as expenditures that are depreciable over two or more years.

A SERIES OF EQUAL LUMP SUM EXPENDITURES OCCURRING AT PERIODIC INTERVALS SEVERAL YEARS APART

Frequently a series of lump sum expenditures is scheduled for intervals of several years. For instance, painting may be needed every 4 years, new HVAC units every 12 years, and so on. The annual costs equivalent to these lump sums may be found as illustrated in the following example.

Example 8.7   (*Given P, periodic lump sum payments, F, i, n, find NAW*)

A civic center with an expected life of 40 years is proposed at an estimated cost of \$7,000,000 including \$1,000,000 for the land, and \$6,000,000 for construction of the center. The following maintenance costs are anticipated.

paint every 4 yr	\$40,000/each painting
replacement air conditioner system every 10 yr	\$500,000/each replacement
replacement wood floors and seats every 20 yr	\$1,500,00/each replacement

Resale value of salvageable components at the end of 40 years is \$500,000.

The civic center will be financed with 40-year revenue bonds at 9 percent, paid out in equal annual installments. How much annual revenue must be obtained to finance the project? The land value remains constant at $1,000,000 (a common assumption in engineering cost analysis studies). This problem may be solved by any of several approaches.

SOLUTION 1   SINKING FUND METHOD
Set up a savings account (sinking fund) for each major cost. The first cost of the center is $6,000,000 and the annual equivalent payment to pay off in 40 years is A_1. In addition, interest payments on a loan to buy the land are provided by A_2. Then an equal amount ($A_3 + A_4 + A_5$) is deposited annually so that $40,000 will accumulate every 4 years for painting, $500,000 will accumulate every 10 years for replacing the air conditioning unit, and $1,500,000 will accumulate every 20 years for replacing the wood floors and seats. This takes care of the expenses as they arise, but also gives a balance in the account at the end of 40 years ready to be spent for new paint, air conditioning, and floors and seats just when the structure is scheduled for demolition. This balance may then be treated as additional salvage value and counted as equivalent annual income in A_6.

first cost of $A_1 = \$-6,000,000(A/P, 9\%, 40)$ $= -\$558,000/\text{yr}$
civic center

interest on $A_2 = -1,000,000 \times 0.09$ $= -\quad 90,000$
value of land

paint $A_3 = -40,000 \, (A/F, 9\%, 4)$ $= -\quad 8,750$
$ 0.2176$

air
conditioning $A_4 = -500,000 \, (A/F, 9\%, 10)$ $= -\quad 32,900$
$ 0.0658$

wood floors $A_5 = -1,500,000 \, (A/F, 9\%, 20)$ $= -\quad 29,250$
and seats $ 0.0195$

salvage value $A_6 = +(500,000 + 40,000 + 500,000$
plus balance $ + 1,500,000)(A/F, 9\%, 40)$ $= +\quad \underline{7,620}$
in accounts $ 0.0030$

net annual worth $A_1 + A_2 + A_3 + A_4 + A_5 + A_6$ $= \underline{\underline{-\$711,280/\text{yr}}}$

An annual revenue of $-\$711,280$ is needed to finance the project. This will result in a debt of $2,540,000 just before liquidation of the last accounts and salvage value, which are needed to finally balance the account.

SOLUTION 2   AMORTIZE OVER THE PERIODIC INTERVAL
Each periodic lump sum amount may be separated from the original cost of the project and amortized over each interval between periodic payments.

The advantage of this method is that all accounts (except salvage value) are paid down to zero at the end of 40 years. Assume the contract for construction is divided among four contractors. The painting contractor gets $40,000, the air conditioning contractor gets $500,000, the wood floors and seats contractor gets $1,500,000, and the remaining $3,960,000 goes to the general contractor. All of the $6,000,000 construction money is borrowed at one time but paid off in separate accounts over the life of each account. For instance, the $40,000 borrowed for painting is paid off over four years. Then another $40,000 is borrowed every fourth year to pay for the repainting required at the end of the fourth year, the eighth year, and so on. At the end of 40 years the painting loan account is again paid down to zero as it has been at the end of every fourth year.

paint	$A_1 = \$-40,000(A/P, 9\%, 4)$ 0.3087	$= -\$\ 12,350/\text{yr}$	
interest on land	$A_2 = -1,000,000 \times 0.09$	$= -\ \ 90,000$	
air conditioning	$A_3 = -500,000(A/P, 9\%, 10)$ 0.1558	$= -\ \ 77,900$	
wood floors and seats	$A_4 = -1,500,000(A/P, 9\%, 20)$ 0.1095	$= -\ \ 164,250$	
remainder of civic center	$A_5 = -3,960,000(A/P, 9\%, 40)$ 0.0930	$= -\ \ 368,280$	
salvage value	$A_6 = +500,000(A/F, 9\%, 40)$ 0.0030	$= +\ \ \ \ \underline{1,500}$	
net annual worth	$A_1 + A_2 + A_3 + A_4 + A_5 + A_6$	$= -\$711,280/\text{yr}$	

GIVEN THE ANNUAL SERIES, FIND THE GRADIENT

Many practical situations occur where the present level of income is not adequate to cover the payments required to amortize the capital cost. Whenever the present worth is greater or less than the equivalent annual series a gradient may be found to make up the difference. This type problem is illustrated in the following example.

Example 8.8 (*Given P, A, G, periodic lump sum payments, F, i, n, find NAW, G*)

A 100-unit university married housing project (that uses state bond funds at about 6%) is expected to incur the following costs and income.

cost new	$1,000,000
maintenance	$100,000/yr increasing $2,000/yr
major overhaul every 10 yr	$200,000
salvage value at the end of 40 yr	$100,000

a. What annual income is required for this project to break even?
b. The initial annual rental income is $1,500 per unit per year ($150,000/yr for the entire project). An annual increase in rent is contemplated, since this initial income level is not sufficient to cover all costs. What annual arithmetic gradient increase will be necessary every year so that income over the 40-year life will be sufficient to cover all costs (provided the books need not be balanced until the end of the 40 years).

SOLUTION

cost new $\quad A_1 = \$1,000,000(A/P,6\%,40) \qquad = -\$\ 66,500$
$$\qquad\qquad\qquad\qquad 0.0665$$

maintenance $\quad A_2 = \qquad\qquad\qquad\qquad\qquad -\ 100,000$

maintenance $\quad A_3 = 2,000(A/G,6\%,40) \qquad = -\quad 24,720$
gradient $\qquad\qquad\qquad 12.359$

major overhaul $\quad A_4 = 200,000(A/F,6\%,10) \qquad = -\quad 15,180$
$$\qquad\qquad\qquad\qquad 0.0759$$

balance
in overhaul
account
at EOY 40 $\quad A_5 = (200,000+100,000)(A/F,6\%,40) = +\quad 1,950$
plus salvage $\qquad\qquad\qquad 0.0065$

net annual worth $\qquad\qquad\qquad\qquad = -\$204,450$

a. The annual income required for breakeven is $-\$204,450$.
b. The required gradient is found as:

$-204,450$ annual income required
$+150,000$ actual income
$-\ 54,450$ deficit per year

required gradient $G = 54,450\,(G/A,6\%,40) = \$4,406$
$$\qquad\qquad\qquad 1/12.359$$

The project could be supported with an initial annual income of $150,000 the first year and an increase of $4,406 each year thereafter.

COMBINED VARIETY OF PAYMENT TIMES AND SEQUENCES CONVERTED TO EQUIVALENT ANNUAL WORTH

A wide variety of payment times and sequences can be combined to produce an equivalent annual series. In each case the problem is divided into basic components. The components are then converted into terms of an equivalent series over the life of the project and then summed. Care must be taken when summing that income is plus and cost is minus.

Example 8.9 (*Given P, A, G, F, ETZ≠PTZ, find NAW*)

A backhoe is expected to incur the following costs and income:

cost new	−$60,000
annual O & M	−$7,000 the first year
with increases of	−$1,000 each year
major overhaul after 5 yr	−$20,000
extra repair costs in years 7–10	−$1,500/yr
resale value after 10 yr	+$10,000 (income)

What must be the equivalent annual income to offset these cost and income items if $i=20$ percent?

SOLUTION

cost new $A_1 = -\$60,000(A/P, 20\%, 10)$ $= -\$14,310$

annual O & M $A_2 = -7,000/\text{yr}$ $= -\ 7,000$

annual increase $A_3 = -1,000(\underset{2.0739}{A/G, 20\%, 10})$ $= -\ 3,074$

major overhaul $A_4 = -20,000(\underset{0.4019}{P/F, 20\%, 5})(\underset{0.2385}{A/P, 20\%, 10})$

                              $= -\ 1,917$

extra repair $A_5 = -1,500(\underset{2.5887}{P/A, 20\%, 4})(\underset{0.3349}{P/F, 20\%, 6})$
cost years
7–10 $\times (\underset{0.2385}{A/P, 20\%, 10})$

                              $= -\ 310$

resale value $A_6 = +10,000(\underset{0.0385}{A/F, 20\%, 10})$ $= +\ \underline{1,385}$

annual income required to break even with costs, ΣA_{1-6} $= \ \underline{\underline{\$26,226}}$

SELECTING THE BETTER ALTERNATIVE BY COMPARISON OF EQUIVALENT ANNUAL WORTHS

As discussed in Chapter 7, the comparison of alternatives by various methods should yield the same answer regardless of the method used. Thus the discussion in Chapter 7 dealing with types of alternatives and criteria for mutually exclusive alternatives is equally relevant here.

    Annual worth comparisons are often the quickest and easiest to perform of all the techniques, because the problem of unequal lives is *automati-*

cally handled without further mathematical manipulation in that *equal replacement costs are automatically implicit.* The following example illustrates this point.

Example 8.10 (*Given two alternatives with P, A, F, i, n, find and compare NAW*)

The problem statement is given in Example 7.3 (Chapter 7).

SOLUTION
In this example the life of type A pavement is ten years while the life of type B pavement is five years. In comparing the equivalent annual costs in typical textbook problems, frequently no adjustment is made for escalating replacement costs at the end of the shorter life. This implies that the replacement cost is identical to the first cost. Since real-world problems often involve escalating replacement costs, numerous problems later in this text will illustrate methods of dealing with escalating replacement costs.

Type A

A_1 = annualized purchase cost $= -20,000 \underset{0.1359}{(A/P, 6\%, 10)} = -\$2,718$

A_2 = annualized maintenance cost $= -1,000$

A = annualized salvage $= 2,500 \underset{0.0759}{(A/F, 6\%, 10)} = +\underline{\quad 190}$

net annual worth of type A $= A_1 + A_2 + A_3$ $= -\$3,528$

Type B

A_1 = annualized purchase cost $= -5,000 \underset{0.2374}{(A/P, 6\%, 5)} = -\$1,187$

A_2 = annualized maintenance cost $= -2,000$

net annual worth of type B $= A_1 + A_2$ $= -\$3,187$

The answer is to *choose type B* because it has the lowest net annual worth of costs.

This is the same decision reached when working the problem by the present worth method (Example 7.3) even though the unequal lives were not considered in the annual worth method (or were they?). To check, the present worth of costs calculated in Example 7.3 are converted to annual worth over the *ten*-year period for *both* types as follows.

$NAW \text{ (type A)} = -25,964 \underset{0.1359}{(A/P, 6\%, 10)} = -\$3,529$

$NAW \text{ (type B)} = -23,456 \underset{0.1359}{(A/P, 6\%, 10)} = -\$3,188$

These are the same answers (within roundoff error) achieved by the AW technique. Therefore, the conclusion is that unequal lives were accounted for by assuming equal replacements.

The computational advantages to the annual worth technique becomes obvious when unequal lives such as 7 and 5 years or 10 and 11 years are being considered. In the first case, the lowest common multiple of 7 and 5 is 35. In the second case the lowest common multiple of 10 and 11 is *110*. Calculations based on present worth would be very tedious and often time consuming. The ease in annual worth comparisons is illustrated in the following example.

Example 8.11 *(Given three alternatives with $P, A, F, i, n_1 \neq n_2 \neq n_3$, find and compare NAW)*

Three different artificial turfs are available for covering the playing field in a college football stadium. The costs associated with each are tabulated as follows (assume $i = 15\%$).

	TURF KING	TURF EASE	TURF MAGIC
Cost new (installed) ($)	− 540,000	−608,000	−467,000
Annual maintenance cost ($)	− 2,300	− 1,600	− 2,500
Expected life (yr)	12	15	10
Salvage value ($)	+ 54,000	+ 57,000	+ 40,000

SOLUTION

Since different lives are involved the lowest common multiple of 10, 12, and 15 would be 60. Therefore, to simplify the calculations, use the AW technique.

$$NAW\,(\text{King}) = -540,000\,(A/P, 15\%, 12) - 2,300$$
$$\qquad\qquad\qquad\qquad 0.1845$$

$$+\ 54,000\,(A/F, 15\%, 12) \qquad = -\$100,070$$
$$\qquad\qquad 0.0345$$

$$NAW\,(\text{Ease}) = -608,000\,(A/P, 15\%, 15) - 1,600$$
$$\qquad\qquad\qquad\qquad 0.1710$$

$$+\ 57,000\,(A/F, 15\%, 15) \qquad = -\$104,370$$
$$\qquad\qquad 0.0210$$

$$NAW\,(\text{Magic}) = -467,000\,(A/P, 15\%, 10) - 2,500$$
$$\qquad\qquad\qquad\qquad 0.1993$$

$$+\ 40,000\,(A/F, 15\%, 10) \qquad = -\$93,600$$
$$\qquad\qquad 0.0493$$

The results indicate that *Turf Magic* is the better selection since it represents the lowest net annual worth.

Remember that the decision is strongly dependent on the value of interest used. A different value of i may result in a different answer.

In the preceding problem the assumption of equal replacement costs was assumed valid. If this assumption is incorrect, then the problem must be worked differently, as follows.

Example 8.12 (*Given two alternatives with P, rising replacement costs, i, $n_1 \neq n_2$, find and compare NAW*)

Two roofs are under consideration for a building needed for 20 years. Their anticipated costs and lives are:

	ROOF C	ROOF D
Cost new ($)	50,000	25,000
Replacement cost ($)	—	rise 10%/yr
Life of roof (yr)	20	10
Salvage value @ 20 yr ($)	0	0
Interest rate (%)	12	12

SOLUTION

Roof C

$$A_1 = -\$50,000 \ (A/P, 12\%, 20)$$
$$0.1339 \qquad\qquad = -\$6,695/\text{yr}$$

Roof D

$$A_2 = -\$25,000 \ (A/P, 12\%, 20)$$
$$0.1339 \qquad\qquad = -\$3,348/\text{yr}$$

$$A_3 = -\$25,000 \ (F/P, 10\%, 10) \ (P/F, 12\%, 10) \ (A/P, 12\%, 20)$$
$$2.5937 \qquad\quad 0.3220 \qquad\quad 0.1339$$
$$= - \ 2,796/\text{yr}$$

net annual worth roof $D = A_2 + A_3$ $\qquad\qquad = -\$6,144/\text{yr}$

The two $25,000 roofs still cost less since the replacement cost is rising at a slower rate (10%/yr) than the interest (12%) that would be tied up in the investment in the second 10-year half of the 20-year life of the $50,000 roof. In the practical situation, account should also be taken of possible business interruption and other inconveniences associated with installing a new roof.

The previous examples dealt with costs only. The same procedures are involved if incomes as well as costs occur.

Example 8.13   *(Given two alternatives with P, income, A, F, i, $n_1 \neq n_2$, find and compare NAW)*

A consulting firm has a choice of two word processing machines. Each machine will permit a reduction in clerical staff. Estimated costs, lives, and salary savings are tabulated below. Assume $i = 15$ percent.

	PROCESSOR A	PROCESSOR B
Cost new ($)	−8,000	−17,000
Salary savings/yr ($)	+3,500	+ 4,800
Life (yr)	5	6
Salvage value ($)	0	+ 4,000

SOLUTION

Whenever the "do nothing" alternative is possible, it should always be considered. In this case the procedure is to calculate the net annual worth for $i = 15$ percent and select the alternative with the greatest net annual worth (select "do nothing" if *NAW* of the two alternatives is negative).

$$NAW(A) = - 8,000 \underset{0.2983}{(A/P, 15\%, 5)} + 3,500 = +\$1,114$$

$$NAW(B) = - 17,000 \underset{0.2642}{(A/P, 15\%, 6)} + 4,800$$

$$+ 4,000 \underset{0.1142}{(A/F, 15\%, 6)} \qquad = \ +\$765$$

The results indicate that *processor A* has the greater positive net annual worth.

SUMMARY

Methods are developed for transfering a variety of different types of cash flows into an equivalent series of periodic payments. Since many alternative engineering investment decisions can be represented by their respective cash flows, these alternatives can be reduced to equivalent series of periodic payments and selections made of the better alternatives on a simple numerical basis. For alternatives involving costs *only*, the least equivalent annual cost alternative is selected. For alternatives involving incomes as well as costs, the alternative with the highest net annual worth (*AW* of incomes minus *AW* of costs) is selected. Where circumstances permit, the "do nothing" alternative is considered, and selected if none of the proposed alternatives yield a positive net annual worth at the stated value of *i*.

Annual worth analysis often involves fewer calculations than present worth analysis if differing lives are under consideration, because annual worth implicitly assumes equal replacement conditions (unless otherwise noted) and the least common multiple of the different lives without extra calculations. Annual worth analysis is often more easily understood, especially where rates or unit costs are being determined. For example, calculation of the annualized cost of a toll bridge provides the information necessary to calculate the toll rate for vehicles using the toll bridge.

The use of the word *annual* is merely for convenience in considering periodic cash flow conditions. The periods of time involved may be monthly, quarterly, daily, or any other. Remember that the compounding period for interest is the same as the period used to measure n, the life of the proposal.

PROBLEMS FOR CHAPTER 8, ANNUAL WORTH ANALYSIS

Problem Group A

Given three or more of the variables P, A, F, i, n, find the annual equivalent cost or income. Where alternatives are proposed, compare and indicate which is the better.

A.1 Plans are under consideration to acquire a public boat ramp and park at a lake. There is an immediate need for 5 acres with a need in another ten years for an additional 5 acres. The first 5 acres may be acquired for about $40,000 and the second 5 acres may be purchased for an additional $40,000. If purchased when needed, ten years from now, it will probably have increased in value about $1,500 each year between now and then.
 a. If money can be borrowed at 6 percent, which is the most economical alternative?
 b. What intangible considerations might cause you to recommend the other alternative?

A.2 a. A parking structure is proposed for on-campus use. It is estimated to cost $2,500 per space plus $50 per year for maintenance. Annual parking decals will be sold to pay off the structure. With $n=20$, $i=6$ percent, what must be the annual charge for a decal?
 b. As an alternative, an hourly parking fee is proposed. In addition to the above costs, an attendant is required at $50 per space per year. Assuming an average of ten hours of parking per day, 220 days parking per year, 80 percent occupancy rate, what must be the hourly parking rate? (Calculate cost on an annual basis and divide by number of hours of use per year.)

A.3 A "giant slide" used by the recreation department is in need of major repairs. The slide cost $1,200 10 years ago. To overhaul now and extend its life another 6 years would cost about $400. A new slide will cost $1,400 and have an estimated life of 12 years. Annual maintenance on the existing slide will be about $100 per year if overhauled, while the new slide will cost about $50 per year to maintain. Resale value of both slides is estimated at about $200 regardless of age. With $i=15$ percent, which should be selected?

A.4 A steel bridge on a county highway near the ocean cost $350,000 12 years ago. Although the average annual maintenance costs of $7,200 per year have seemed excessive to the county supervisors, these costs, mainly for painting, have been necessary to prevent severe corrosion in this particular location.

A consulting structural engineer proposes to the county supervisors that he be employed to design and to supervise the construction of a reinforced concrete bridge to replace this steel bridge. He estimates the total first cost of the concrete bridge to be $420,000 and the net salvage from the steel bridge either now or any time in the next 38 years to be $20,000. The required net outlay is therefore $400,000. Since the county receives $500,000 a year from state gas tax funds to be used for county highway improvement, the engineer points out that this bridge could be financed without borrowing. His economic comparison is as follows.

ANNUAL COST OF PROPOSED BRIDGE

Depreciation (based on 50-yr life)=$400,000÷50	=$8,000	
Annual maintenance cost	= 200	
Total annual cost	=$8,200	

ANNUAL COST OF PRESENT BRIDGE

Depreciation (based on 50-yr life)=$350,000÷50	=$ 7,000	
Annual maintenance cost	= 7,200	
Total annual cost	=$14,200	

Comment on the engineer's analysis. Show the calculations and results you would use to present the comparative annual costs of the two bridges. Use $i=6$ percent.

A.5 Solar hot water heaters are available for $600 installed. No maintenance is expected for the first 15 years. After 15 years they are expected to have a value of $100. A developer asks your advice on whether to install the solar heaters in all the new houses in a development in place of conventional gas hot water heaters costing $100 installed. Fuel costs are expected to average $8 per month at the start, with an increase of $0.50 per year over the 15-year period. Maintenance costs on the gas heaters is expected to be negligible, with a resale value after 15 years of $10 ($i=12\%$). Assume end-of-year payments.

a. Which is the most economical alternative for the home buyer, if the house with the solar heater costs an extra $500?

b. In your opinion, do you think customers will pay an extra $500 for a house with the solar heater? Why?

A.6 A commercial parking firm is considering construction of a parking lot as an interim use of a 4-acre tract in Metro City's central business district (CBD). The firm uses a 15 percent minimum acceptable rate of return. The tract can be leased for ten years at $8,000 per year. Construction costs are estimated to be $40,000; 20 percent of the cost will be from available funds, the remaining 80 percent will be financed by a 10 percent, five-year loan. At termination of the interim use as a parking lot, all improvements must be removed; estimated cost $6,000. Operating costs are expected to be $30,000 per year. What is the uniform equivalent annual cost?

A.7 An elementary school crossing guard is currently employed at $3,600 per year. The guard can be replaced by a pedestrian actuated traffic light costing $12,000 initially plus $300 per year for maintenance. Assume a life of ten years and no salvage value for the light, and $i=6$ percent. How does the annual cost of the light compare with the annual cost of the guard? List other benefits of having a crossing guard on duty. List other benefits of the light. Which are you inclined to recommend?

A.8 A front end loader can be had for $10,000 under either of two plans. Which results in the lowest annual cost?

a. Borrow $10,000 at bank and repay in equal annual installments over five-year period at 8 percent on unpaid balance.

b. Borrow $2,000 down payment and repay the $2,000 in equal annual payments over five years at 12 percent. Pay off remaining $8,000 balance owed dealer at $2,000 per year for five years.

Problem Group B

This group involves unit costs, $n_1 \neq n_2$, $ETZ \neq PTZ$.

B.1 Compare the equivalent annual net income from the two draglines ($i=8\%$).

	DRAGLINE A	DRAGLINE B
Full price	$50,000	$80,000
Cash down payment	10,000	50,000
Percent on unpaid balance (loan pd=5 yr)	7	6
*Est. annual production @ $0.25/yd^3	200,000 yd^3	250,000 yd^3
Est. resale value	$8,000/8 yr	$16,000/10 yr
Average annual maintenance and operating cost	$40,000	$45,000

*Use this figure to find gross income.

B.2 An earthmoving machine with an estimated life of ten years will need a scheduled major overhaul at the end of the fifth year, costing $10,000. If $i=10$ percent, what is the equivalent annual cost over the ten-year life?

B.3 The repair costs on an earth moving machine are expected to average $5,000 at end of years 1 through 4 and $8,000 at end of years 5 through 7. If $i=10$ percent, what is the equivalent uniform annual cost over the seven-year period?

B.4 A delivery van was purchased three years ago for $8,000. The expected life was seven years at which time the salvage value was expected to be $600; annual maintenance and operating costs were expected to be $1,700 per year. As result of an accident, $2,500 in repairs are needed now. If repaired, the van will last out the original seven-year period but will have a salvage value of $300. However, annual maintenance and operating expenses have been found to be $1,900 per year rather than the $1,700 estimated prior to purchase.

A new van can be purchased for $10,000. Its estimated life is eight years; its expected salvage value $900; maintenance and operating expenses are estimated to be $1,600. A trade-in of $2,200 will be allowed on the existing van in its unrepaired condition.

Should the existing van be repaired and returned to service or should a new van be purchased if 10 percent return on investment is used?

B.5 A company is considering the following three investments. If their $MARR*$ is 15 percent, which investment, if any, should they select?

*$MARR$=minimum attractive rate of return, the minimum acceptable rate of return.

	P	Q	R
Initial investment	$-\$40,000$	$-\$30,000$	$-\$20,000$
Annual net income	$+\$7,400$	$+\$6,300$	$+\$3,200$
Life (yr)	20	15	20
Salvage value	0	0	0

B.6 A toll bridge is being considered for design and construction. It is estimated the bridge will cost $73,000,000 and require annual operating and maintenance costs of $2,000,000. If 110,000 vehicles are expected to cross the bridge each day, what toll rate should be charged if $i=7$ percent?
  a. Assume 50-year life with no salvage value.
  b. Assume perpetual life.

B.7 Same as Problem B.6 except that $i=10$ percent.

Problem Group C

Arithmetic or geometric gradients.

C.1 A proposed civic center is estimated to cost $10,000,000 and have a life of 45 years. The maintenance costs are estimated to be:

paint and normal repairs	$20,000/yr (billed and paid at the end of the first year and each succeeding year) and increasing 10%/yr over the 45-yr life.
replace air conditioner, floor tile, and seat upholstery every 15 yr	the cost 15 yr from now is estimated at $2,000,000 and the cost is expected to double for the replacement required at the end of the thirtieth year.

After 45 years it is estimated the entire civic center will have a salvage value of $500,000.
  a. If the civic center will be financed with 45-year revenue bonds at 8 percent to be paid off in equal annual installments, how much annual revenue is required to pay off the bonds plus paint, repairs, and replacements as listed above?
  b. A group of citizens feel that revenue from the civic center itself should amount to $200,000 the first year (use EOY receipt) and increase 10 percent per year thereafter for the 45-year life. Under the 8 percent, 45-year financing of (a), how much annual subsidy is needed to pay for the P&I payments, maintenance, and other costs listed above?

C.2 A contractor has just borrowed $100,000 from the bank at $i=1$ percent per month, and purchased a dragline. He expects to keep it for 120 months. The following operating and maintenance costs are anticipated.

monthly operator's salary, maintenance, lube & filters	$2,000/month
expected gradient increase in monthly costs each month	100/month
replace cables and cutting edges at the end of every 12 months	3,000/yr
complete overhaul of engines and tracks at the end of every 30 months	10,000/each time

At the end of the 120-month period the dragline should sell for about $20,000 in "as is" condition, without a final overhaul or replacement of cables. The dragline is expected to move 50,000 yd^3 the first month. Due to continued wear, age, and down time, the amount will *decrease* by 100 yd^3 per month.

How much should the contractor charge in dollars per cubic yard to cover all his dragline costs plus 10¢ per cubic yard for overhead and profit?

C.3 A proposed airport terminal is expected to cost $4,000,000 with a life of 40 years. The operating and maintenance costs are estimated as follows.

operations   1–40 yr, begin at $100,000/yr, rising $5,000/yr

maintenance   1–20 yr, begin at $50,000/yr, rising 10% the second
               year and 10% more each year thereafter

end of year   10 yr, $20,000 major rehabilitation

maintenance   21–40 yr, begin @ $70,000/yr rising $1,000/yr
               the twenty-second year and $1,000 more each year
               thereafter

All costs of the terminal including construction will be financed at 6 percent and will be repaid by charging each passenger a fee.

a. Assuming 1,000,000 passengers per year, what fee would be charged per passenger?

b. Assuming 70,000 passengers the first year and an added 100,000 passengers each year thereafter, what fee should be charged?

C.4 Same as Problem C.3 except the 1–20 year maintenance increases at the rate of $5,000/yr/yr.

C.5 Operating and maintenance costs for a package sewage treatment plant are expected to be $10,000 per year the first year and increase $1,000 per year thereafter for five years. If $i = 10$ percent, what is the equivalent annual cost? All costs are charged at the *end* of the year in which incurred.

C.6 A large recreation center is proposed at an estimated $10,000,000 cost and a life of 40 years. The following maintenance costs are anticipated.

clean and paint every 4 yr $60,000/each painting

replace mechanical systems every 10 yr $800,000/each replacement

replace wood floors and seats every 20 yr $1,000,000/each
                                         replacement

The resale value of salvageable components at the end of 40 years is estimated at $2,000,000. The recreation center will be financed with 40-year bonds at 6 percent paid out in equal annual installments. How much annual revenue is needed to finance this project?

C.7 A bridge is needed at a certain location and designs in both steel and concrete are available with the following cost estimates. The steel bridge is estimated to cost $200,000 now with a salvage value of $30,000 at the end of 20 years.

Maintenance costs on the steel bridge are estimated at $1,000 per year for the first 3 years, and then increase by $200 per year for the next 17 years ($1,200 the fourth year, $1,400 the fifth year, etc.).

The concrete bridge is estimated to cost $180,000 with a negative salvage value of $20,000. (It will cost $20,000 to demolish the structure.)

Maintenance costs on the concrete bridge are estimated at $2,000 per year for the full 20 years.

Compare the annual costs of the bridges, using $i=7$ percent. Assume that maintenance costs for both bridges are paid in advance at the beginning of each year. For example, the maintenance cost for the first year is paid out at the *beginning* of year 1.

C.8 A contractor has a contract to construct a 12,000-ft-long tunnel in 30 months. She is trying to decide whether to do the job with her own forces or subcontract the job. She asks you to calculate the equivalent *monthly cost* of each alternative. Her $i=1.75$ percent per month. Production under both alternatives will be 400 ft per month.

Alternative A: Buy a tunneling machine and work with contractor's own forces.

cost of tunneling machine	$500,000 paid now
salvage value of machine @ EOM 30	$100,000
cost of labor and materials	$80/ft for the first 10 months, increasing by 0.5%/month at EOM each month thereafter ($80.40/ft at EOM 11, $80.802/ft at EOM 12, etc.)

Alternative B: Subcontract the work. Cost is $200 per foot of advance of the tunnel heading, for the full 12,000 ft.

C.9 Same as Problem C.8 except that cost of labor and materials increases by $0.50 per month at EOM each month thereafter ($80.50/ft at EOM 11, $81.00/ft at EOM 12, etc.).

C.10 Same as Problem B.6 except maintenance costs will increase by $200,000 per year ($2,000,000 EOY 1, $2,200,000 EOY 2, etc.).

Chapter 9
Future Worth Method of Comparing Alternatives

KEY EXPRESSION IN THIS CHAPTER

NFW = Net future worth, the equivalent lump sum value at a specified future date after summing the equivalent future values of all costs, incomes, and savings

COMPARING VALUES AT SOME FUTURE DATE

Future worth (FW) comparisons of alternative investments or projects are frequently used when the owner or manager expects to sell or otherwise liquidate the investment at some future date and wants an estimate of net worth at that future date. The future worth is simply the amount of cash or equivalent value the owner will have at that future date.

Future worth also is employed to estimate future population or other nonmonetary levels of activity such as traffic count, consumer demand for water, electricity and other products, goods, and services.

The application of the future worth method is similar to the present worth and annual worth methods and values are readily convertible between all three by application of appropriate equivalence factors. In comparing alternatives the lives must be of similar length, or an adjustment must be made so that the alternatives are compared over similar time spans.

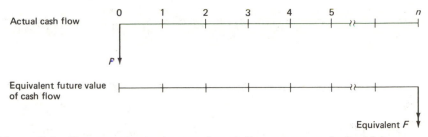

Figure 9.1   Diagram of actual present cash flow versus equivalent future cash flow.

Summary of Three Types of Cash Flow Converted to Equivalent Future Value

Cash flows previously have been classified into the following three basic types:

1. Single payment lump sum
2. Periodic series of equal payments
3. Periodic series of nonequal payments
   a. Arithmetic gradient
   b. Geometric gradient
   c. Irregular amounts and/or times

The future value of each of the above cash flows depends upon the time of occurrence of the flow, as well as the interest rate. Each type is discussed in turn in the following paragraphs.

1. *Single payment lump sum, converted into an equivalent future value.*
   a. Lump sum payment occurs at some time prior to the point in time for which the future value is calculated. Interest at rate i is earned on the payment from the time the payment is made. See Figure 9.1.

Typical problem solution equation is

$$F = P(F/P, i, n)$$

Example 9.1   (*Given P, i, n, find F*)

If $10,000 is invested now at 6 percent compounded, how much will accumulate (principal and interest) in 12 years?

SOLUTION

$$F = 10,000(F/P, 6\%, 12) = \underline{\underline{20,120}}$$
$$\underset{2.0122}{}$$

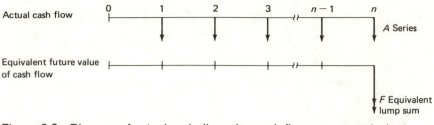

Figure 9.2   Diagram of actual periodic series cash flow versus equivalent future cash flow.

2. *Periodic series of equal payments, converted into an equivalent future value.*
    a. The periodic series ends at the same point in time for which the future value is calculated. Interest at rate i is earned on all payments made. See Figure 9.2.

Typical problem solution equation is

$$F = A(F/A, i, n)$$

Example 9.2   *(Given A, i, n, find F)*

If $1,000 per year is invested at the end of each year at 6 percent for 12 years, what will be the total amount of principal and interest accumulated at the end of the twelfth year after the twelfth payment?

SOLUTION
The cash flow diagram (Figure 9.3) for this example is similar to that above.

$$F = 1,000(F/A, 6\%, 12) = \$16,870$$

b. The periodic series ends at some point in time *other* than the point in time for which the future value is calculated. See Figure 9.3.

Typical problem solution equation is

$$F = A(P/A, i, n_1)(F/P, i, n_2)$$

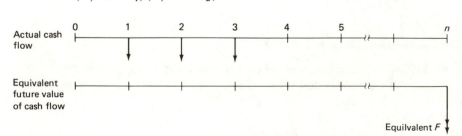

Figure 9.3   Diagram of a short series cash flow versus equivalent future cash flow.

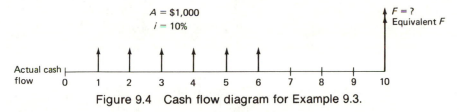

Figure 9.4 Cash flow diagram for Example 9.3.

Example 9.3 (*Given A, i, n, ETZ $\neq$ PTZ, find F*)

An income of $1,000 per year is expected from a certain enterprise at the end of years 1 through 6. These funds will be invested at 10 percent as soon as received and will remain invested at that interest until the end of year 10. How much will the account be worth at the end of year 10?

SOLUTION
The cash flow diagram for this problem is shown in Figure 9.4.

The problem can be solved as follows.

$$F = A(P/A, i, n_1)(F/P, i, n_2)$$
$$F = \$1,000 \, (P/A, 10\%, 6) \, (F/P, 10\%, 10)$$
$$\qquad\qquad\quad 4.3553 \qquad\qquad 2.5937$$
$$= \$11,300 \text{ future value of the series}$$

3. *Periodic series of nonequal payments, converted into an equivalent future value.* If the nonequal payments form an arithmetic series or a geometric gradient series, the conversion to future value proceeds very similarly to the equal payment series described in paragraph 2 above. (Remember, however, that the arithmetic gradient first payment occurs at the end of the second period rather than the first as in the geometric gradient and the equal payment series.) See Figure 9.5.

Typical problem solution equation, arithmetic gradient is,

$$F = G(P/G, i, n_1)(F/P, i, n_2)$$

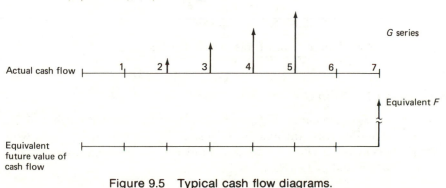

Figure 9.5 Typical cash flow diagrams.

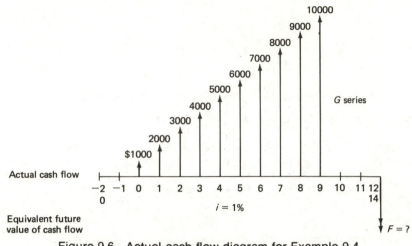

Figure 9.6   Actual cash flow diagram for Example 9.4.

Example 9.4 (*Given G, i, n, ETZ≠PTZ, find F*)

A contractor is arranging financing for a project he is about to construct. He needs to borrow $1,000 at the beginning of the first month, $2,000 at the beginning of the second month, and increase the amount borrowed by $1,000 per month through the beginning of the tenth month. He will not need to borrow any more, and expects payment for the entire job from the owner at the end of the twelfth month. The bank agrees to loan the money at 1 percent interest compounded monthly on the amount borrowed. How much will the contractor have to repay at the end of the twelfth month?

SOLUTION

See Figure 9.6. The problem can be solved by shifting the *ETZ* back to −2, and using

$$F = G(P/G, i, n_1)(F/P, i, n_2)$$

$$F = \$1,000\ (P/G, 1\%, 11)\ (F/P, 1\%, 14)$$
$$50.8068 1.1495$$

$$= \underline{\$58,400}\ \text{future value of this gradient series}$$

Example 9.5 (*Given C, r, i, n, find F*)

Make a series of ten annual deposits beginning with $1,000 at the end of the first year and increase the amount deposited by 10 percent per year, with $i = 10$ percent on all sums on deposit. Find the accumulated principal and interest at the end of 12 years.

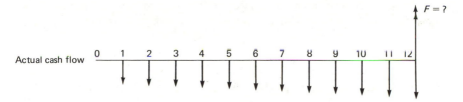

Figure 9.7   Actual cash flow diagram for Example 9.5.

SOLUTION

See Figure 9.7. The problem can be solved as follows.

$$\text{since } i = r, \qquad P = \frac{Cn}{1+r}$$

$$F = \frac{Cn}{1+r}(F/P, i, n)$$

$$F = \frac{\$1,000 \times 10}{1.10}(F/P, 10\%, 12)$$

$$= \$28,530 \text{ future value of this gradient series}$$

Comparison of Alternatives

As with present worth and annual worth, comparisons between alternatives may be made using future worth equivalents. Two types of alternatives commonly occur: those involving costs only and those involving incomes (or benefits) as well as costs. For problems involving costs only, the alternative with the lowest equivalent future worth (cost) is usually preferable. For alternatives involving incomes as well as cost the alternative with the greatest net future worth (NFW) is usually better, with due consideration being given the "do nothing" alternative. The following example illustrates the method.

Example 9.6   *(Given competing alternatives with P, A, i, $n_1 \neq n_2$, find and compare the NFW)*

A construction company is considering the purchase of one of the two new front end loaders whose data are listed below.

	FRONT END LOADER A	FRONT END LOADER B
First cost ($)	− 110,000	− 40,000
Annual net income ($)	+ 16,000	+ 13,000
Useful life (yr)	10	5
Replacement cost escalation	NA	10%/yr compounded
Salvage value ($)	+ 10,000	+ 10,000

1. For an i of 8 percent, which loader, if any, should be selected?

$$NFW(A) = -\ 110,000\ (F/P,8\%,10) + 16,000\ (F/A,8\%,10) + 10,000$$
$$2.158914.4866$$

$$= +\ \underline{\$4,307}$$

$$NFW(B) = -\ 40,000\ (F/P,8\%,10) - 40,000\ (F/P,10\%,5)\ (F/P,8\%,5)$$
$$2.15891.61061.4693$$
$$+\ \ 13,000\ (F/A,8\%,10) + 10,000 + 10,000\ (F/P,8\%,5)$$
$$14.48661.4693$$

$$= +\ \underline{\$22,005}$$

SOLUTION 1

Loader B, with the greatest positive NFW, is preferred.

2. If $i = 20$ percent, which loader, if any, should be selected?

$$NFW(A) = -\ 110,000\ (F/P,20\%,10) + 16,000\ (F/A,20\%,10) + 10,000$$
$$6.191725.958$$

$$= -\underline{\$255,759}$$

$$NFW(B) = -\ 40,000\ (F/P,20\%,10) - 40,000\ (F/P,10\%,5)\ (F/P,20\%,5)$$
$$6.19171.61062.4883$$
$$+\ \ 13,000(F/A,20\%,10) + 10,000 + 10,000\ (F/P,20\%,5)$$

$$= -\underline{\$35,637}$$

SOLUTION 2

Since neither alternative results in a positive NFW, the preferred solution is "do nothing," providing that alternative is available.

THE "DO NOTHING" ALTERNATIVE

When considering an investment involving income, frequently the investor has the option of either investing in one of several alternatives or not investing at all (known as the "do nothing" alternative). However, in some instances the "do nothing" alternative is not available even when income is expected. The investment under consideration may be a part of a larger operation that would not function properly without this investment as an integral part. For instance, a contractor who owns a construction firm may set up a separate equipment company to own and maintain his heavy construction equipment. This equipment company may not make any profit on its own but could contribute significantly to the profitability of the contractor's construction firm. Similarly a department store may operate a lunch counter or a branch post office at a loss in order to attract more customers into the store. Thus, even when income is involved the "do nothing" alternative may or may not be available depending upon the particular proposal under consideration.

In part (2) of Example 9.6, the "do nothing" alternative is the better selection (if it is available) since the company can make more money by investing the purchase price of either loader in an investment earning 20 percent [if the company is borrowing at $i = 20\%$, then just repaying some of their debts represents an investment that saves (yields) 20% rate of return] rather than in investing in either piece of equipment.

This leads naturally to the question of what rate of return will these two possible alternatives yield the company? Obviously they both will yield greater than 8 percent since in part (1) the NFW's are positive at 8 percent, and, they both yield less than 20 percent since in part (2) the NFW's are negative at 20 percent. The methodology of finding the actual rate of return for each alternative is presented in the next chapter.

SUMMARY

In this chapter, the methods for finding and comparing future worths of a variety of cash flow combinations are presented. These methods are similar in many respects to the methods used for finding present worths and annual worths. Future worth is useful for finding the future value of an investment or for comparing alternatives on the basis of net future worth at some specified rate of return. As with present worth, the lives of various alternatives *must be equal* in order for the comparison to be valid.

PROBLEMS FOR CHAPTER 9, FUTURE WORTH

Problem Group A

Given three or more variables, find F; one-step operation or reasonably simple.

A.1 Assume that 30 years ago a trust fund was set up for your benefit with an initial investment of $10,000. The trustees had to choose between investing the funds in (a) a cross section of common stock generally following the value of the Dow Jones Industrial Average (DJIA), (b) a selected mutual fund, or (c) a savings account bearing interest at 5 percent compounded annually. Which alternative would have yielded the greatest total amount of principal plus interest by now, 30 years later.

  a. The increase in value of the investment in the DJIA stocks may be estimated from the following approximate values.

DATE	DJIA
30 years ago	200
now	890

In addition, dividends averaged 4.1 percent of the current stock value and are assumed to have been reinvested as received. (*Hint:* Find i for increase in

value of DJIA and add 4.1% to it to obtain a new i to use in finding the
value of $10,000.)

b. The selected mutual fund advertises that $10,000, if invested in their average
fund 30 years ago, would amount to $94,008 by now, 30 years later.

A.2 A developer client of yours purchased some property for $50,000. The terms
were $10,000 down and a note for $40,000, payable at the rate of $5,000 per
year with $8\frac{1}{2}$ percent interest on the unpaid balance.

a. When is the last payment due?

b. How much is the last payment?

A.3 Traffic on First Street is currently 1,500 cars per day and increasing at the rate
of 6.3 percent per year. When the traffic reaches 4,000 cars per day, you intend
to recommend that new traffic signals be added. How soon will that be if the
present rate of increase continues?

A.4 The main trunk line of a sanitary sewer is being planned for an area experiencing a steady rate of growth. The initial flow through the pipe is expected to be
10,000 gallons per day for the first year. As the pipe ages, a small amount of
infiltration is expected to occur, amounting to about 5 percent of the sanitary
sewage flow each year. Additional development of the area is expected to add
about 300 gallons per day for each new dwelling unit, and about 50 new
dwellings per year are expected on the line every year for the next ten years. At
the end of ten years the development will be completed, but the infiltration is
expected to continue at the same 5 percent compounded per year. For simplicity, assume all new dwellings are connected at EOY.

a. What is the expected flow at the end of 20 years?

b. How much of the flow at the end of 20 years is sanitary sewage, and how
much is infiltration?

Problem Group B

Arithmetic gradient, comparisons.

B.1 A retirement plan requires a 5 percent annual salary contribution from an
employee aged 25 who expects to retire at age 60. He now makes $12,000 per
year and expects annual raises averaging $1,000 per year from now until
retirement. The retirement 5 percent of his salary $(0.05 \times \$12,000 = \$600$ the
first year) is placed in a trust fund at the end of every year ($n=35$) along with a
matching 5 percent from his employer and the fund draws 4 percent interest
compounded. The plan is voluntary, and as an alternative he realizes he can
invest his own 5 percent (no employer matching funds) in mortgages and
securities at an estimated 12 percent return. At age 60, how much would there
be in each alternative fund?

B.2 A friend of yours purchased some land ten years ago for $10,000. At the end of
the first year she paid taxes of $120. Every year thereafter the taxes have
increased by $20 per year each year. Four years ago she had some drainage and
fill work done that cost $1,500. A potential buyer is interested in the property
now and your friend wants to know what selling price to ask in order to realize
15 percent compounded on all cash invested in the property.

B.3 An engineer uses a car extensively for company business and has a choice
between using the company car or buying his own car and receiving a mileage

allowance from the company. Compare the equivalent future value of each alternative and find which is preferred. Assume all payments are credited at EOY, and $i = 10$ percent.

EOY	0	1	2	3	4
Purchase price	$10,000				
Fuel (cents/mi)		8.0	8.9	9.8	10.7
Tires, repairs (cents/mi), maintenance, insurances, etc.		5.0	5.6	6.2	6.8
Resale value					$3,000
Mileage reimbursement paid by the company (cents/mi)		22.0	24.0	26.0	28.0

a. Assume the car is driven 10,000 miles per year.

b. Assume the car is driven 20,000 miles per year.

B.4   Deposit $1,000 at the end of the second year, $2,000 at the end of the third, and increase the amount deposited by $1,000 per year thereafter until the end of the tenth year with $i = 10$ percent. Find the accumulated principal and interest at the end of ten years.

Problem Group C

Geometric gradient.

C.1   A graduating class of civil engineers decides they will present the engineering college with a gift of $100,000 at their twenty-fifth reunion. There are 50 in the class and they can invest at 8 percent. They plan to start out giving a small amount at the end of the first year and increase their gifts to the class fund by 10 percent each year. Find the amount each graduate would donate at the end of the first year in order to reach their class goal of $100,000 at the end of 25 years.

Problem Group D

Both arithmetic and geometric gradients, comparison.

D.1   A contractor is considering two alternate plans to set aside a certain amount every year to replace a large tractor. Plan 1 is to set aside $1,000 per year beginning at the end of the second year and continue increasing the amount $1,000 per year until the tractor is scheduled for trade-in at the end of the tenth year (nine payments).

Plan 2 is to set aside $2,000 now, at the *beginning* of the first year and increase this amount by 25 percent every year until the end of the ten years (11 payments). In both cases the funds are invested at 10 percent interest. How much would be in each fund at the end of the tenth year?

D.2   A building project is brought to you for analysis. The costs and income are estimated as follows.

		END OF YEAR
Construction	$1,000,000	0
O & M	$80,000/yr	1 through 20 increasing $5,000 the second year and $5,000/yr each year thereafter
Gross income	$200,000/yr	1 through 20 increasing 8% the second year and 8% more each year thereafter
Resale value	$800,000	20

Find the net gain or loss for the project at the end of 20 years by summing the future values of all costs and income. Assume the owner can lend and borrow at 10 percent interest.

D.3 A contractor client of yours set up a sinking fund (savings account) to finance purchases of new equipment. The fund now has $5,758 in it and the contractor is earning a nominal $7\frac{1}{2}$ percent interest compounded monthly on the balance in the account. He intends to set aside $0.10 for every cubic yard of earth moved in order to build up the fund. He is currently moving 50,000 yd^3 of earth a month and makes deposits at the end of every month.
a. At the end of another 24 months, how much will be in the account?
b. If the deposits are increased by 10 percent per month (geo grad) how much will the account total in 24 months?

Problem Group E

PTZ ≠ ETZ.

E.1 If $1,000 per year is deposited at the end of years 1 through 5 at 6 percent and $2,000 per year is deposited at the end of years 3 through 7 at 8 percent, how much will the investments be worth at the end of year 10 if each deposit continued to bear interest compounded at the same rates as obtained when deposited?

E.2 As a consulting engineer you plan to build your own office complex in a few more years for your own occupancy and use. You have just heard of a choice lot in a new office park that is available at a bargain price of $30,000 cash payable right now. You call your local banker and he offers to loan you the money at 12 percent (annual rate) compounded monthly. No repayments need be made until the building is constructed and a mortgage taken out.

You plan on starting construction 35 months from now on a building that will cost $100,000. To finance the construction your banker offers to loan an additional $20,000 per month for five months with interest also at 12 percent (annual rate) compounded monthly until repaid.

At the end of the five-month construction period you expect to take out a mortgage in an amount sufficient to repay the banker the entire principal plus accumulated interest to date on the original $30,000 for the land, plus the $100,000 for the building. The mortgage will have a 9 percent interest rate and will be repaid in 240 equal monthly payments.
a. How much do you owe the banker at the end of the construction period?
b. How much are the monthly payments on the mortgage?

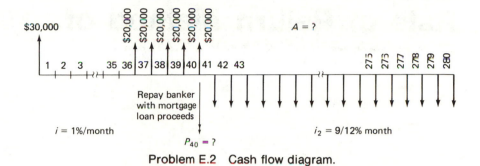

Problem E.2 Cash flow diagram.

E.3 Your client the utility board asks that you determine the rate to charge customers in a new residential development in order to pay for a new waste water treatment plant over the next 20 years. The following data are provided.

Plant cost is $1,000,000. This cost will be repaid to the utility board through a special annual end-of-year assessment to each customer. Number of customers in the development is 1,000, and interest on unpaid balance, $i=7$ percent. To encourage the developer to build in their area, the utility board agreed to charge as little as possible the first year, and then increase the rate by 4 percent per year over the balance of the 20-year period.

a. Find the annual charge per customer that must be charged the first year.

b. Find the annual charge per customer charged for the twentieth year.

Chapter 10
Rate of Return Method of Comparing Alternatives

BASICS OF *ROR*

The rate of return (*ROR*) on an investment is a concept that is already familiar to most people. A return of $6 interest per year on a deposit of $100 is easily understood to imply a rate of return of 6 percent. Therefore, those charged with decision-making responsibilities usually appreciate the ease of comprehending a study of alternate investments when the results are given in terms of rates of return. The project with a return of 18.2 percent obviously is more desirable than an alternative project with a return of only 9.3 percent, all other factors being approximately equal. Because the results are easily understood, the rate of return method together with the companion incremental rate of return method are powerful tools for comparing alternatives and providing decision makers and their supporters with the information required to obtain better investment and allocation of resources of all types.

In order to apply the rate of return method each alternative investment must have a numerically measurable return of income or some equivalent value. Then the rate of return is calculated as the percent interest at which the present worth of the cost equals the present worth of the income. This of course is also the percent interest at which the equivalent annual cost equals the annual income, as well as the percent interest at which future worth of cost equals future worth of income. In more general terms, if income is defined as benefits, and disbursements are defined as costs, then the rate of return is the interest rate for which the benefits are equivalent to the costs. The cash flow analysis may be equated in any of the following forms.

$$PW \text{ benefits}/PW \text{ costs}=1$$

$$PW \text{ benefits}-PW \text{ costs}=\text{net present worth }(NPW)=0$$

$$\text{annualized benefits}-\text{annualized costs}=\text{net annual worth }(NAW)=0$$

$$\text{future worth benefits}-\text{future worth costs}=\text{net future worth}=NFW=0$$

Example 10.1

A $1,000 cost resulted in a lump sum payment of $2,500 at the end of ten years. What is the rate of return (ROR) on this investment?

SOLUTION
The solution is obtained by formulating the income and cost into *equivalent* amounts (in this case the present worth is the simplest) and solving for the value of i which makes the $NPW=0$.

$$PW \text{ benefits}=2,500(P/F, i, n)=2,500\frac{1}{(1+i)^{10}}$$

$$PW \text{ costs}=1,000$$

$$NPW=2,500\frac{1}{(1+i)^{10}}-1,000=0$$

$$\frac{1}{(1+i)^{10}}=\frac{1}{2.5}, \qquad (1+i)^{10}=2.5$$

$$i=\sqrt[10]{2.5}-1=0.096 \quad \text{or} \quad \underline{\underline{9.6\%}}$$

The problem could also be easily solved using NFW, or NAW.

STEP-BY-STEP PROCEDURE FOR FINDING *ROR*

The foregoing simple example could be solved directly because there was only one unknown value of i. More commonly the unknown value, i, appears several times, and the most practical general method of solution is by successive approximation (trial and error). The following step-by-step procedure is recommended.

  Step 1. Make a guess at a trial rate of return.
  Step 2. Count the costs as negative ($-$) and the income or savings as positive ($+$). Then find the equivalent net worth of all costs

and income. Use either present worth or annual worth (or future worth, if preferred).

Step 3. If the equivalent net worth is positive, then the income from the investment is worth more than the cost of the investment, and the actual percent return is higher than the trial rate, i.

Step 4. Adjust the estimate of the trial rate of return, and proceed with steps 2 and 3 again until one value of i is found that results in a positive ($+$) equivalent net worth, and another higher value of i is found with a negative ($-$) equivalent net worth.

Step 5. Solve for the correct value of i by interpolation.

An example involving two lump sum cash flows as well as a periodic series cash flow will serve to illustrate the procedure.

Example 10.2

Assume a bond sells for $950. The bond holder will receive $60 per year interest as well as $1,000 (the face amount of this bond) at the end of ten years. Find the rate of return.

SOLUTION

Find the interest rate at which the present worth of the income ($60/year for 10 yr, plus $1,000 at the end of 10 yr) equals the present worth of the cost ($-$950). The steps recommended above are applied as follows.

Step 1. Make a trial selection of 7 percent for a trial rate of return. (The yield is $60/yr on an investment of $950, plus an extra $50 at the end of 10 yr, so the i probably will be a little above $60/950 = 0.063$ or 6.3%.)

Step 2. Find the present worth of all costs and income using $i = 7$ percent. The income consists of $60 per year for ten years, plus $1,000 at the end of year 10. The cost is one $950 payment due right now.

$$\text{income } P_1 = \$60/\text{yr} \underset{7.024}{(P/A, 7\%, 10)} = \quad \$421.44$$

$$\text{income } P_2 = \$1,000 \underset{0.5083}{(P/F, 7\%, 10)} = \quad 508.30$$

$$\text{cost } P_3 \qquad\qquad\qquad\qquad = - \quad 950.00$$

$$\text{net present worth} \qquad\qquad = - \quad \$20.26$$

Step 3. The net present worth is negative so the trial rate, 7 percent, is too high. The present worth of income ($929.74) is less than the present worth of costs ($950.00) or, at 7 percent you are offered an opportunity to pay $950 for an income worth $929.74.

Step 4. Adjust the estimate downward to, say, 6 percent. Then, the net present worth of the income minus cost is

$$P_1 = \$60/\text{yr} \ (P/A, 6\%, 10) = \$60 \times 7.360 = \$441.60$$
$$P_2 = \$1000 \ (P/F, 6\%, 10) = \$1000 \times .5584 = \ \ \ 558.40$$
$$P_3 = \ -950.00$$
$$\text{net present worth} \ \ \ \ \ \ \ \ \ \ \ \ \ = +\$50.00$$

Two values of i have been found, each resulting in a different sign $(+ \text{ or } -)$ thus bracketing the value of $NPW=0$. An interpolation to the correct value of i is found as follows.

i VALUE	NPW
6%	+50
ROR i	0
7%	−20.26

$$\text{rate of return, } i = 6\% + \frac{50.00}{50 + 20.26} \times 1\% = 6.7\%$$

The 6.7% return is called the bond's "yield to maturity" because it includes redemption income when the bond matures. (The "current yield" is $\$60/\$950 = 6.3\%$ and is frequently used in financial circles with reference to bonds whose maturity date is fairly distant. Bonds are discussed in more detail in Chapter 17.)

CONCURRENT INCOME AND COSTS

The rate of return method can be used where both costs and income occur during the same periods. To save a step, use the net payment for those periods as in the following example.

Example 10.3

An entire fleet of earthmoving equipment may be purchased for $2,600,000. The anticipated income from the equipment is $920,000 per year with direct expenses of $420,000 per year. The market value after five years is expected to be $1,000,000. What is the rate of return?

SOLUTION

Step 1. Assume $i = 10$ percent for first trial.
Step 2. Find present worth of all income and costs.

$$\text{income} = \ \ \$920,000/\text{yr}$$
$$\text{expenses} = -420,000/\text{yr}$$
$$\text{net annual income} = \ \ \$500,000/\text{yr}$$

The present worth of this net income for five years is

$500,000/yr $(P/A, 10\%, 5)$ = $1,985,500 present worth of income
                         3.791

Market value in five years is $1,000,000. The present worth of $1,000,000 income expected five years from now is

$1,000,000 $(P/F, 10\%, 5)$ = $620,900 present worth of market value
                        0.6209

total present worth of all income at 10% = $1,985,500 + 620,900

= $2,516,400 total

This means that if you desire a 10 percent return you can only invest $2,516,400 cash in this project and accept $500,000 per year for five years plus $1,000,000 at the end of the fifth year. Actually $2,600,000 is required to purchase this fleet, so:

Step 3. *Conclusion.* The income (with a *PW* of $2,516,400 at 10%) is worth less than the cost (with *PW* of −$2,600,000), resulting in a negative *NPW*, so the trial estimate is too high.

Step 4. Revise the trial to 8 percent and repeat steps 1 through 3. The *PW* at 8 percent is found as

net income $500,000/yr $(P/A, 8\%, 5)$ = $1,996,500
                        3.993

market value in 5 yr $1,000,000 $(P/F, 8\%, 5)$ =   680,600
                              0.6806

total *PW* of all income at 8%   = $2,677,100

Conclusion. Since the *PW* of income at 8 percent is higher ($2,677,100) than the *PW* of cost (−$2,600,000), resulting in a positive *NPW*, the rate of return is higher than 8 percent. Interpolation will give a final figure.

8% *PW* income= $2,677,100   10% *PW* income= $2,516,400
   PW cost   = −2,600,000   *PW* cost   = −2,600,000
               +$ 77,100   −$ 83,600

rate of return = 8% + $\dfrac{77,100}{160,700}$ × 2% = 8.96% = 9.0%

This problem can also be worked by using net annual worth. To illustrate:

net annual worth = annualized benefits − annualized costs
NAW = 500,000 − 2,600,000$(A/P, i\%, 5)$ + 1,000,000$(A/F, i\%, 5)$ = 0
try 10%   (0.2638)   (0.1638) = −22,077
try 8%   (0.2505)   (0.1705) = +19,269

By interpolation i = 8.93 or 8.9 percent (the slight difference in this answer is due to round off).

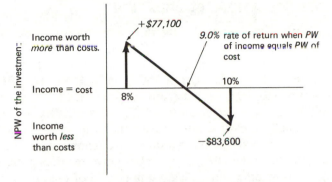

Figure 10.1 Interpolation to find rate of return.

Income and/or Costs Include Gradients

Gradients present no unusual problems when determining rate of return. Simply find present worth (or annual worth or future worth) of each item of cost and each item of income and find the interest rate of return at which the two sums are equal.

Example 10.4

A bus company is for sale for $150,000. The net income this year is $50,000 but is expected to drop $6,000 per year next year and each year thereafter. At the end of ten years the franchise will be terminated and the company assets sold for $50,000. What is the interest rate of return on this investment?

SOLUTION
Try $i=15$ percent.

$$P_1 \hspace{6.5cm} = -\$150,000$$

$$P_2 = -G(P/G, 15\%, 10) = 6{,}000 \times 16.979 = -\$101,890$$

$$P_3 = A(P/A, 15\%, 10) = 50{,}000 \times 5.0187 = +\$250,950$$

$$P_4 = F(P/F, 15\%, 10) = 50{,}000 \times 0.2472 = +\$\ 12,360$$

$$P_{TOT} = P_1 + P_2 + P_3 + P_4 \hspace{2cm} +\$\ 11,420 \text{ for } i = 15\%$$

Try $i=20$ percent.

$$P_{TOT} \hspace{6cm} = -\$\ 9,648 \text{ for } i = 20\%$$

Interpolating,

$$i = 15\% + \frac{11,420}{11,420 + 9,648} \times 5\% = \underline{\underline{17.7\%}}$$

OTHER CONSIDERATIONS BESIDES *ROR*

Some organizations have employed the rate of return as the sole measure of desirability when comparing alternatives. Actually the alternative with the higher *ROR* may *not* be the most profitable. (An investment of $1 now that returns $2 at EOY 1 yields a *ROR* of 100%, but only $1 of profit, whereas an investment of $10,000 now that returns $12,000 at EOY 1 yields a *ROR* of only 20% but a much larger profit of $2,000. Assuming the two investments are mutually exclusive alternatives and cannot be duplicated, the *lower ROR* probably is the better investment for minimum attractive rate of return (*MARR*) less than 20%.) If one alternative requires a higher initial investment than the other, an evaluation is needed of the rate of return on the increment of initial investment (the difference between the two initial investments). The return on this increment of investment is called the incremental rate of return (*IROR*) and is discussed in Chapter 11.

MULTIPLE RATES OF RETURN

For most practical problems involving interest rate of return there is only one solution for the *ROR*. However, there are a few instances where more than one *ROR* is possible. *This possibility is indicated when the cumulative cash flow changes signs more than once.* In the following example, a running total of the value of the cumulative cash flow is shown in the right-hand column. Note that it changes from negative to positive (first change of sign) and then from positive back to negative (second change of sign). This multiple change of sign signals the possibility (but not certainty) that more than one solution may occur.

Example 10.5

An investment consists of the following expenditures ($-$) and incomes ($+$). Find the rate of return.

END OF YEAR	NET CASH FLOW FOR THIS YEAR	CUMULATIVE CASH FLOW
0	$-$$ 500	$-$$500
1	$+$ 1150	$+$ 650
2	$-$ 660	$-$ 10

SOLUTION

The cash flow diagram is shown in Figure 10.2. The problem may be solved in the usual manner by setting the *PW* of the cost equal to the *PW* of the income, as follows.

$$PW \text{ of cost} = 500 + \frac{660}{(1+i)^2} = PW \text{ of income} = \frac{1150}{1+i}$$

Figure 10.2   Cash flow diagram with reversed flow.

This equation can be solved either algebraically or by trial and error. Solving algebraically gives the following results:

$$500(1+i)^2+660-1150(1+i)=0$$

$$[(1+i)-1.1][(1+i)-1.2]=0$$

This yields *two* roots,

$i=0.1$ or   10%   and $i=0.2$ or   20%

Solving by the quadratic equation yields the same results, as does solution by trial and error.

A graph of the rate of return versus the PW of the total investment is plotted in Figure 10.3. Inspection of the graph indicates that if the initial expenditure were increased by $1 (to −$501), there would be only one solution, the tangent point of the parabola with the x axis at $i=15$ percent. Further, if the initial investment were raised to more than $501, there would be no solution, since all total PW's of cost plus income would be negative regardless of the trial i employed.

In connection with the above discussion, the following observations may prove helpful.

1. Only one change or no change in the sign of the cumulative cash flow indicates there is only one solution i.

2. It is theoretically possible to have any number of solutions, providing that the sign of the cumulative cash flow changes more than once. For example,

$$[(1+i)-1.05][(1+i)-1.10][(1+i)-1.15]=0$$

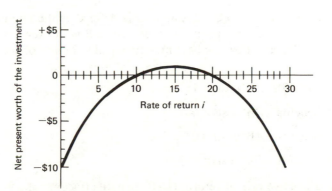

Figure 10.3   Graph of rate of return versus PW using a two-solution example.

This "solution" will solve for $i = 5$ percent, $i = 10$ percent, and $i = 15$ percent. This "solution" may be multiplied out to find the equivalent cash flow, as follows.

$$+ (1+i)^3 - 3.30(1+i)^2 + 3.63(1+i)^2 - 1.33 = 0$$

Any convenient multiple of these numbers will yield to the same three interest rate solutions.

3. Multiple rates of return, intuitively, seem contrary to reason and common experience. However, multiple rate solutions do occur due to the implicit assumption that *all* the cash flow values (both positive income and negative costs) will earn the solution interest rate (the derived rate, i, that makes the net equivalent worth equation equal to zero). Actually, any income must be invested at some externally available market rate, i, and earn interest until the income is returned to the project to pay costs when the signs change. Since there is a practical limit to externally available rates, i, that practical limit represents a cap on the actual choice from the multiple solutions. Thus, if the maximum available external rate is $i = 15$ percent, and a multiple internal rate solution is $i = 10$ percent and 20 percent, then the 20 percent alternative is disqualified as being higher than the external rate of 15 percent. More detailed treatments of multiple interest rates are found in a number of texts, such as the following.

1. Eugene L. Grant, W. Grant Ireson, and Richard S. Leavenworth, *Principles of Engineering Economy*, Sixth Edition, Ronald Press Company, New York, New York, 1976, pp. 543–562.
2. Donald G. Newnan, *Engineering Economic Analysis*, Engineering Press, San Jose, California, 1976, pp. 128–138.
3. George A. Taylor, *Managerial and Engineering Economy*, Third Edition, D. Van Nostrand Company, New York, New York, 1980, pp. 150–151.

SUMMARY

The rate of return (ROR) on investments is defined as the interest rate for which benefits from the investment are *equivalent* to the costs of the investment. The cash flow analysis may be equated in any of the following forms:

PW benefits$/PW$ costs $= 1$

PW benefits $- PW$ costs $= 0$

AW benefits $- AW$ costs $= 0$

FW benefits $- FW$ costs $= 0$

For a positive rate of return, ($i > 0$) benefits (income or savings) must exceed costs. Since there are often several terms in the cash flow equation involving the unknown interest rate, i, a trial and error approach is generally used.

Step 1. Make a guess at a trial rate of return.

Step 2. Solve one of the above forms of the cash flow equations.

Step 3. Compare the equivalent worth of benefits to costs. If the worth of income is higher than the worth of costs (equivalent net worth is positive), the trial value of i is too low.

Step 4. Choose another trial rate of return and proceed with steps 2 and 3 until the value of i representing a net worth of zero is bracketed. Then solve for i by interpolation.

When comparing alternatives several factors must be considered. They are:

1. Comparisons must be on the basis of equivalent costs and benefits. Therefore, the alternatives must be made to have the same lives.
2. If different initial investments are involved between alternatives, the alternative with the highest *ROR* may or may not be the most profitable. Whether or not it is the most profitable depends upon the *ROR* of the *incremental* investment, which is presented in Chapter 11.

Once in a while problems occur with more than one reversal of signs in the cash flow, and consequently, more than one i value solution to the cash flow equation is possible.

PROBLEMS FOR CHAPTER 10, RATE OF RETURN

Problem Group A

Given three or more of the variables P, A, F, n, find the rate of return, i, for each proposal.

A.1 A shopping center costing $2,000,000 will have an estimated net annual income of $200,000. In ten years it is estimated the center can be sold for $1,500,000. What is the expected rate of return?

A.2 A parking lot is proposed for downtown. The land, clearing, and development costs are $150,000. The land can be sold in ten years for $100,000. Parking revenue is expected to net $15,000 per year. What is the rate of return?

A.3 A parking garage for 160 cars is proposed for West University Avenue opposite the campus at a total cost of $250,000. The net annual income (after expenses) is expected to be $150 per car space per year for 20 years. At the end of 20 years the estimated value of the property is $200,000. What percent return would this yield on the investment?

A.4 A mortgage note for $10,000 at 6 percent calls for payment of *interest only* at the end of every year ($600/yr) for ten years. At the end of the tenth year the $10,000 will be paid off, together with the $600 interest for that year. The mortgage note is offered for sale at $7,000. What is the *actual* interest rate at this price?

A.5 Find the percent return on the investment of $10,000 cash down payment on a $60,000 power shovel. Estimated gross income is $20,000 per year. Estimated operating and maintenance expenses are $8,000 per year not including payments. Payments on the $50,000 balance may be made over a 10-year period at 7 percent on the unpaid balance. The shovel can be sold in 12 years for an estimated $10,000. (*Hint:* Try 50%.)

A.6 According to a recent issue of *Florida Trend* magazine, a bank in Winter Park pays interest on certificates of deposit at the time of deposit rather than waiting $3\frac{1}{2}$ years. A depositor of $10,000 gets a check for $2,275 at the time of deposit and $3\frac{1}{2}$ years later the original $10,000 is returned. Assume monthly compounding.
   a. Calculate the nominal annual rate of return on this system.
   b. What is the nominal annual rate of return if the $2,275 is paid at the end of $3\frac{1}{2}$ years instead of the beginning?

A.7 A new bus route is proposed with the following estimated costs and income.

$$2 \text{ new buses @ } \$50,000 \text{ each} = \$100,000$$

$$\text{operating costs} = \$1/\text{mi}/2 \text{ buses (or } \$0.50/\text{mi}/\text{bus})$$

$$\text{average speed including stops} = 10 \text{ mph}$$

$$\text{operating hours} = 6 \text{ AM–9 PM, 7 days/week}$$

$$\text{overhead charged to this route} = \$10,000/\text{yr}$$

average total number of riders
on this route

$$\text{(includes both buses) daily} = 900 \text{ @ } \$0.25 \text{ each}$$

$$\text{resale value of buses after 10 yr} = \text{negligible}$$

What is the rate of return if these estimates prove correct?

A.8 The public works department requests permission to purchase a new mower for $10,000. They estimate it will save $1,200 per year over the 13-year life of the machine. No salvage value is expected at the end of that time.
   a. What rate of return would this savings represent on the investment of $10,000?
   b. Funds are available for the purchase from a new $7\frac{1}{2}$ percent bond issue. Would you recommend that the request to purchase be approved? Why?

A.9 The cost of heavy construction dredging equipment is rising at the rate of 6 percent compounded every year. A sum of $400,000 is now on hand. Two alternatives are available: (a) One $400,000 unit if purchased now should serve the needs of a growing contractor for the next 10 years. (b) One $250,000 unit may be purchased now and another identical unit purchased in five years at a price increase as outlined above. At what interest rate must the extra $150,000 left on hand be invested in order to provide the needed cash in five years? Neglect obsolescence costs and comparative operating costs.

A.10 An investor friend of yours purchased some stock six years ago and kept a cash flow record of all expenditures and income. What rate of return did the investment yield?

EOY		
0	$3,476	purchase stock (including commission)
1	142	dividend
2	215	dividend
3	222	dividend
4	375	dividend
5	350	dividend
6	4,013	sale of stock (net after commission)

Problem Group B

Involves arithmetic gradients.

B.1 A contractor asks your advice on the purchase of a new dump truck which is available for $40,000. You estimate that the net profit earned by the truck the first year will be $10,000. However, with increasing maintenance and repair costs, and decreasing productivity you estimate that the net profit will decline by $600 per year each year. Thus you estimate $9,400 the second year, $8,800 the third, and so on. The contractor expects to keep the truck for ten years and sell it for $5,000 at the end of the tenth year. Estimate the rate of return for the contractor. (*Hint*: Try 15%.)

B.2 A dragline is available for $75,000. Net income from available work is estimated at $25,000 this year, but due to rising costs, competition, and obsolescence, is expected to decrease $3,000 per year in each succeeding year. At the end of ten years the dragline should be worth about $25,000. What is the rate of return on this investment? (*Hint*: Try between 15% and 20%).

B.3 A private utility company is considering the feasibility of servicing a large new subdivision. You are provided with the following income and cost projections and are requested to find the rate of return, *i*. (*Hint*: Try 10%.)

EOY		EOY CAPITAL COST	EOY NET OPERATING INCOME
0	Begin design	$ 50,000.00	
1	Begin construction	$300,000.00	
2	Complete construction	$500,000.00	
	Commence operation		
3	Operations underway,		− $100,000.00
	subdivision in growth stages		
4			− 60,000.00
5			− 20,000.00
6			+ 20,000.00
7			+ 60,000.00
8			+ 100,000.00
9	Subdivision fully developed		+ 140,000.00
	(change of gradient)		
10			+ 145,000.00
11			+ 150,000.00
.			(Increase $5,000.00
.			per year through
.			thirty-fifth year)
35			+ 270,000.00
35	Salvage value		+ 500,000.00

Problem Group C

Involves geometric gradients.

C.1 A small shopping center can be constructed for $500,000. The financing requires $50,000 down, and a $450,000, 30-year mortgage at 10 percent with equal annual payments. The gross income for the first year is estimated at $60,000, and is expected to increase 5 percent per year thereafter. The operating and maintenance costs should average about 40 percent of the gross income and rise at the same 5 percent annual rate. The resale value in 30 years is estimated at $1,000,000. Find the rate of return on this investment over the 30-year life. (*Hint*: Try between 10% and 20%.)

Problem Group D

ETZ≠PTZ

D.1 A 20-year loan to the owners of flood-damaged property provides for no interest to be paid for the first four years and for 2 percent interest to be paid from the fifth through the twentieth year. Loans in Needeep County amounted to a total of $50,000,000. The total interest payments are scheduled at $1,000,000 per year payable at the end of the fifth through the twentieth year, with the $50,000,000 principal also due at the end of the twentieth year.
a. What is the true interest rate received by the lending agency?
b. If this lending agency is paying 6 percent for the money it loans out, what is the *PW* of the subsidy?

D.2 Your client is interested in buying land and building an office park on it and operating the office park as an investment for a period of about ten years. She asks you to determine the before-tax rate of return on the investment using the following estimates of cost and income. (*Hint*: Find *i* between 10% and 20%.)

EOY		COST	INCOME
		(ALL FIGURES ARE ×1,000)	
0	Purchase land	− $100	$ 0
1	Complete the design	− 80	0
2	Complete construction	− 1100	0
3	Operating expenses and income	− 200	+ 100
4	Operating expenses and income	− 210	+ 150
5	Operating expenses and income	− 220	+ 200
6	Operating expenses and income	− 230	+ 250
7	Operating expenses and income	− 240	+ 300
8	Operating expenses and income	− 250	+ 350
9	Operating expenses and income	− 260	+ 400
10	Operating expenses and income	− 270	+ 450
10	Sell entire complex		+ 3,900

Chapter 11
Incremental Rate of Return (*IROR*) on Required Investments

KEY EXPRESSIONS IN THIS CHAPTER

IROR = Incremental rate of return, the interest rate of return on each increment of added initial investment.

MARR = Minimum attractive rate of return, the minimum acceptable rate of return. Investment opportunities yielding *less* than the *MARR* are considered *not* worthwhile.

RATE OF RETURN ON THE INCREMENT OF INVESTMENT

Circumstances, often beyond our control, frequently require us to make sizable expenditures that do *not* produce income or add to profit. For instance, new environmental regulations may require a steel mill to install pollution control devices, a construction contractor may receive orders from OSHA to buy new safety equipment, or a private homeowner may need a new replacement roof for his or her home. None of these expenditures produce income by themselves, but all are necessary expenditures required for the continued operation of some important function.

Competitive alternatives usually exist for fulfilling most needs in our relatively free economy. The steel mill can consider a number of alternative methods of pollution control, the contractor can select from competing safety equipment, and the homeowner can choose from a wide selection of roof materials. Even though there is no direct dollar profit or "return" resulting from these investments, the incremental rate of return ($IROR$) method is well suited and often used to find the most economical selection.

Instead of calculating the income earned by *every* dollar of investment as in the rate of return method, the $IROR$ method determines the amount earned by each *additional* dollar of investment above the required minimum. Thus the $IROR$ is the prospective return on each optional incremental dollar of capital investment under review.

The problem now assumes the following form. If $1 *must* be invested with no measurable return expected, and an additional (incremental) $1 *may* be added to this investment with a positive return expected on the second $1, should the second $1 be invested in addition to the first $1? The following example illustrates this situation.

Example 11.1

Assume a steel mill must select one pollution control device from two suitable types available having the characteristics listed below.

POLLUTION CONTROL DEVICE	A	B
Cost new ($)	1,000,000	1,200,000
Operating costs ($)	300,000/yr	270,000/yr
Life, n (yr)	20 yr	20 yr
Salvage value ($)	50,000	60,000

Find the incremental rate of return ($IROR$).

SOLUTION

Note: The rate of return on each alternative is negative. Therefore there is no need to determine the ROR and the only consideration is which alternative yields the best *incremental* rate of return.

Comparing the two alternatives, the lowest initial investment is required by alternative A, at $1,000,000. Alternative B requires $1,200,000 capital outlay ($200,000 more than A). This raises the basic question "since a minimum of $1,000,000 must be spent on pollution control anyway, should an extra $200,000 be invested so that B may be purchased instead of A." Will the lower operating cost and higher salvage value of B more than compensate for the extra $200,000 initial cost of alternative B? To answer these questions an incremental rate of return ($IROR$), or rate of return on the increment of investment, is calculated. As the name implies, the method is similar to that used for rate of return, but only the *increments* of cost and saving (or income) need be considered. The $IROR$ can be found by the following procedure.

1. Write a present worth equation for each alternative (annual or future worth equations may be used equally well, but present worth is used more commonly).
2. Find the equation for the present worth of the increments (incremental equation) by subtracting the equation for the alternative with the lower first cost from the equation of the higher first cost alternative (in determining which alternative is lower cost, only the first cost is considered).

For Example 11.1, the present worth equation for each alternative is written as follows.

alternative B: $\quad P_B = -1,200,000 - 270,000(P/A, i\%, 20)$

$\quad\quad\quad\quad\quad\quad\quad + 60,000(P/F, i\%, 20)$

alternative A: $\quad P_A = -1,000,000 - 300,000(P/A, i\%, 20)$

$\quad\quad\quad\quad\quad\quad\quad + 50,000(P/F, i\%, 20)$

The incremental equation (for the present worth of the increments) is derived by subtracting the equation for the alternate with the lower first cost (alternate A) from the equation for alternate B as follows.

$P_B - P_A = \Delta P = -200,000 + 30,000(P/A, i\%, 20)$

$\quad\quad\quad\quad\quad + 10,000(P/F, i\%, 20)$

3. Then the incremental equation is set equal to zero and solved for i, the *IROR*. The interest rate found is the rate of return on the incremental investment, which in this example is the extra $200,000 required for alternative B over A. This interest rate, i, is found by the normal procedure for rate of return, summarized for review as follows.
   a. Assume an i value.
   b. Solve the incremental equation for the net present worth at the assumed i value.
   c. If the resulting present worth is positive (+), make a second trial at a higher i. Continue with different trial i values until the sign (+ or −) of the next resulting present worth changes.
   d. Interpolate to find the actual value of i representing the *IROR* for a zero present worth of the incremental equation.

For the steel mill pollution control of Example 11.1, the incremental rate of return on the additional $200,000 investment in alternative B over A is found as

$\Delta P = -\$200,000 + \$30,000(P/A, i, 20) + \$10,000(P/F, i, 20)$

Substituting trial i values, assume $i = 13$ percent.

$$\Delta P = -\$200,000 + \$30,000(7.0248) + \$10,000(0.08679) = +\$11,612$$

The result is a positive present worth, so in order to find the i at which $\Delta P = 0$, the next trial i should be higher; try $i = 14$ percent.

$$\Delta P = -\$200,000 + \$30,000(6.6232) + \$10,000(0.07277) = -\$576$$

The resulting value of ΔP now has changed from a positive to a negative value, indicating that for $\Delta P = 0$, i lies between 13 and 14 percent. Interpolating yields

$$i = 13\% + \frac{11,612}{11,612 + 576} \times 1\% = 13.9\%$$

Thus, the extra $200,000 required for alternate B over alternate A earns a rate of return of 13.9 percent through lower operating costs and higher salvage value. Assuming the firm has the additional $200,000 ready for investment, the decision makers must now consider whether other alternative investments would yield more than the 13.9 percent return available through investing the incremental $200,000 in alternate B. In response to this question a minimum attractive rate of return ($MARR$) is usually established. Only investment opportunities with a rate of return equal to or higher than the $MARR$ are considered as worthwhile investments. Opportunities to invest at a rate *lower* than the $MARR$ are usually rejected as not sufficiently profitable. If a lot of good opportunities occur to invest at the $MARR$ rate or higher, and investment funds run short, the $MARR$ can be raised to suit the current supply and demand for the firm's investment funds. (In addition to the rate of return, all investment opportunities should be evaluated for risk and other characteristics as well.)

MARR, THE MINIMUM ATTRACTIVE RATE OF RETURN

By establishing the $MARR$, the following assumptions are made.

1. Sufficient funds are available to purchase the highest cost alternative under consideration, should the investment be justifiable.
2. If one of the lower cost alternatives is selected, the excess funds (the difference in cost between the selected lower cost alternative and the highest cost alternative) are assumed to be invested elsewhere earning the $MARR$.

Thus the $MARR$ involves consideration of how best to invest the *total available* funds, and not just the funds utilized in any particular alternative. The following paragraphs and examples illustrate these points.

For the steel mill pollution control example the incremental rate of return is compared to the current $MARR$. If the $MARR$ is 13.9 percent or above, then why invest the $200,000 in B at 13.9 percent when it can be invested at the higher *MARR* rate? More profit can be obtained by purchasing the least costly alternate that will fulfill minimum requirements for only $1,000,000. On the other hand if the *MARR* is less than 13.9 percent it is obviously more profitable to invest the extra $200,000 in alternate B.

Amount of investment
$1,200,000 _____

	Optional $200,000 for higher cost pollution control equipment	Opportunity to invest an additional $200,000 @ +13.9% *IROR*
$1,000,000 _____		
	Required $1,000,000 investment for pollution control equipment	Required $1,000,000 yields negative rate of return

COMPARING MULTIPLE ALTERNATIVES WHEN (1) THE *MARR* IS KNOWN, AND (2) THE "DO NOTHING" ALTERNATIVE IS NOT AVAILABLE

Frequently more than two alternatives occur and these can be compared systematically by the *IROR* method as follows.

1. List the alternatives in order of *ascending* first cost. The alternative with the lowest first cost is placed at the top of the list.
2. Set up one equation for each alternative in terms of present worth. (Annual cost or future value may be used equally well.)
3. Subtract the equation of each alternative from each of those lower on the list (higher in first cost), and determine the resulting *IROR* value of *i*. Set up a matrix of values.
4. With the known *MARR*, examine each *IROR* value and cross out as "rejects" *IROR* values that are less than the *MARR*.
5. Begin with the top row of the matrix and select the highest *IROR* value in this row noting the alternative in whose column it lies. Temporarily accept this new alternative (in whose column the highest *IROR* lies) as the better one.
6. Proceed down the matrix to the row of the new alternative and find the highest *IROR* in that row. Note which alternative heads that column, temporarily accept that alternative, and proceed downward to the row of that alternative. Repeat this process until reaching the lowest row containing at least one alternative with an *IROR* value higher than the *MARR*. The alternative accepted last is now the best of all those listed.

Example 11.1 *(Continued)*

For example, suppose that the steel mill in Example 11.1 had five alternatives instead of the two listed. Assume further that each had the first costs listed below, and that the "do nothing" alternative is not a viable option in this case.

ALTERNATIVE	FIRST COST
A	$1,000,000
B	$1,200,000
C	$ 900,000
D	$1,150,000
E	$1,400,000

SOLUTION

The first step is to list the alternatives in order of ascending first cost, as follows.

ALTERNATIVE	FIRST COST
C	$ 900,000
A	$1,000,000
D	$1,150,000
B	$1,200,000
E	$1,400,000

Then, assume that additional information on operating costs, salvage values, and so forth are provided (not shown). From these values the ten *IROR* values are derived (calculations not shown) and listed in matrix table form as follows:

COST OF ALTERNATIVE	ALTERNATIVE	IROR			
		A	D	B	E
$ 900,000	C	24.1%	14.1%	20.8%	11.2%
$1,000,000	A		5.2%	13.9%	7.9%
$1,150,000	D			10.1%	8.2%
$1,200,000	B				9.6%

The table is read from left to right, so that alternatives in the column on the left are compared to other alternatives in the top row. Values of *IROR* are shown only where a lower cost alternative in the left-hand column can be compared to a higher cost alternative in the top row. For example the lower cost alternative C at the top of the left-hand column is compared in turn to each higher cost alternative in the top row, A, D, B, E. On the first $100,000 increment of A above C ($1,000,000 − $900,000 = $100,000) an *IROR* of 24.1 percent can be obtained. On the $250,000 increment of D above C an *IROR* of 14.1 percent can be earned, and so on.

COST OF ALTERNATIVE	ALTERNATIVE	A	D	B	E	This line shows *IROR* of A,D,B, E, compared to C
$ 900,000	C	24.1%	14.1%	20.8%	11.2%	
$1,000,000	A		5.2%	13.9%	7.9%	
$1,150,000	D			10.1%	8.2%	
$1,200,000	B				9.6%	

The process of selecting the better alternative begins with the lowest cost alternative C at the top row of the matrix. The *IROR* values on the row to the right of C are examined and the largest one selected. This is the 24.1 percent which can be earned by investing $100,000 in addition to the required $900,000 expenditure for minimum first cost pollution control device C. If the *MARR* is less than 24.1 percent then alternative A is the better selection over C. If the *MARR* is greater than 24.1 percent then C is the better alternative and none of the other alternatives or rows need be examined. *Whenever the MARR exceeds the IROR between the lowest cost alternative and any other alternative (MARR >the highest IROR in the top row of the matrix) the lowest cost alternative is the better investment.* Assume that in Example 11.1 the *MARR* <24.1 percent and alternative A is temporarily accepted. Then is it profitable to invest more than the $1,000,000 needed for A?

If *MARR* <24.1%, select A temporarily, and proceed to line A. If *MARR* >24.1%, select C and end search. Proceed to row A and compare the *IROR* values to the right of A. The highest value is 13.9 percent, which indicates that an additional investment of $200,000 over the $1,000,000 required for A would yield an *IROR* of 13.9 percent. For an *MARR* >13.9 percent (but less than 24.1 percent found in row C) proceed no further, invest in A at $1,000,000.

For an *MARR* <13.9 percent alternative B should be accepted temporarily. The same procedure is followed when comparing B to E.

The results may be summarized as follows:

for $MARR > 24.1\%$, use alternative C
for $24.1\% > MARR > 13.9\%$, use alternative A
for $13.9\% > MARR > 9.6\%$, use alternative B
for $MARR < 9.6\%$, use alternative E

There are no values of $MARR$ under which alternative D is the better investment.

Graphically the selection process may be viewed as follows. Since an investment in pollution control equipment of at least $900,000 is required, would you recommend an additional investment

of $100,000 in A which yields an $IROR$ of 24.1%
of $250,000 in D which yields an $IROR$ of 14.1%
of $300,000 in B which yields an $IROR$ of 20.8%
of $500,000 in E which yields an $IROR$ of 11.2%

Amount of investment

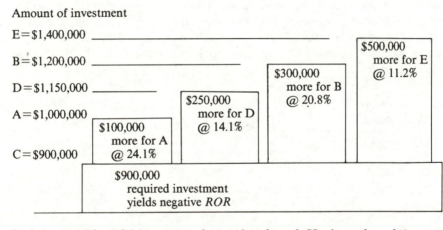

If the $MARR$ is <24.1 percent, then A is selected. Having selected A, at a cost of $1,000,000 would you now recommend an additional investment of

$150,000 in D which yields an $IROR$ of 5.2%
$200,000 in B which yields an $IROR$ of 13.9%
$400,000 in E which yields an $IROR$ of 7.9%

Amount of investment

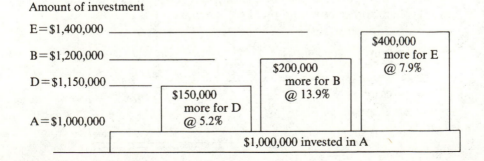

If the *MARR* is <13.9 percent, then alternative B is the logical selection. Having selected B, at a cost of $1,200,000, would you now recommend an investment of an additional $200,000 in E that yields an *IROR* of 9.6 percent?

Amount of investment

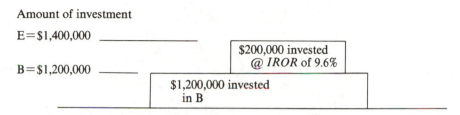

If the *MARR* <9.6 percent then E should be selected. The total investment pattern would now appear graphically as follows.

Amount of investment

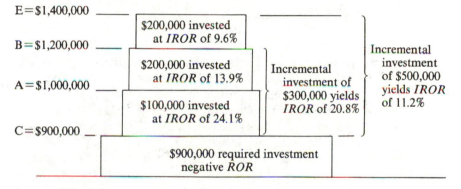

SOLUTION BY ARROW DIAGRAM

Where a number of alternatives are available, a graphical solution using an arrow diagram may simplify the selection process. The diagram is drawn as a closed figure with the same number of corners as there are alternatives. Thus the five alternatives of Example 11.1 are drawn as a pentagon, six alternatives are drawn as a hexagon, and so on. The first corner drawn always represents the alternative requiring the lowest capital investment (C in Example 11.1). Proceeding *clockwise* around the diagram, each successive corner represents the alternative with the next higher capital cost. The sides between the corners are tipped with arrows pointing toward the next higher cost corner and labeled with the corresponding return on incremental investment (*IROR*) as shown in Figure 11.1.

To graphically determine the better alternative, start at the corner representing the lowest capital investment, in this case C. Select the line with the highest rate of return leading from that starting corner (in this case, 24.1%). If the *MARR* is higher than this rate (*MARR* >24.1%), then the

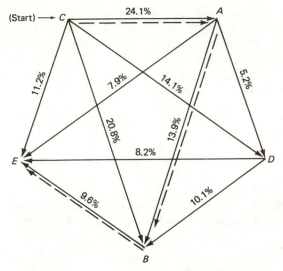

Figure 11.1   Arrow diagram for Example 11.1.

lowest cost capital investment C costing $900,000 is the better one. Assume for Example 11.1 that the *MARR* is lower (*MARR* < 24.1%); then alternate A at the other end of the CA line is the better choice. Then proceed to select the highest rate of return leading away from corner A, that is 13.9 percent leading to corner B. From B only one arrow leads away and that goes to E. No arrows lead away from E, indicating that E is the alternative with the highest capital investment. Reviewing the diagram, if the *MARR* is greater than 24.1 percent, select the (lowest cost) alternative C. If the *MARR* is between 24.1 and 13.9 percent, then A is the better selection; if the *MARR* is between 13.9 and 9.6 percent, then B is the better selection; and if the *MARR* is less than 9.6 percent, then E is the better selection.

OPPORTUNITY FOR "DO NOTHING" (NO INVESTMENT)

In Example 11.1 if the steel mill is relieved of the need to install pollution control devices then another alternative is available, the opportunity for "no investment." The pollution control devices produce no income but instead increase the costs, which in turn results in a negative rate of return when each alternative is considered independently. Thus the "no investment" opportunity is now added to the arrow diagram as the lowest first cost option, at corner 0. The comparative returns on incremental investment are as indicated along the appropriate sides of Figure 11.2. The graphic solution now begins at corner 0 and all arrows away from 0 yield negative rates of return. The diagram simply indicates that if the *MARR* is higher than −5.8 percent then no investment from this group of alternatives is recommended.

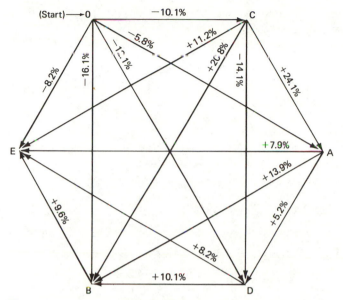

Figure 11.2 Arrow diagram with "do nothing" alternative.

COMPARING DIFFERENT LIFE SPANS

If the alternatives have different life spans the *IROR* may be found by several methods.

1. *Assume multiple equal replacements.* Find the present worth of enough future replacements of each alternative to reach a common life span.
2. *Use equivalent net annual worth.* Formulate the equation for each alternative in terms of equivalent annual worth instead of present worth [this differs from (1) above only by the appropriate A/P equivalence factor].

Example Illustrating Multiple Replacements

The next example, 11.2, illustrates how *IROR* is found by multiple replacements when life spans differ between alternatives.

Example 11.2

A firm of engineers is in need of additional office space. A landlord offers them suitable space with three alternative leasing plans from which they may select: (1) a one-year lease for $5,000 paid in advance, with option to renew for five subsequent one-year leases under the same terms; (2) a two-year lease for $9,000 paid in advance, with option to renew two more times; and (3) a three-year lease for $13,000 paid in advance, with option to renew one more time.

    a. Find which ranges of *MARR* favor which lease terms.
    b. If the firm can invest its working capital elsewhere in the business at a 20 percent pretax rate of return, which is the better selection of lease terms?

SOLUTION
The one- and two-year leases are compared over the least common multiple of life spans, in this case a two-year period.

$$\text{two 1-yr leases,} \quad P_1 = -\$5,000(P/F, i, 1) - \$5,000$$

$$\text{one 2-yr lease,} \quad P_2 = -\$9,000$$

The present worths of the two plans are equated and solved for i.

$$\$5,000 + \$5,000(P/F, i, 1) = \$9,000$$

$$P/F = \frac{\$9,000 - \$5,000}{5,000}$$

$$\frac{1}{(1+i)^1} = 0.8$$

$$i = 0.25 = 25\%$$

In comparing the one- and two-year leases the firm has an opportunity to invest an extra $4,000 now instead of $5,000 a year from now. The effect is the same as investing $4,000 now and receiving a $5,000 return (benefit) in one year. Thus the rate of return on the incremental investment (*IROR*) is $1,000/$4,000=25 percent for the one-year period.

    The one- and three-year leases are compared over the least common multiple, in this case a three-year period, in the same manner as above.

$$\text{three 1-yr leases,} \quad P_1 = -\$5,000 - \$5,000(P/A, i, 2)$$

$$\text{one 3-yr lease paid in advance,} \quad P_3 = -\$13,000$$

$$(P/A, i, 2) = \frac{13,000 - 5,000}{5,000} = 1.60$$

The equations yield factors listed below which are then interpolated to find i.

i	$(P/A, i, 2)$
16%	1.6053
IROR	1.6000
17%	1.5853

$$IROR = \frac{1.6053 - 1.6000}{1.6053 - 1.5853} \times 1\% + 16\% = \underline{\underline{16.3\%}}$$

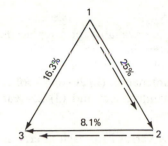

Figure 11.3   Arrow diagram for Example 11.2.

In comparing the two-year lease to the three-year lease the shortest common time period is six years. The *IROR* between these alternatives is found as follows.

$$P_2 = -\$9,000 - \$9,000(P/F, i, 2) - \$9,000(P/F, i, 4)$$
$$P_3 = -\$13,000 - \$13,000(P/F, i, 3)$$

Subtracting P_2 from P_3 yields

$$\Delta P = -\$4,000 + \$9,000(P/F, i, 2) - \$13,000(P/F, i, 3)$$
$$+ \$9,000(P/F, i, 4)$$

Solving for $\Delta P = 0$, select trial i values. Try i at 8 percent.

   find $\Delta P = +\$11.41$

Try $i = 9$ percent.

   find $\Delta P = -\$87.48$

Interpolation yields

$$i = 8\% + \frac{11.41}{11.41 + 87.48} \times 1\% = \underline{\underline{8.1\%}}$$

The arrow diagram method (see Figure 11.3) provides a graphic solution process for this example as follows.

   for *MARR* > 25%, use the 1-yr leases
   for 25% > *MARR* > 8.1%, use 2-yr leases
   for *MARR* < 8.1%, use 3-yr leases

Example Illustrating Annual Costs

When determining the *IROR*, annual costs are easier to use than present worth in cases where different life spans occur and the lowest common multiple happens to be a large or awkward figure to work with. The following example illustrates this situation.

Example 11.3

Three grades of roofing are available for a replacement roof for a residence and you are asked to determine which is the more economical. Data on the roofs are presented as follows: (1) 15-year roof costs $1,800 plus $100 per year for end of year maintenance; (2) 20-year roof costs $2,200 plus $50 per year for end of year maintenance; and (3) 25-year roof costs $3,000 and requires no maintenance.

    a. Plot a graph of A versus i for each alternative.
    b. On the graph show which ranges of i favor which alternatives.

Discussion: The least common multiple for the n values in this problem is 300 years, which renders the present worth method of finding *IROR* cumbersome and tedious. The annual cost method is much simpler.

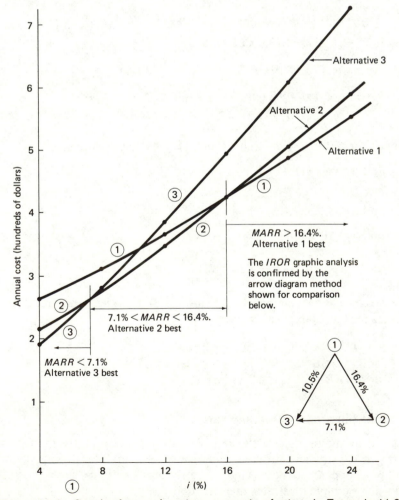

Figure 11.4   Graph of annual costs versus rate of return in Example 11.3.

SOLUTION

Find *i* for the *IROR* in terms of annual costs.

Alternative 1

$$A = -\$100 - \$1,800(A/P, i, 15)$$

Alternative 2

$$A = -\$50 - \$2,200(A/P, i, 20)$$

Alternative 3

$$A = -\$3,000(A/P, i, 25)$$

Solution by the usual method [find the *IROR* for each pair, e.g., Alternative 2 − Alternative 1 = $-2,200(A/P, i, 20) + 1,800(A/P, i, 15) - 50 + 100 = 0$] yields the following matrix

	2	3
1	16.4%	10.5%
2		7.1%

A graph of the values of annual costs versus *i* is shown in Figure 11.4

GRADIENTS

Changes in costs with the passage of time are a normal occurrence in real life and a reasonable mathematical simulation usually can be formulated by use of gradients. The following example illustrates the use of gradients in *IROR* problems.

Example 11.4

The state DOT has decided to construct a new road to bypass a congested area. They ask you to find the better of the three alternative routes from the data provided below.

ROUTE	A	B	C
Initial cost ($)	$1,500,000	$1,300,000	$1,600,000
Maintenance costs ($)	100,000/yr	120,000/yr	130,000/yr
Gradient maintenance costs ($)	12,000/yr/yr	12,000/yr/yr	12,000/yr/yr
Savings per vehicle (seconds)	25 s	15 s	40 s
Salvage value ($)	500,000	200,000	700,000
Estimated life (yr)	15	20	25

Each of the three routes are estimated to carry 5,000 vehicles per day for the first year, with a gradient of 500 vehicles per day for each year thereafter. The value of time for each vehicle is estimated at $3 per hour.

1. Find the incremental rate of return between the lowest cost route and the next lower cost alternative route.

2. The incremental rates of return for the other alternatives are as listed below. Draw an arrow diagram showing the three alternatives including appropriate directions for each arrow between alternatives.

	A	C
B	find	16.7%
A		13.1%

3. Draw dashed arrows on the arrow diagram to illustrate the method of reaching the better alternative of the three if the *MARR* is 15 percent.

SOLUTION 1

Since the alternatives have different life spans, the equations are more easily formulated in terms of equivalent annual costs. Route B is the lowest cost and is subtracted from the next higher cost alternative route A, and the result solved for the *IROR* value. Since savings are given in terms of seconds (s) rather than dollars, the first step is to find the annual dollar value of savings in time for each alternative route, calculated as

savings route A: $\quad A_A = \$3/hr \times 5{,}000 \text{ vehicles/day} \times 365 \text{ days/yr}$
$$\times 25 \text{ s/vehicle} \div 3{,}600 \text{ s/hr} = \$38{,}021/yr$$

savings route B: $\quad A_B = \$3/hr \times 5{,}000 \text{ vehicles/day} \times 365 \text{ days/yr}$
$$\times 15 \text{ s/vehicle} \div 3{,}600 \text{ s/hr} = \$22{,}813/yr$$

Then the gradient in savings per year is found as

savings route A: $\quad G_A = \$3{,}802/yr/yr$

savings route B: $\quad G_B = \$2{,}281/yr/yr$

The total equivalent annual costs plus savings is then formulated for each route as follows (all quantities are $\times\$1{,}000$).

A: $\quad A_A = -1{,}500(A/P, i, 15) + 500(A/F, i, 15) - 100$
$$- 12(A/G, i, 15) + 38.021 + 3.802(A/G, i, 15)$$

B: $\quad A_B = -1{,}300(A/P, i, 20) + 200(A/F, i, 20) - 120$
$$- 12(A/G, i, 20) + 22.813 + 2.281(A/G, i, 20)$$

Then trial *i* values are substituted into the equations until a positive and a negative result of $A_A - A_B$ are obtained.

FOR $i = 15\%$		FOR $i = 20\%$
A_A	$= -345.421$	-408.317
A_B	$= -355.070$	-406.469
$A_A - A_B =$	$+ \quad 9.649$	$- \quad 1.848$

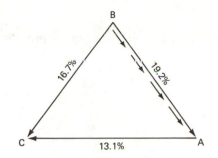

Figure 11.5   Arrow diagram for Example 11.4.

Thus by interpolation

$$IROR = i = 15\% + 5\% \times \frac{9.649}{9.649 + 1.848} = 19.2\%$$

If problems of this type are encountered often, it is well to write or acquire a computer program to save time in finding solutions and ranging variables.

SOLUTION 2
The three alternatives are represented by a three-sided polygon (triangle) with the lowest first cost alternative at the top (see Figure 11.5). The next higher first cost alternatives are represented by each corner in sequence proceeding clockwise around the polygon. From every corner except the last, arrows are drawn from lower first cost to higher first cost.

SOLUTION 3
    a. Begin at the lowest cost alternative B, and
    b. Find the highest *IROR* leading from this point (*IROR* = 19.2%).
    c. Compare *IROR* with *MARR* (*IROR* = 19.2% > 15% = *MARR*), so choose *IROR* and temporarily accept alternative A.
    d. Find highest *IROR* leading from this point (*IROR* = 13.1%).
    e. Compare *IROR* with *MARR* (*IROR* = 13.1% < 15% = *MARR*), so reject alternative C, stay with A.

Conclusion. With the *MARR* at 15 percent, route A should be constructed for $1,500,000 rather than route B for $1,300,000, since the extra $200,000 will earn an *IROR* of 19.2 percent. Route C should be rejected since the extra $100,000 required over route A will only earn an *IROR* of 13.1 percent.

COMPARING ALTERNATIVES BY THE *ROR* VERSUS THE *IROR*

The following example illustrates some of the problems encountered if only the *ROR* method is used for comparing alternatives. In this example the relative rankings for the preferred alternatives reverse themselves as the *i* values are ranged from one value to the next.

Example 11.5

The following information has been obtained on two alternative machines:

	MACHINE X	MACHINE Y
Initial cost ($)	200	400
Annualized income ($)	75	120
End life salvage value ($)	50	150
Life (yr)	6	12

Which alternative should be selected?

Note: There are unequal lives involved, therefore you must either (a) assume multiple equal investments and benefits of machine X until the 12-year period is concluded, or (b) assume some different salvage value for machine Y after 6 years. Remember, considering each alternative on the basis of equivalent annualized worth *automatically* assumes multiple equal investments and benefits.

SOLUTION

If we make assumption (a) above and calculate net annualized worth, we obtain

Machine X

$$A = 75 - 200(A/P, i\%, 6) + 50(A/F, i\%, 6) = 0$$

| try 30% | 0.37839 | $0.07839 = +3.24$ |
| try 40% | 0.46126 | $0.06126 = -14.189$ |

by interpolation: $i = 31.9\%$

Machine Y

$$A = 120 - 400(A/P, i\%, 12) + 150(A/F, i\%, 12) = 0$$

| try 30% | 0.31345 | $0.01345 = -3.364$ |
| try 25% | 0.26845 | $0.01845 = +15.39$ |

by interpolation: $i = 29.1\%$

If the maximum *ROR* were the correct criterion, then the decision would be to choose machine X. But, if we check the net annual worth of each machine at a specific value of interest, then we may, or may not, arrive at the same solution.

To illustrate, let us take a value for i of 20% and determine the net annual worth of each machine.

Net annual worth of machine X

$$A(X) = 75 - 200 \underset{0.30072}{(A/P, 20\%, 6)} + 50 \underset{0.10072}{(A/F, 20\%, 6)} = 19.9$$

Net annual worth of machine Y

$$A(Y) = 120 - 400 \underset{0.225262}{(A/P, 20\%, 12)} + 150 \underset{0.025262}{(A/F, 20\%, 12)} = 33.7$$

In this case machine Y has the greater net annual worth and thus would be selected. But this is a different answer than when the rate of return was examined, so we must conclude that utilizing the net annual worth for a specific interest rate can yield results very different from those found by simply calculating the *ROR* of each investment. Different decisions can be reached, and thus some further analysis is needed. At this point, the major conclusion to be drawn is that choosing among multiple alternatives involves *more* than simply choosing the alternative with the highest *ROR*. A return of 20 percent on $10,000 yields a lot greater profit ($2,000) than a return of 100 percent on a $1 investment ($1 profit) (assuming the investments are mutually exclusive and nonduplicative). If machine X is chosen (earning 31.9% on this $200), then the remaining money available ($400 − 200 = $200 in this case) is assumed to earn some *MARR* (for example 20%). Now the complete question to be answered is whether $200 earning 31.9 percent plus $200 earning 20 percent is better or worse than $400 earning 29.1 percent (machine Y). Calculation of the net annual worth at 20 percent reveals that machine Y is the better investment, assuming an *MARR* of 20 percent. The *IROR* method considers the *differences* between the alternatives over the same number of years. Figure 11.6 shows the cash flows extended to provide a comparison over the same total life span.

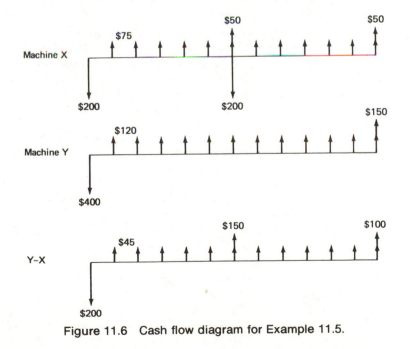

Figure 11.6 Cash flow diagram for Example 11.5.

$$NPW = -200 + 45(P/A, i\%, 12) + 150(P/F, i\%, 6) + 100(P/F, i\%, 12)$$

try 30%	3.1903	0.20718	0.04293

$$NPW = -21.02$$

try 25%	3.7252	0.26215	0.06872

$$NPW = +13.83$$

by interpolation $IROR = 27.0\%$

This means the \$200 *additional* investment earns 27 percent. If the *MARR* is 20 percent then machine X would earn 31.9 percent and the unused additional investment would presumably earn 20 percent, while machine Y (\$400 investment) would earn 29.1 percent. A \$400 investment earning 29.1 percent is better than a \$200 investment at 31.9 percent plus another \$200 investment at 20 percent (the *MARR*).

The answer (decision) depends upon the *MARR*. For this example, if the *MARR* is less than 27.0 percent (the *IROR* in this case), then machine Y is the best alternative. If the *MARR* is between 27.0 and 31.9 percent, then machine X is the best alternative. If the *MARR* is greater than 31.9 percent, then *neither* alternative should be selected.

COMPARING MULTIPLE ALTERNATIVES WHEN (1) THE *MARR* IS KNOWN, AND (2) THE "DO NOTHING" ALTERNATIVE IS AVAILABLE

Often there are a number of alternatives from which to choose, each being mutually exclusive and none required (the "do nothing" alternative is available). When this occurs, the time-consuming calculation of every alternative's *ROR* and the *IROR* for *every* difference between all possible combinations of alternatives becomes unnecessary *if* (1) the *MARR* can be established, and (2) the "do nothing" alternative is available. To analyze this type of multiple alternative, the following steps should be followed.

Step 1. List the alternatives *in order* of ascending first cost. The alternative with the lowest first cost is placed at the top of the list (or at the left column).

Step 2. Calculate the *ROR* for each alternative. Discard any alternative whose *ROR* is less than the *MARR*. (In those cases where there are no incomes and the "do nothing" alternative is not a possible choice, Step 2 can be omitted because the *ROR* for all alternatives is less than zero and the decision must be based on the *IROR* values.)

Step 3. Regroup the alternatives after discarding any unacceptable alternatives based on *ROR*.

Step 4. Stepwise determine the *IROR* for each difference in alternatives beginning with the lowest cost alternative, discarding either (a) the lower cost alternative if the *IROR* is greater than the *MARR*, or (b) the higher cost alternative if the *IROR* is less than the *MARR*.

Step 5. Select the single remaining alternative after analyzing all alternatives.

The following example illustrates this method.

Example 11.6

An engineering firm is considering five different small computers to aid the firm in running complicated and repetitive calculations. Analysis of the various machines, when applied to the firm's business, indicates the following cash flow information.

	F	G	H	I	J
First cost ($)	25,000	18,000	22,500	28,400	32,000
Annual net income ($)	5,000	3,800	4,800	6,000	6,600
Useful life (yr)	8	8	8	8	8

Which computer should be selected if the *MARR* is 12%?

SOLUTION

First, arrange the alternatives in ascending order of first cost, to wit:

	G	H	F	I	J
First cost ($)	18,000	22,500	25,000	28,400	32,000
Annual net income ($)	3,800	4,800	5,000	6,000	6,600

Second, calculate the *ROR* of each alternative by setting the $NPW=0$.

Computer G

$$-18,000+3,800(P/A, i\%, 8)=0$$

$$(P/A, i\%, 8)=18000/3800=4.7368$$

$$\text{try } 13\%, \quad (P/A, 13\%, 8)=4.7988$$

$$\text{try } 14\%, \quad (P/A, 14\%, 8)=4.6389$$

by interpolation, $ROR=13.4\%$

Computer H

$$-22,500+4,800(P/A, i\%, 8)=0$$

$$(P/A, i\%, 8)=4.6875$$

by interpolation, $ROR=13.7\%$

Computer F

$$-25{,}000+5{,}000(P/A, i\%, 8)=0$$
$$(P/A, i\%, 8)=5.000$$

by interpolation, $ROR=11.8\%$

Computer I

$$-28{,}400+6{,}000(P/A, i\%, 8)=0$$
$$(P/A, i\%, 8)=4.7333$$

by interpolation, $ROR=13.4\%$

Computer J

$$-32{,}000+6{,}600(P/A, i\%, 8)=0$$
$$(P/A, i\%, 8)=4.8485$$

by interpolation, $ROR=12.7\%$

Third, regroup the alternatives, after discarding those alternatives whose *ROR* does not meet the *MARR* (computer F in this case would be discarded).

	G	H	I	J
First cost ($)	18,000	22,500	28,400	32,000
Annual net income ($)	3,800	4,800	6,000	6,600
ROR (%)	13.4	13.7	13.4	12.7

Fourth, stepwise determine the *IROR* of each difference in two alternatives discarding the least desirable of each two.

Thus computer G can be considered as the comparison between computer G and the "do nothing" alternative. As the *IROR* of computer G over the "do nothing" is 13.4 percent, which is greater than the *MARR* of 12 percent, discard the "do nothing" alternative. Comparing H over G, the cash flow becomes

	H−G
Incremental first cost ($)	4,500
Incremental net income ($)	1,000
$-4{,}500+1{,}000(P/A, i\%, 8)=0$	
$(P/A, i\%, 8)=4.500$	

By interpolation, the *IROR* = 14.9 percent > 12 percent *MARR*, therefore discard G. Next, compare I over H.

	I−H
Incremental first cost ($)	5,900
Incremental net income ($)	1,200
$-5{,}900+1{,}200(P/A, i\%, 8)=0$	
$(P/A, i\%, 8)=4.9167$	

By interpolation, $IROR = 12.3$ percent > 12 percent $MARR$, therefore, discard H. Next, compare J over I.

	J–I
Incremental first cost ($)	3,600
Incremental net income ($)	600
$-3,600 + 600(P/A, i\%, 8) = 0$	
$(P/A, i\%, 8) = 6.000$	

By interpolation, $IROR = 6.9$ percent < 12 percent $MARR$, therefore, discard J. The one remaining alternative, computer I, should be selected.

Note that if the net present worth of each alternative is calculated at the $MARR$ of 12 percent, the best alternative would be computer I, as determined using $IROR$ analysis. And the same answer would be obtained using net annual worth or net future worth. To illustrate:

	F	G	H	I	J
First cost ($)	25,000	18,000	22,500	28,400	32,000
Present worth of annual net income					
$A(P/A, 12\%, 8)$ ($)	24,838	18,878	23,845	29,806	32,787
Sum	-162	878	1,354	1,406	787

The natural question is why expend the effort calculating $IROR$ when NPW or NAW will usually yield the answer with much less effort. There are generally two reasons for calculating $IROR$. They are:

1. When the $MARR$ is not known and thus each alternative must be examined to determine over what range (if any) of i it will be the best alternative.
2. The client (Board of Directors, City Council, etc.) wishes to know the anticipated rates of return on the proposed investment and incremental investment.

SUMMARY

The incremental rate of return ($IROR$) is defined as the rate of return for an *additional* initial investment when comparing against a lower cost investment. Two types of investment decisions occur. The first type involves costs only. Under such conditions the rate of return for each investment is obviously negative and, thus, need not be calculated. The question to be answered is which alternative yields the lowest equivalent cost? To determine this the following procedure should be followed.

Step 1. List the alternatives *in order* of ascending first cost.
Step 2. Stepwise determine the $IROR$ for each difference in alternatives, beginning with the lowest cost alternative. Compare each

IROR with the minimum attractive rate of return (*MARR*), if known, immediately discarding either (a) the lower cost alternative if the *IROR* is greater than the *MARR* or (b) the higher cost alternative if the *IROR* is less than the *MARR*.

If the *MARR* is not known, then all *IROR*'s should be calculated and analyzed.

PROBLEMS FOR CHAPTER 11, INCREMENTAL RATE OF RETURN

Problem Group A

Given a series of alternative investments, each with one or more P, A, F values, but all alternatives have the same n values.

A.1 A profitable company is considering the following three investments. Using incremental rate of return analysis techniques, which alternative, if any, should they select, (a) if their *MARR* is 15 percent; (b) if their *MARR* is 20 percent?

	P	Q	R
Initial investment ($)	40,000	30,000	20,000
Annual net income ($)	7,400	5,700	3,200
Salvage value ($)	0	0	0
Life (yr)	20	20	20

A.2 Same as Problem A.1 except the life of each alternative is 30 years.

A.3 A sanitary sewer can be constructed using either a deep cut or a lift station, with the following costs estimated. Using incremental rate of return analysis techniques determine which alternative should be selected for an *MARR* of 7 percent.

	DEEP CUT ALTERNATIVE	LIFT STATION ALTERNATIVE
Installation cost of pipe ($)	275,000	235,000
Installation cost of lift station ($)	—	15,000
Annual maintenance costs ($)	400	1,800
Expected life (yr)	40	40

A.4 Same as Problem A.3 except the expected life of the lift station is only ten years.

A.5 A friend asks your advice on which of two possible investments he should choose. Prepare recommendations based on an *IROR* analysis. Clearly state the assumptions needed in order to make your recommendation.

Alternative A. Place $10,000 in an eight-year certificate of deposit yielding 8 percent per year compounded quarterly.

Alternative B. Invest $12,000 in a business venture for which the company promises to pay your friend $550 per quarter for eight years.

A.6 Same as Problem A.5, except in alternative B the company promises to pay $260 per quarter for eight years, then pay back the initial $12,000 investment.

A.7 A contractor asks your advice on selecting the best of three alternative financing methods for acquiring company pickups for use by her foremen. The alternatives are listed as follows.

1. Purchase the pickups now for $5,000 each. Sell after four years for $1,000 each.
2. Lease the pickups for four years at $1,200 per year paid in advance at the beginning of each year. Contractor pays all operating and main-tenance costs, just as if she owned them.
3. Purchase the pickups on time payments, with $750 down payment and $1,320 per year at the end of each year. Sell after four years for $1,000 each.

a. Plot a graph of *PW* versus *i* for each alternative. (*i* is the pretax rate at which the contractor can invest available funds.)
b. On the graph, show which ranges of *i* favor which alternatives.

A.8 A consulting firm finds it must provide reprints of a large number of items such as plans and specifications for its customers at a fixed nominal cost. It narrows the alternative possible means of providing this service for the next five years down to the following list of three.

1. Outside contract. A $1,000 deposit (returnable at EOY 5) is required to reserve these facilities. In addition, the fixed fee from customers is not enough to cover the cost of the work. Loss to the consultant amounts to $100 per year.
2. Lease the reproduction equipment, brand B. A deposit of $4000 is required by the lessor, returnable upon termination of the lease. A cash flow profit of $320 per year for the reproduction work is assured under this plan.
3. Purchase brand C with guaranteed buy-back. This firm sells the reproduc-tion equipment for $10,000 and guarantees to buy it back for 50 percent of the purchase price in five years. Calculations indicate this plan would yield a positive cash flow of $2,000 per year (not counting the purchase and resale cash flows).

In each case the study period is five years.
a. Find the rate of return for each alternative.
b. Find the incremental rate of return for each pair.
c. Sketch a graph of present value of each alternative versus *i* for each alternative.
d. Analyze the curves and tell which alternative investment is best suited for each range of *i*.
e. If the consulting firm is earning 20 percent on capital invested in the firm, but must provide the reproduction service, which is the best alternative?
f. If the consulting firm has the option of *not* providing the reproduction service, and they are earning 20 percent on capital, which is the best alternative?

A.9 Your client, a contractor, needs some new trucks to complete a project currently under contract. He asks you to analyze the three alternative types of trucks on the basis of the information listed below, and advise which is the best buy.

TYPE OF TRUCK	COST NEW	ANNUAL O&M COST	ANNUAL INCREASE IN O&M COSTS	RESALE VALUE AT EOY 4
A	$25,000	$0.85/mi	$0.080/mi/yr	$5,000
B	28,000	0.80	0.080	7,500
C	32,500	0.75	0.045	7,500

The trucks will average 20,000 miles per year and will be sold at the end of four years for the resale values listed.

a. Find the incremental rate of return between A and B. (*Hint*: Try i for B–A between 30 and 40%.)

b. Assume the *IROR* of C–A is 22.3%, and of C–B is 24.8%. Draw an arrow diagram showing the *IROR* values and tell what ranges of *MARR* favor which alternatives.

c. Show which alternative is recommended if the contractor's *MARR* = 15 percent.

A.10 An OSHA inspector has just ruled that the lighting fixtures in your client's plant are inadequate to meet minimum standards. He must install new lighting in order to stay in business and asks you for a recommendation. You develop three alternative installations with figures as tabulated below. In order to assist your client in making the best selection, develop the following information.

a. Find the incremental rate of return between A and B.

b. Assume the *IROR* of C–A is 17.1%, and C–B is 32.4%. Draw an arrow diagram showing which alternative to choose for an *MARR* = 5 percent.

c. Show the break points (if any) and tell what ranges of *MARR* favor which alternatives.

TYPE	COST NEW	ANNUAL OPERATING COST	ANNUAL INCREASE IN OPERATING COST	SALVAGE VALUE AT EOY 20
A. Mercury	$30,000	$14,000/yr	$2,000/yr/yr	$ 3,000
B. Fluorescent	45,000	13,000	1,900	13,000
C. Metal halide	54,000	10,000	1,800	3,000

(*Hint*: Find i, B–A between 10 and 20%; C–A between 10 and 20%; C–B between 30 and 40%.)

A.11 Assume that you are a consulting engineer with your own office. Your company car has just been totaled in a wreck, but the insurance company promptly wrote you a check for $3,000 to cover the loss. You are now selecting a replacement car. Any of the three listed below is acceptable from a functional point of view.

a. Find the incremental rate of return for each pair of alternatives.
b. Draw an arrow diagram showing which alternative to choose if the *MARR* $=15$ percent.
c. Show the break points in the *MARR* (if any), and tell what ranges of *MARR* favor which alternatives.

TYPE	COST NEW	ANNUAL O&M COST	ANNUAL INCREASE O&M COST	RESALE VALUE AT EOY 4
A. Electra	$5,000	$0.17/mi	$0.014/mi/yr	$1,000
B. Monaco	5,700	0.16	0.016	1,500
C. Montego	6,500	0.15	0.009	1,500

The car will be driven about 20,000 miles per year and sold at the end of the fourth year.

(*Hint*: Try *i* for B–A between 10 and 20%; C–A between 20 and 30%; C–B between 20 and 30%.)

A.12 A friend of yours is considering the purchase of a home. The seller offers a choice of two options.

	OPTION A	OPTION B
Purchase price	$40,000	$39,000
Down payment	5,000	6,000
Mortgage	35,000	33,000
Interest rate on the mortgage	9%	9%
Payout period for mortgage	25 yr	25 yr

a. What is the incremental rate of return on the difference between the two options? (*Hint*: Try between 15 and 25%.)
b. Which option do you recommend and with what conditions (if any)?

Problem Group B

The *n* values of each alternative differ.

B.1 A subscription to a professional magazine is available at three different rates:

1 yr $10
2 yr $17.50
3 yr $25.00

Assume you intend to subscribe probably for the rest of your professional career. Find the before-tax rate of return on:
a. The two-year subscription compared to the one-year.
b. The three-year subscription compared to the one-year.
c. If you can invest your money elsewhere at 20 percent, before taxes, which alternative is the best selection?

B.2 A professional society whose dues are normally $100 per year is offering life memberships for $1,250. A young engineer of 24 is considering the life membership offer. If she stays active in the society until age 65 (41 years), what rate of return will the investment in life membership yield?

a. Compared to the $100 per year membership dues paid in advance, with the first year's dues payable right now.

b. Compared to the $100 per year membership dues paid in advance, but the dues are expected to increase by $5 per year every year ($100 the first year, $105 the second, etc.).

B.3 The daily newspaper is available at the news stand for $0.15 a copy seven days a week. As an alternative, the paper may be taken on a subscription basis for $52 per year.

a. Find the rate of return on the subscription price compared to the daily rate, using a 365-day year.

b. Assume that the Sunday edition at the daily rate is $0.35 a copy instead of $0.15, and there are 52 Sunday editions per year. What is the rate of return now for the subscription at $52 per year?

B.4 The school board asks your help in determining which is the better alternative for obtaining portable classrooms.

1. One-year renewable leases, with $100 per month paid in advance, at the first of every month.

2. Purchase with guaranteed buy-back. The portable classrooms are purchased for $10,000 each, and the vendor agrees to repurchase them back from the school board at the end of three years for $5,700, if the school board should want to sell them at that time.

3. Three-year lease for $3,000 paid in advance.

Find the rate of return of:
a. Alternative 2 compared to 1.
b. Alternative 3 compared to 1.
c. If a selection of the most economical alternative is made when funds may be borrowed at 6 percent, does the selection change if interest rates rise to 9 percent?

B.5 Three grades of roofing are available for a proposed office building, and you are asked to determine which is the most economical. Data on the roofs are presented as follows. (Find i in terms of A rather than PW.)

1. Fifteen-year roof costs $18,000 plus $1,000 per year for end of year maintenance.

2. Twenty-year roof costs $22,000 plus $500 per year for end of year mainte-nance.

3. Twenty-five-year roof costs $30,000 and requires no maintenance.

a. Plot a graph of A versus i for each alternative. (i is the pretax rate at which the building owner can invest available funds.)

b. On the graph, show which ranges of i favor which alternative.

B.6 Your firm has a large fleet of cars, and tires are a major expenditure. You are asked to determine which of the three alternatives listed below is the most economical. The vehicles average 2,000 miles per month. Assume i is the monthly rate of interest and is compounded monthly.

1. 20,000-mile tire costing $21 each, wholesale
2. 30,000-mile tire costing $29 each.
3. 40,000-mile tire costing $38 each.

a. Plot a graph of A versus i for each alternative.
b. On the graph, show which ranges of i favor which alternatives.

Problem Group C

Formulate your own problem.

C.1 Formulate a problem involving alternative methods of acquisition of goods or services of interest to you. Find the better alternative by the *IROR* method, and determine the ranges of i for which each alternative is feasible.

Chapter 12
Break-Even Comparisons

THE BREAK-EVEN POINT

Engineers working with private enterprise companies frequently are dealing with capital investments intended to return a profit from the subsequent sale of goods or services produced as a result of the capital investment. Costs for these companies usually occur in two basic categories—the initial capital investment and the operating and maintenance (O&M) costs. Some financial managers use the terms:

1. *Fixed costs* for those capital costs (including construction, acquisition, or installation costs) that remain relatively constant regardless of changes in the volume of production.
2. *Variable costs* for those costs that vary directly with production output.

These two types of costs can be portrayed (in simplified form) on a graph as shown in Figure 12.1. Since the fixed costs are the costs of providing plant and equipment in a standby ready-to-serve condition before starting production, the fixed costs are shown as a horizontal line on the graph, indicating the same constant annual cost regardless of the number of units

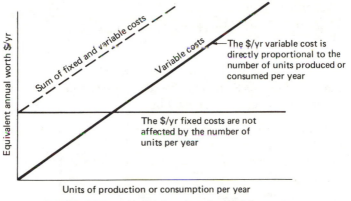

Figure 12.1   Linear fixed and variable costs.

produced. On the other hand, variable costs are usually considered as the operating costs of actually turning out a product. If each unit of the product is assumed to cost a constant amount of dollars per unit to produce, then the variable cost versus output plots on the graph as an inclined straight line.

When comparing alternative investments, often there is an inverse relationship between fixed capital cost and the variable operating expenses. In many instances a higher initial capital cost will purchase more efficient equipment which in turn results in lower operating and production costs. Engineers are frequently confronted with decisions between higher capital cost for more efficient equipment with lower operating costs, and the converse. For example, is it more economical to purchase a new large dump truck for $30,000 with O & M costs of $0.10 per ton-mile or buy a used truck with the same total capacity for $10,000 but with O & M costs of $0.15 per ton-mile? The answer depends on how many ton-miles will be hauled. For a low number of ton-miles per year, the used truck with the lower capital costs will be least expensive. As the number of ton-miles per year increases, the lower O & M costs of the new unit becomes a significant factor so that at high ton-miles per year the new unit becomes the more economical. At some point in between, called the *break-even point*, the costs will be equal. If this point is known, the owner who expects to haul few ton-miles can benefit from the economy of the used unit in the low ton-mile range, while the owner with expectations of hauling more than the break-even amount can maximize profit by purchasing the new unit.

The break-even point may be found by following the logical three-step procedure, as follows:

1. Find the annual equivalent of the capital costs.
2. Find the independent variable (ton-miles in this case) and set up an equation for each alternative cost combination. The equation usually

takes the form of

total annual cost = annual equivalent capital cost

$$+ (cost/variable\ unit)(number\ of\ variable\ units/year)$$

This is an equation that graphs as a straight line.
3. Find the break-even points. These break-even points are the points of intersection of the graphed lines.

A graphical solution can be obtained by plotting the lines on a graph, and scaling the points of intersection. An algebraic solution can be obtained by solving the equations simultaneously. These procedures are illustrated below in Example 12.1.

Example 12.1

A contractor is thinking of selling his present dump truck and buying a new one. The new truck costs $30,000 and is expected to incur O&M costs of $0.10 per ton-mile. It has a life of 15 years with no significant salvage value. The presently owned truck can be sold now for $10,000. If kept it will cost $0.15 per ton-mile for O&M, and have an expected life of five years, and no salvage value. Use $i = 10$ percent. Find the break-even point in terms of ton-miles per year.

SOLUTION

The annual equivalent to the capital investment cost is

$$A_{(new\ truck)} = \$30,000\ (A/P, 10\%, 15) = \$3,944/yr$$
$$0.13147$$

$$A_{(present\ truck)} = \$10,000\ (A/P, 10\%, 5) = \$2,638/yr$$
$$0.26380$$

The present truck "costs" $10,000 because the contractor gives up $10,000 cash if he keeps it, just as he gives up $30,000 if he decides to purchase the new one.

The total annual equivalent cost for each year for each alternative is simply the annual equivalent capital cost plus the annual O&M cost, as follows:

$$total\ annual\ cost_{(new\ truck)} = A_1$$

$$A_1 = \$3,944/yr + (\$0.10/ton\text{-}mi)(X\ ton\text{-}mi/yr)$$

$$total\ annual\ cost_{(present\ truck)} = A_2$$

$$A_2 = \$2,638/yr + (\$0.15/ton\text{-}mi)(X\ ton\text{-}mi/yr)$$

where X is the number of ton-miles per year. The break-even value of X may

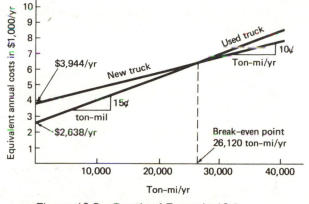

Figure 12.2   Graph of Example 12.1.

be found by solving the equations simultaneously (finding the intersection point by one point common to both equations).

	general form	$y=mx+b$
	for new truck	$y=0.10X+3,944$
	for present truck	$y=0.15X+2,638$

$$X=\frac{3,944-2,638}{0.15-0.10}=26,120 \text{ ton-mi/yr}$$

This is the break-even point. If the new truck carries an average load of 10 tons, the annual mileage at break-even is 26,120 miles per year. The graphical solution is illustrated in Figure 12.2.

From the graph it is evident that the break-even point occurs where the difference in slopes of the two lines overtakes the initial gap between the two lines. This point can be determined by the following process.

1. Find the amount of the gap between the two lines at the vertical axis ($1,306/yr at 0 ton-mi/yr);
2. Divide this amount ($1,306/yr) by the difference in the slope of the lines ($0.15−$0.10=$0.05/ton-mi) to obtain 26,120 ton-miles per year for the break-even point.

BREAK-EVEN INVOLVING INCOMES (BENEFITS) AND COSTS

Often the purpose of a capital investment is to produce a product with a direct dollar market value, and the capital investment becomes the direct source of a cash flow income stream.

If the product can be sold for a certain number of dollars per unit produced then the income can be depicted on the graph as another inclined

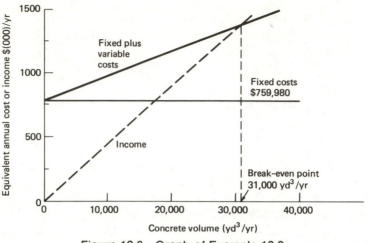

Figure 12.3   Graph of Example 12.2.

straight line representing dollars of income as a function of output volume, as shown in Figure 12.3. To illustrate the concept the following example is given.

Example 12.2

The purchase of an existing ready mixed concrete plant in good operating condition requires a capital investment of $3,700,000 (plant, trucks, property, etc.). Calculation of operating and maintenance costs of producing concrete are $20.52 per cubic yard of concrete (this includes material costs for cement and aggregate as well as O&M costs for the equipment). If the concrete is sold for $45.00 per cubic yard, what annual volume of concrete must the company sell to break even if their before-tax *MARR* is 20 percent? Assume a 20-year plant life with no salvage value.

SOLUTION

The annual fixed cost of the investment is

$$A = 3,700,000 \, (A/P, 20\%, 20) = \$759,830$$
$$0.20536$$

The various components of costs and income are portrayed in Figure 12.3.

From the graph, the break-even point of about 31,000 yd³ of concrete sales each year can be found.

To solve algebraically, simply set the costs equal to the income.

$$\$759,830 + \$20.52 B = \$45 B$$

$$B = \frac{759,830}{24.48} = 31,040 \text{ yd}^3$$

This type of analysis is helpful in determining break-even production outputs, and estimating the marginal profit or loss resulting from various production volumes. In the previous example, for every cubic yard of concrete over the break-even 31,040 yd³, the company will earn a profit of $45.00 − 20.52 = $24.48. And conversely for every cubic yard under 31,040 yd³ they will lose $24.48.

FOUR TYPES OF FIXED/VARIABLE COST COMBINATIONS

Four commonly encountered examples of fixed and variable cost combinations are depicted in simplified form in Figure 12.4. The first graph illustrating practically no fixed costs can be exemplified by a casual day-laborer who offers his services as a manual laborer on an hourly basis to

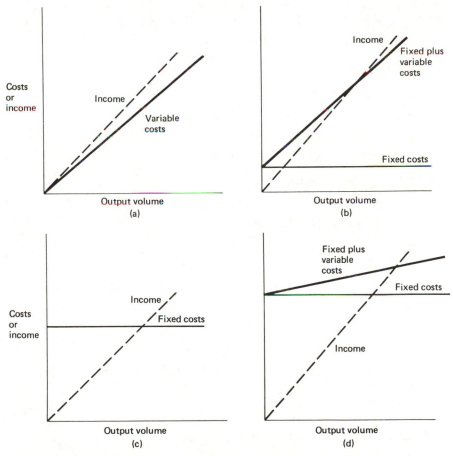

Figure 12.4   (a) Zero fixed costs. (b) Low fixed costs, high variable costs. (c) Zero variable costs. (d) High fixed costs, low variable costs.

contractors and others. He carries no tools and thus has no fixed costs except the cost of maintaining himself in good health. He gets paid only for hours worked and thus his income varies in direct proportion to those hours worked. The second graph showing variable costs could picture the owner of a piece of land who is receiving royalties for the removal of oil from the land. Her fixed costs are the costs of the land and her income is derived from the oil removed. The third and fourth graphs illustrate the more commonly encountered example of combined fixed and variable costs situations. Many managers spend considerable time and effort looking for ways to lower the slope of the variable cost line, and often succeed by raising the level of the fixed cost line. Example 12.3 (as well as most other examples in this chapter) illustrates this relationship.

Algebraic Relationships for Break-Even Analysis

Since many realistic investment situations can be approximated by straight line equivalence, the following simple algebraic relationships may prove useful in finding break-even points.

Let

B = break-even output (units/period)

F_d = Fixed cost per period

R = Total revenue (income) per period

V = Variable cost per unit

M = Profit per period (or loss)

C = Total costs per period

N = Output units (or time periods)

S = Selling price (income/unit)

$$R = NS \tag{12.1}$$

$$C = NV + F_d \tag{12.2}$$

At break-even, $R = C$. By letting $N = B$ (number of units to break even)

$$BS = BV + F_d$$

$$B = F_d/(S - V) \tag{12.3}$$

The quantity $(S - V)$ is sometimes referred to as the "contribution," because it is the contribution of the net profit per unit (or selling price per unit minus variable cost per unit) toward paying off the fixed costs.

$$M = R - C = N(S - V) - F_d \tag{12.4}$$

Further Applications of Linear Break-Even

The following example illustrates a break-even analysis for hiring additional help.

Example 12.3

A small, but profitable, consulting engineering firm is considering hiring another engineer for $30,000 salary+$30,000 (for fringe benefits, technical support, etc.)=$60,000 per year. Currently their fixed costs total $740,000 per year and their variable costs average $11,200 per engineering contract. Their income per engineering contract averages $19,300.

1. If the new engineer is hired and his costs are all fixed costs, how many *more* contracts must be acquired for the company to break even?
2. If the addition of the new engineer will permit the company to reduce its subcontracting efforts to other engineers to the point where the average cost per engineering contract is reduced to $10,400 and they expect approximately 100 contracts next year, should the new engineer be hired?

SOLUTION 1
The existing break-even is

$$B(\text{existing}) = \frac{\$740,000}{(19,300 - 11,200)} = 91.4 \text{ contracts}$$

The new break-even is

$$B(\text{new}) = \frac{740,000 + 60,000}{(19,300 - 11,200)} = 98.9 \text{ contracts}$$

additional contracts $= 98.8 - 91.4 = 7.4$ or   8 contracts

SOLUTION 2
Here the decision would depend on the overall profit generated by the 100 contracts.

existing profit, $M = 100(19,300 - 11,200) - 740,000 = +\$70,000$

new profit, $M = 100(19,300 - 10,400) - 800,000 = +\$90,000$

The answer is to hire the new engineer in anticipation of an additional $20,000 profit.

Comment. The additional fixed cost (a new engineer in this case) reduced the variable costs per unit V (by $800) and thus raised the "contribution" $(S - V)$ that could be applied to the fixed costs. With the large number of contracts expected, the overall profit was increased. A graph illustrating this example is shown in Figure 12.5. Examination of the graph reveals that *if* the estimate of 100 contracts is *not* realized, the company could end up losing money by hiring the engineer.

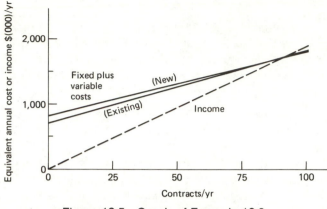

Figure 12.5   Graph of Example 12.3.

LIMITATIONS OF BREAK-EVEN ANALYSIS

Problem areas commonly encountered when applying break-even analysis in actual practice include the following.

1. Difficulty in allocating all costs into purely fixed and purely variable components. Most costs have a fixed component and a variable component. For example, the unit labor cost for erecting concrete forms may include (a) the cost of a supervisor who will be employed regardless of whether or not any forms are built and (b) the cost of carpenters and helpers which may be totally variable. Over-simplification of the costs can result in erroneous data and conclusions.

2. The relationship between costs and outputs are often not linear. To assume that the variable cost of an activity is directly proportional to the quantity involved sometimes neglects important cost reductions available at higher production volumes, such as the discounts achieved through volume purchasing, shipping, handling, and so on.

3. The break-even chart may be valid for a limited time duration only because the data may change significantly with time. Therefore, break-even analysis (like many other methods) should be used with caution, particularly on long-lived investments. The real world contains many examples of preliminary studies that indicated one course toward a profitable activity, only to have circumstances change drastically, resulting in the indicated course leading to a loss rather than profit. For example, in 1977 to 1979 U.S. automakers decided to continue producing large, expensive, and inefficient automobiles because of potentially larger unit profits ("contribution"). But, market conditions changed abruptly when gasoline costs began escalating. Sales of large cars decreased rapidly and the automakers who were depending on sales of large cars found themselves in serious financial difficulty because they could not achieve their break-even volume.

Used knowledgeably, break-even analysis is a useful tool for examining relationships between fixed and variable costs and incomes. It is also an aid in making rational decisions involving capital investments.

RISING RATES (GRADIENT)

Unfortunately, future expenditures usually do not remain at conveniently level rates, but typically rise at a rate the estimator must attempt to predict. The following example illustrates the effect of rising rates by comparing the break-even points using (a) constant labor rates versus (b) rising labor rates.

Example 12.4

A water and sewer utility department is considering the acquisition of a "bell holer" which will dig small bell holes for bell and spigot pipe joints. The equipment costs $5,000 with no predictable salvage value after ten years. The machine can dig five holes an hour and requires one operator at $8 per hour. Fuel, maintenance, insurance, and overhead adds another $2 per hour.

At present the bell holes are hand dug by laborers at $5 per hour averaging two holes per person per hour.

1. How many holes per year is the break-even point if $i = 10$ percent?
2. If labor costs increase $0.50 per hour per year for both laborers and equipment operators, what is the break-even point using equivalent annual cost of the gradient, if $i = 10$ percent?

SOLUTION 1

Annual capitalization cost of bell holer,

$$A = \$5,000 \, (A/P, 10\%, 10)$$
$$\qquad\qquad 0.1627$$
$$\quad = \$813.50$$

There is zero salvage value in ten years. Per hole operating costs for bell holer are as follows.

operator	$ 8/hr
fuel, etc.	2
total	$10/hr ÷ 5 holes/hr = $2/hole

total cost for bell holer/hole = $2/hole + ($813.50/yr)/(Y holes/yr)

total cost for hand labor/hole = $5/hr ÷ 2 holes/hr = $2.50/hole

break-even point = $2.50 = $2.00 + $813.50/$Y$

$$Y = 1627 \text{ holes/yr}$$

SOLUTION 2

Annual equivalent of increased labor costs,

$$A = \$0.50 \underset{3.755}{(A/G, 10\%, 10)} = \$1.86$$

Per hole operating costs for bell holer are as follows.

operator $\quad\quad\quad\quad\quad$ $\$8 + \$1.86 = \$\ 9.86$

fuel, etc. $\quad\quad\quad\quad\quad\quad\quad$ $=\quad\underline{2.00}$

$\quad\quad\quad\quad$ $\$11.86 \div 5$ holes/hr $= \$2.37$/hole

total cost for hand labor/hole $= (5 + 1.86)/2$ holes/hr

$$= \$3.42/\text{hole}$$

break-even point $= \$3.43 = \$2.37 + \$813.50/Y$

$$Y = 813.50/1.06 = \underline{\underline{767 \text{ holes/yr}}}$$

Comment. In this case the rising labor rates significantly decrease the number of holes per year required for the break-even point.

BREAK-EVEN COSTS, FIND UNKNOWN LIFE

In ordinary algebra, as long as the number of independent equations exceeds the number of unknowns, the problem can usually be solved. Break-even problems occur under a wide variety of circumstances, and the unknown variable could be any one of these variables involved in the problem. An example of a break-even problem with an unknown service life follows.

Example 12.5

A dredge is in need of a pump that will pump 15,000 gal/min at a 10 ft total dynamic head. The specific gravity of the dredged material will average 1.20. There are two good pumps available with data as shown ($i = 7\%$). How long must the Pumpall last to have a cost equivalent to the Gusher?

	PUMPALL	GUSHER
Cost new ($)	− 15,600	− 43,200
Efficiency (%)	70	78
Installation costs ($)	− 4,200	− 3,900
Maintenance and repair costs ($/yr)	− 3,850	− 3,850
Salvage value ($)	None	+ 10,000
Estimated life (yr)	Find	12

The pump is expected to operate 310 days per year, 24 hours per day. Power is available at $0.042/kWh. Note:

1 horse power (hp) $= 0.746$ kW

$$\text{hp output required} = \frac{\text{dynamic head} \times \text{gal/min} \times \text{specific gravity}}{3,960}$$

SOLUTION

The break-even point in terms of the estimated life of the Pumpall pump is found by setting the costs of the two pumps equal. The only unknown variable is the life of the Pumpall pump.

Pumpall

cost in place, $\quad A_1 = -(15,600 + 4,200)(A/P, 7\%, n)$

$$= -\$19,800(A/P, 7\%, n)/\text{yr}$$

operating cost, $\quad A_2 = -\dfrac{\begin{bmatrix} 10 \text{ ft} \times 15,000 \text{ gal/min} \times 0.746 \text{ kW/hp} \\ \times 1.20 \times 310 \text{ days/yr} \times 24 \text{ hr/day} \\ \times 0.042/\text{kWh} \end{bmatrix}}{3,960 \times 0.70}$

$$A_2 = -\$15,137/\text{yr}$$

Gusher

cost in place, $\quad A_3 = -(43,200 + 3,900)\underset{0.1259}{(A/P, 7\%, 12)}$

$$= -\$5,930/\text{yr}$$

salvage, $\quad A_4 = +10,000 \underset{0.0559}{(A/F, 7\%, 12)} = +\$559/\text{yr}$

operating cost, $\quad A_5 = -\dfrac{\begin{bmatrix} 10 \text{ ft} \times 15,000 \text{ gal/min} \times 0.746 \text{ kW/hp} \\ \times 1.20 \times 310 \text{ days/yr} \times 24 \text{ hr/day} \\ \times \$0.042/\text{kWh} \end{bmatrix}}{3,960 \times 0.78}$

$$A_5 = -\$13,585/\text{yr}$$

total annual cost of Gusher $= A_3 + A_4 + A_5 = -\$5,930 + 559 - 13,585$

$$= -\$18,956$$

Setting the Pumpall cost equal to the Gusher

$$-19,800(A/P, 7\%, n) - 15,137 = -18,956$$

$$(A/P, 7\%, n) = 0.19288$$

Checking the tables yields:

$n = $ less than 7 yr

The Pumpall would need a seven-year life to equal the cost of the Gusher.

KELVIN'S LAW, COSTS THAT VARY DIRECTLY AND INVERSELY WITH ONE VARIABLE

Engineers sometimes encounter problems where an increase in the independent variable involves an increase in one type of cost and a decrease in another. For instance, in 1881, Lord Kelvin pointed out that the capital cost of conducting electricity through a wire increases as the area of the wire increases, but the cost of transmission losses due to resistance decreases with the same increase in wire area. He proposed that the best solution was the lowest total of capital costs plus transmission costs, obtained as a solution to an equation in the general form of

$$Y = AX + \frac{B}{X} + C$$

where

$X =$ independent variable, the area of the wire in this case

$AX =$ costs that increase with the increases in the wire area
(capital costs)

$B/X =$ costs that decrease with increases in the wire area
(operating or energy costs)

$C =$ the fixed costs incurred regardless of the wire area

The equation plots as a curve, in this case concave up, with a dip low point. The minimum point may be determined by finding the first derivative (slope equation) and setting equal to zero (slope at tangent to low point) as follows.

$$\frac{dY}{dX} = A - \frac{B}{X^2} = 0$$

$$X = \sqrt{\frac{B}{A}}$$

Thus, the cross-sectional area at the minimum cost point is the square root of the indirectly varying costs divided by the directly varying costs. This wire area is then substituted back into the equation to determine the total cost of the installation. An example follows.

Example 12.6

A 200-ft length of copper wire is required to conduct electricity from a transformer to an electrically powered pump at a municipal water plant.

Find the most economical size wire, assuming the following data.

pump motor size (hp)	100
pump motor voltage (V)	440
pump motor operating efficiency (%)	80
operating hours per year	4,000
length of wire conductor needed (ft)	200
installed cost of copper wire	$0.80/lb + $200 for fittings
estimated life (yr)	20
salvage value ($/lb)	0.40
electrical resistance of copper (ohms in²/ft length)	8.17×10^{-6}
conversion W/hp @ 100% efficiency	745.7
cost of energy ($/kWh)	0.05
interest rate, i (%)	7
copper unit weight (lb/in.³)	0.32

SOLUTION

The amperage required by the electric pump motor is:

$$(100 \text{ hp} \times 745.7 \text{ W/hp})/0.8 \text{ eff} \times 440 \text{ V}) = 211.8 \text{ A}$$

The electrical resistance is inversely proportional to the area of the wire cross section (a in.²). The energy loss in kilowatts hours (kWh) due to resistance on the wire is calculated as

$$\frac{I^2 R \times \text{number of hr/yr}}{1,000 \text{ W/kW} \times a \text{ in.}^2} = \frac{211.8^2 \times 8.17 \times 10^{-6} \times 200 \times 4,000}{1,000 \times a}$$

$$= 293.2 \text{ kWh}/a \text{ in.}^2$$

Thus, the energy loss due to resistance in the wire amounts to 293.2 kWh/(a in.²) each year. The annual cost of this lost energy is

$$A_1 = \$0.05/\text{kWh} \times 293.2 \text{ kWh}/(a \text{ in.}^2/\text{yr}) = \$14.66/a \text{ in.}^2/\text{yr}$$

Annual equivalent costs: The cost of the wire is given in terms of dollars per pound, and the weight is proportional to the area, a. Therefore, the cost is calculated as

$$P = 200 \text{ ft} \times 12 \text{ in./ft} \times 0.32 \text{ lb/in.}^3 \times a \text{ in.}^2 \times \$0.80/\text{lb} = \$614.4a$$

The annual equivalent value is

$$A_2 = \$614.4a \underset{0.0944}{(A/P, 7\%, 20)} = \$58.00a/\text{yr}$$

The salvage value is $0.40 per pound, or 307.2a$. The annual equivalent is

$$A_3 = \$307.2a\,(A/F,7\%,20) = \$7.50a/\text{yr}$$
$$0.0244$$

The fixed cost of installation and fittings is $200, regardless of the area, so the annual equivalent is calculated as

$$A_4 = \$200\,(A/P,7\%,20) = \$18.88/\text{yr}$$
$$0.0944$$

Summing, the total of all annual equivalent costs is

$$A_{\text{TOT}} = \$14.66/a + \$58.00a - \$7.50a + \$18.88$$

Optimum cross section, a, for the lowest net cost: The equation for total cost derived above may be graphed as a sag curve of dollars versus area, a. The lowest cost point on the curve is found by taking the first derivative of the equation and setting equal to zero (finding the point of zero slope). As previously explained, the low point is found at

$$X = \sqrt{\frac{B}{A}} \quad \text{or} \quad a = \sqrt{\frac{14.66}{58.00 - 7.50}} = 0.539 \text{ in.}^2$$

For circular wire, the diameter is

$$\underline{d = 2\sqrt{a/\pi} = 0.828 \text{ in. diameter}}$$

The weight of the wire is

$$0.539 \text{ in.}^2 \times 0.32 \text{ lb/in.}^3 \times 12 \text{ in./ft} = 2.07 \text{ lb/ft}$$

The equivalent annual cost of the entire installation is

$$A = \frac{14.66}{0.539} + (58.00 - 7.50)0.539 + 18.88 = \underline{\underline{\$73.30/\text{yr}}}$$

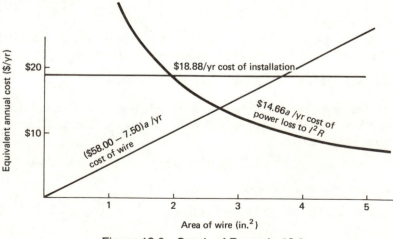

Figure 12.6   Graph of Example 12.6.

From an economics viewpoint, the lowest installation is a 0.828-in. diameter copper cable weighing 2.07 lb/ft. Of course, electrical manufacturers' catalogues show the nearest size available and electrical codes must be checked to ensure the resulting wire size complies with code requirements. Figure 12.6 illustrates the component costs of this problem.

Applications to Span Length (Column or Pier Spacing) Determination

Other problems involving a single variable with both inverse and direct relationships include the determination of column or pier spacing and the consequent span length. The longer spans save on the number of columns or piers but increase the size and cost of the spanning member. The following example illustrates this type of problem.

Example 12.7

Assume a river is to be spanned by a highway bridge that will measure 2,000 ft from abutment to abutment, and a determination of pier spacing and span length is needed. The following data are provided. The piers are estimated to cost $100,000 each regardless of the span design. The end abutments will cost $250,000 each, and no pier is required at the abutments (of course). The weight of the span dead load per foot is estimated as

$$w = (10s + 200) \text{ lb/ft}$$

where s represents the spacing between pier centerline in feet. The cost of the span is estimated as $0.60 per pound.

SOLUTION
The number of piers required is the number of spans less one, since no piers are required at the ends of the bridge where the abutments are located. This number can be expressed as $2,000/s - 1$. Therefore, the total cost of the piers is the number of piers times the cost per pier or $(2,000/s - 1)\$100,000$. Likewise, the weight of all the spans is $(10s + 200)$ lb/ft $\times 2,000$ ft. Therefore, the cost of these spans is this weight times $0.60 per pound. The total cost of the bridge is found by summing these parts as

$$T = (10s + 200)(2,000)(\$0.60) + \left(\frac{2,000}{s} - 1\right)(\$100,000) + 2(\$250,000)$$

This is reduced to

$$T = 12,000s + \frac{2 \times 10^8}{s} + 640,000$$

Since this is the equation for the cost, the minimum cost span is found by taking the first derivative of this equation and setting it equal to zero.

$$\frac{dT}{ds} = 12,000 - \frac{2 \times 10^8}{s^2}$$

$$s = \sqrt{\frac{2 \times 10^8}{12,000}} = 129.1 \text{ ft}$$

The theoretical minimum span occurs using 15.5 spans of 129.1 ft. Since the number of spans must equal a whole number, either 15 or 16 may be used. The cost of each alternative is found by substituting in the equation above.

NUMBER OF SPANS	SPAN LENGTH	$12,000s + \dfrac{2 \times 10^8}{s} + 640,000 = $ TOTAL COST
15	133.3	$1.6 \times 10^6 + 1.5 \times 10^6 + 640,000 = \$3.74 \times 10^6$
16	125.0	$1.5 \times 10^6 + 1.6 \times 10^6 + 640,000 = \$3.74 \times 10^6$

Since these points are equidistant from the minimum point on the curve, the resulting costs are the same in either case.

The same solution technique may be adapted for other similar type problems, such as floor spans in buildings, distance between booster pumps on a pipeline, and many other problems involving costs that vary directly with a variable X, together with costs that vary inversely with the same variable, and fixed costs.

SUMMARY

Often an increased first investment cost will result in decreased periodic costs (such as operation and maintenance costs). These periodic costs are usually directly related to the use rate of the investment. Different initial investments can be compared on the basis of their break-even volume—the volume at which two alternatives yield the same overall cost. Knowing the break-even volume, or output, decisions can be made concerning which alternative will result in the lowest overall equivalent cost.

Both income and cost can be expressed in terms of the output volume. Under such circumstances break-even volumes can be determined on the basis of expected income versus fixed and variable costs.

Suggested Readings

H. G. Theusen et al., *Engineering Economy*, Fourth Edition, McGraw-Hill Book Company, New York, New York; pp. 207–211.

N. N. Barish and S. Kaplan, *Economic Analysis for Engineering and Managerial Decision Making*, Second Edition, McGraw-Hill Book Company, New York, New York, 1978, pp. 665–682.

James Riggs, *Engineering Economics*, McGraw-Hill Book Company, New York, New York, 1977, pp. 50–67.

PROBLEMS FOR CHAPTER 12, BREAK-EVEN COMPARISONS

Problem Group A

Break-even problems between two or more alternatives involving costs only. Given sufficient data to determine annual costs (at zero number of units/yr) and unit costs as the number of units per year increase, find break-even costs.

A.1 A tree spade is available for $20,000 with zero salvage value in ten years. Overhead and maintenance costs are estimated at $10 per hour. It can dig up and transplant an average of one tree per hour. The current method is by hand labor, with a crew costing $12 per hour to dig up and transplant an average of one tree every 1.5 hours. How many trees per year must be moved to justify the tree spade expenditure ($i=6\%$)?

A.2 A trencher is available for $30,000 with an estimated resale value in ten years of $5,000. The O&M costs are estimated at $10 per hour increasing each year by $1 per hour. It is capable of digging a trench at an average rate of about 50 ft/hr. The alternate method is to buy a backhoe at $20,000 with a resale value of $5,000 in ten years and O&M costs of $6 per hour increasing each year by $0.60 per hour. It can dig a trench at an average rate of 20 ft/hr. How many feet of trench per year must be dug to justify the trencher ($i=10\%$)? Plot the results on a graph.

A.3 A bucket truck may be rented, leased, or purchased for sign inspection by the building inspector in a small city. It rents for $5 per hour, or can be leased (including maintenance) for $2,500 per year or can be purchased for $10,000 with guaranteed buy-back of $2,000 in five years. If purchased it can be maintained for about $2 per hour. Find the most economical options with break-even points if any, in terms of hours per year use for each option ($i=6\%$). Plot the results on a graph.

A.4 A decision is required on what type air conditioners should be installed in a residential project your firm is designing. Two units are under consideration with data as follows.

	CARRIER 38 G5 036	CARRIER 38 SE 004
Cost installed	$750	$1,005
Efficiency rating	6.5	8.6
BTU	35,000/hr	37,000/hr

The efficiency rating is the number of BTU's output per watt of electrical consumption. Find the annual cost of each unit under the following assumptions.

a. Assume the unit will last 10 years with no salvage value ($i=8\%$). The unit will be used an estimated ten hours per day, six months of the year. Electricity costs an average of $0.036/kWh.

b. Assume the hours use per year is unknown. Graph the "dollars per year" versus "hours use per year," and find the break-even point.

c. Same as (b) except electrical costs increase by $0.004/kWh each year (first year cost $0.036/kWh, second year cost $0.04/kWh, etc.).

d. What effect would a higher interest rate have on the solution to (b) and (c).

A.5 A friend asks your advice on whether or not to sell his present car and buy a new one. His records indicate that his fixed costs of insurance, tag, and depreciation are currently about $750 per year. His mileage costs, including gas, tires, maintenance, and repairs are costing about $0.082 per mile. Authoritative figures indicate that a new car in the same class as your friend's present car would have a fixed cost of about $1,200 per year and mileage cost of $0.059 per mile.

    a. Find the break-even point in terms of miles per year.

    b. If your friend drives 10,000 miles per year, should he buy or keep for the greatest economy?

A.6 A consumers group needs information on relative costs of each of three systems of hot water heaters for residential use. They provide the information listed below ($i=8\%$).

	INITIAL COST	SERVICE LIFE (YR)	OPERATING EXPENSES	SALVAGE VALUE
Solar water heater	$600	20	1.00/Kg	$50
Electric hot water heater	$100	12	2.00/Kg	0
Gas hot water heater, vented through roof (Kg=1,000 gal)	$200	12	1.60/Kg	$100

Plot a graph of Kg/yr versus $/yr cost and find any break-even points that occur.

A.7 Same as Problem A.6 except that $i=15$ percent.

A.8 A new office building is being planned and you are requested to give an opinion on the alternative window systems listed below.

	DOUBLE PANE	SINGLE PANE
Initial cost	$4.00/ft^2	$2.00/ft^2
Heat loss	10 BTU/hr/°F/ft^2	15 BTU/hr/°F/ft^2

Assume $i=8$ percent; $0.10 per 1,000 BTU cost; $n=40$ years; salvage value $=0$. Graph the results in terms of $/yr/ft^2 versus °F-hr/yr and determine break-even points, if any.

A.9 Same as Problem A.8 except $i=16$ percent.

A.10 A designer finds that a certain building may be wired with either aluminum or copper wire. The aluminum is less expensive per foot but the connections are more expensive. The comparative costs are shown as follows.

	ALUMINUM	COPPER
Cost of wire per foot (cents)	12	15
Cost per connection (cents)	83	27

Find the average length of run between connections at the break-even point between aluminum and copper wire with a connection on one end of the wire only.

A.11 An air conditioner is being selected for a new municipal coliseum. Two are available with the following characteristics ($i=8\%$). What is the break-even point in hours per year of use?

	COLDAIR	FRIGID
Horsepower of compressor	150 hp	150 hp
Cost installed	$160,000	$130,000
Efficiency	90%	87%
Annual O & M costs	24,000/yr	19,500/yr
Life	12 yr	12 yr
Power costs $0.037/kWh	(There are 0.746 kW/hp.)	

A.12 A solar heating system is under consideration for a new office building. The comparative data are:

	INITIAL COST	O & M COSTS EACH YEAR ($/HR)	ESTIMATED INCREASE IN O & M COSTS EACH YEAR	SALVAGE VALUE END OF 10 YR
Solar heat	$20,000	$0.10	0	$15,000
Conventional heat	8,000	1.00	$0.12/hr	0

Using $i = 10$ percent, how many hours per year must the heating system be used to reach the cost break-even point between the two systems?

A.13 Same as Problem A.12 except $i = 18$ percent.

Problem Group B

Kelvin's Law. Find the least cost.

B.1 A short length of electrical conductor is needed to supply a pump at the municipal waste water treatment plant for which you are a consultant. The pump draws 350 A for an estimated 6,500 hours per year. Assume copper wire will be used and the following data have been obtained.

copper weight	0.32 lb/in.^{3}
length of conductor	150 ft
cost of copper wire in place	$0.85/lb plus $2,500
estimated life	25 yr
salvage value	$0.20/lb
electrical resistance	8.22×10^{-6} ohms/in.^{2}/ft of length
power costs	$0.025/kWh

With $i = 8$ percent, find the theoretically most economical size of wire in terms of square inches of cross section.

B.2 Find the most economical span for tower spacing of transmission line towers for high voltage electric lines. The intermediate towers cost $20,000 each, regardless of the spacing; the double braced end towers every 5 mi cost $100,000 and the cost of the high transmission wires is $0.50 per pound in place. The weight of the wire in this case is governed not only by strength requirements but also by conductance. Conductance requirements are met by a minimum weight of 50 lb/ft. Strength requirements are met by a weight per foot of $w = 0.125s + 4$. The average length of a straight line run between double braced end towers is 5 mi (s = span between towers in feet).

Assume $i=8$ percent, and all materials have a salvage value equal to 50 percent of the cost new at the end of a 20-year service life. Include the cost of one double braced end tower for each 5 mi of transmission line.

a. Find the most economical theoretical spacing.

b. Find the nearest actual spacing that will provide a whole number of equal spans.

c. Find the cost per mile (including $\frac{1}{3}$ of an end tower per mile).

B.3 Your client, the city, needs to determine where to locate the new solid waste disposal land fill site, and asks your recommendation. The land farther out from the city is cheaper, but transportation costs increase with distance from the city. You research the problem and come up with the following data.

Land values within a mile of the city limits average $8,000 per acre and drop off in value according to the equation

$$\frac{\$6,000}{M} + \$2,000, \qquad \text{where } M = \text{miles from the city limits } (M > 1)$$

The city now has a population of 50,000 and is growing at a rate of 2,000 per year. Solid waste collections average 4 lb per person per day for 365 days per year. (To simplify calculations, assume the waste is all collected and transported at the end of each year; for 50,000 at the end of the first year, for 52,000 persons at the end of the second year, and so on.)

Transportation costs equal $0.10 per ton per mile of distance between the disposal site and the city limits. The compacted waste weighs 1,000 lb/yd³ average. The disposal site is expected to contain four lifts of compacted waste, with each lift being 3 ft high. Enough land should be purchased to last for ten years.

The city borrows at 7 percent annual rate (1 acre = 43,560 ft²).

a. How much land should be purchased?

b. How far from the city limits should it be located in order to minimize costs?

B.4 Find the optimum thickness of fiberglass insulation to place in a residence in an attic space above the ceiling. The resistance to heat loss is measured in units of BTU-in./hr-°F-ft² (termed R). Fiberglass insulation costs $0.045/ft²-in., and has an R value of 3 BTU-in./hr-°F-ft². Thus the initial installation cost for the fiberglass is

$$P = \$0.045/\text{ft}^2\text{-in.} \times t \text{ in.}, \qquad \text{where } t = \text{thickness of insulation}$$

Assume that oil is used to heat the house and that 1 gal of oil emits 150,000 BTU and costs $1.12/gal the first year with probable increases of $0.0375/gal in each subsequent year. Therefore, the price of heat is calculated as

$$1.12/150,000 \text{ BTU} = \$7.47/10^6 \text{ BTU},$$
$$\text{increasing in price by } \$0.25/10^6 \text{ BTU/yr}$$

Assume that in this particular location there is an average of 200 days per year with average attic temperature 20° below room temperature, yielding a total of

$$200 \text{ days} \times 20°\text{F} \times 24 \text{ hr/day} = 96,000 \text{ degree hr/yr}$$

Thus annual fuel costs are

$$A = 96,000 \text{ °F-hr/yr}(3 \text{ BTU-in./hr-°F-ft}^2)(1/t)(\text{fuel cost/BTU})$$

Assume $i = 10$ percent, and $n = 30$ years.

B.5 Same as Problem B.4 except that $i=15$ percent and $n=40$ years.

B.6 Your client plans to build a multistory apartment building and asks you to determine how many stories it should be. The cost of construction in terms of dollars per square foot increases as the number of floors increases due to the following factors:

1. Increase in vertical services required (elevators, utilities, etc.)
2. Increased cost of foundations and structural supports
3. Decreasing productivity of construction forces with increasing height, increased vertical transport required, and decreasing safety

Assume the cost of construction of a one-story building is $50 per square foot and that this cost increases according to $N^{0.1}$, where N is the number of floors. Therefore the cost of construction for a building with N floors is

$$C_1 = \$50 N^{0.1}$$

On the other hand, the cost of land for each square foot of building decreases with the number of floors. The cost of land for a two-story building is one-half as much per square foot of floor area as it is for a one-story building. (On a two-story building the floor area is double, thus the land cost per square foot of floor area is just half that for a one-story building.) In this case the land cost for the entire lot is $500,000. The building will cover a ground area of 10,000 ft^2. Therefore, the cost of land for each square foot of building floor area is

$$C_2 = \frac{\$500,000}{10,000 N} = \frac{\$50}{N}$$

Maintenance costs also increase with the number of floors at about the same ratio as building costs. Thus the annual maintenance costs will equal

$$C_3 = \$1.00 N^{0.1}/\text{yr} \quad \text{plus an arithmetic gradient of } \$0.15 N^{0.1}/\text{yr}/\text{yr}$$

Use $i=10$ percent, $n=40$ years.
a. Find the theoretical number of floors in the building for least cost.
b. Find the actual number of floors and the present worth per square foot.
c. Find the annual rent in dollars per square foot which must be charged to cover the costs of land, construction, and maintenance, assuming 90 percent occupancy.

B.7 Your client, the Downtown Parking Authority, is considering a new multistory parking structure. Given the following data, find the percent of capacity the garage must operate at in order to break even ($i=15\%$).

	COST ($)	ECONOMIC LIFE (YR)
Land	600,000	infinite
Construction (Capital Improvements)	1,690,000	40
Fixtures	300,000	10
Salvage value	1,000,000	EOY 40
Working cash needed for entire 40-yr period	75,000	(not included in salvage)

The structure will contain 653 spaces and customers will pay $3.65 per space per day for whatever spaces they occupy. Operating expenses are

fixed $230,000/yr

variable $0.50/day for occupied spaces only (total daily cost for the structure varies with the occupancy rate)

The garage will be open for business 250 days per year.

Problem Group C

Break-even analysis for profitable companies.

C.1 A firm manufacturing building supplies has the following fixed costs, variable costs, and income per unit of output. They are considering the purchase of an automated device which will change the costs as indicated. Use $i = 12$ percent.

	EXISTING	AUTOMATED DEVICE ADDITION
Fixed cost ($)	$940,000	$40,000
Life of fixed cost (yr)	17	10
Salvage value ($)	0	0
Variable costs ($/unit)	7.20	6.95
Income ($/unit)	10.20	10.20

a. What annual volume of units must be sold under the existing system to break even?

b. What annual volume of units must be sold to break even if the automated device is added?

c. Prepare a break-even graph for both situations.

d. If 75,000 units are expected to be sold each year, should the automated system be purchased?

e. How many units must be sold each year to justify the purchase of the automated device?

C.2 Same as Problem C.1 except the variable costs with the addition of the automated device are $5.50 per unit.

Chapter 13
Probability Evaluation

PROBABILITY DISCUSSION

A decision to construct any large project is usually preceded by a sequence of increasingly detailed economic studies to determine feasibility. For the preliminary studies only one estimated value is assumed for each element of the problem, for example, assume one value for the life span, one value for the interest rate, and the same for other values represented by P, A, F, and G. Later in the sequence as the studies become more detailed and lifelike it may become evident that some previously assumed values cannot be substantiated with a high degree of certainty but are subject to some statistical variations. Thus, often there is an element of risk or uncertainty associated with each value used. How this inherent variability is handled can critically affect the outcome of the economic study.

The concepts of variability, risk, and uncertainty are part of a large and still growing science discussed in a large body of literature including several good textbooks. Only a brief review of some pertinent aspects are included in this chapter.

The difference in meaning of the terms "risk" and "uncertainty" are significant at this point. Risk concerns those outcomes for which a probability value can be assigned or estimated. The risk involved in such activities as providing fire insurance, bidding on a contract, and selecting a dam spillway height for flooding can be calculated. Uncertainty, on the other hand, refers to outcomes for which a probability value cannot be assigned. Examples include the future price of asphalt or other construction products, future value of the ENR Construction Cost Index, value of inflation one year from today, whether a business venture will be successful or not. Fortunately, many such situations involve values for which probabilities can be estimated. In fact, the concepts of probability permeate many common areas of our world. We hear "there is a 20 percent probability of rainfall today," "the class average on the last quiz was 78," "there is a probability of one in a hundred that a flood will occur at point X on River B," "there is a 50 percent probability of tossing heads with a coin," and so forth. If an engineering alternative involves the purchase of a computer for $50,000 today with a salvage value of $10,000 in seven years, no one would presume the actual salvage value would be exactly $10,000 in seven years. In reality it might be as low as zero or much higher than $10,000, but the best estimate today is given as $10,000. "Best estimates" of such values represented by P, A, F, i, n are usually adequate for preliminary studies, but the elements of probability and statistics can be employed to determine better solutions for more detailed studies.

BASIC PROBABILITY

Probability may be regarded as the relative frequency of events (either discrete or continuous) in the long run. Thus, a 50 percent probability of tossing heads with a coin means that over many tosses of the coin, the average number of heads will closely approximate 50 percent of the total tosses.

Four time-honored axioms regarding probability are:

• *Axiom 1.* The probability (p) of an event (E) occurring [$p(E)$] can be expressed as a number between 0 and 1.0, that is,

$$0 \leqslant p(E) \leqslant 1.0 \tag{13.1}$$

• *Axiom 2.* The sum of the probabilities of all possible mutually exclusive outcomes for a given event equals 1.0. (The probability of an event occurring [$p(E)$] plus the probability of that same event not occurring [$p(\overline{E})$] is equal to 1.0.)

$$p(E) + p(\overline{E}) = 1.0 \tag{13.2}$$

• *Axiom 3.* The probability of either of two *mutually exclusive* events occurring is equal to the sum of the probabilities of the two events, that is,

$$p(E_1 + E_2) = p(E_1) + p(E_2) \tag{13.3}$$

• *Axiom 4.* The probability of two *mutually exclusive* events occurring simultaneously (together) is equal to the product of the probabilities of the two independent events, that is,

$$p(E_1E_2)=p(E_1)\times p(E_2) \tag{13.4}$$

These concepts can be illustrated by responses to questions about the toss of a six-sided die.

1. What is the probability of a number six turning up on the first throw of a die?

Answer: $p(6)=1/6$, because there are six possible outcomes $(1,2,3,4,5$ or 6 on any throw) and the probability of a six on one throw is one out of six possibilities, or $1/6$.

2. What is the probability of *not* throwing a six with one throw of the die?

    Answer: $p(\bar{6})=1-p(6)$ [Axiom 2]

            $=1-1/6=5/6$

3. What is the probability of throwing either a five or six with one throw of the die?

    Answer: $p(E_1+E_2)=p(E_1)+p(E_2)$ [Axiom 3]

            $=1/6+1/6=2/6$

4. What is the probability of throwing two sixes in succession with two throws of the die?

    Answer: $p(E_1E_2)=p(E_1)\times p(E_2)$ [Axiom 4]

            $=1/6\times 1/6=1/36$

Item (4) is analogous to throwing two dice simultaneously and obtaining double sixes. In this case there are 36 possibilities (the sample space) of numbers, of which only one would be a double six.

Expected Value

Expected value (*EV*), or weighted average, is a standard measure for assessing the most probable outcome from an event involving risk (assignable probabilities). Expected value can be determined by multiplying the outcomes by their associated probabilities. The mathematical form is

$$EV=\sum_{j=1}^{m} p_jO_j \tag{13.5}$$

where

p_j = independent probability for event j

O_j = outcomes for event j

$j = 1, 2, 3, \ldots, m$

Note:

$\sum\limits_{j=1}^{m} p_j = 1.0$ (the sum of the probabilities must equal 1.0)

To illustrate, suppose a lottery ticket is sold for $3.00, which if drawn will pay $1,000,000 and there are 500,000 such tickets sold. If you purchase the ticket for $3.00 you will either win $1,000,000 or lose your $3.00. The probability of winning $1,000,000 (outcome O_j = $1,000,000) on any one ticket is 1 in 500,000 ($p_j = 1/500,000$). However the probability of losing the $3.00 cost of the ticket ($O_j = -$3.00$) is $p_j = 499,999/500,000$. Therefore, for any one ticket the expected value is

$$EV = \frac{1}{500,000} \times 1,000,000 + \frac{499,999}{500,000} \times (-3.00)$$

$$EV = \$2.00 - 3.00 = -\$1.00$$

The $-\$1.00 EV$ means that if you bought a large number of tickets you would lose, on the average, $1.00 for every ticket purchased. The expected value to the ticket purchasers for the *whole* lottery is determined as

$$EV = \sum\limits_{j=1}^{m} p_j O_j$$

$$EV = \left[\frac{1,000,000}{500,000} - \frac{499,999}{500,000} \times (-3.00) \right] 500,000 = -\$500,000$$

The numerical value of probabilities (p) may be estimated by two basic methods.

1. Theoretically comparing the number of occurrences to the theoretical number of opportunities for such occurrences. Example: the probability that a flipped coin will come up heads is .5. This can be determined without ever flipping the coin, simply by considering the limited number of possibilities. Also in this category are probabilities of rolling a given combination with a pair of dice, drawing a certain card or combination from a deck, and so forth.

2. Counting the actual number of occurrences of an event and comparing to the actual number of opportunities when a near infinite variety exists. Example: the average ($p = .5$) percent of downtime on a given make and model of construction equipment during its first year of service must be determined experimentally. Or the probability that a 25-year storm ($p = .04$) will produce a maximum runoff of

500 ft^3/s at a given culvert is typically a determination based on actual field measurements.

The following examples typify the practical methods used in practice for deriving probabilities and expected values.

Example 13.1

Assume a certain construction equipment manufacturer's record of the up-time availability of a given type of bulldozer during the first year in service is tabulated below.

NUMBER OF DOZERS REPORTING (a)	PERCENTAGE OF TIME AVAILABLE FIRST YEAR (b)	WEIGHTED DOZERS×% (a·b)
1	100	100%
3	99	297%
5	98	490%
6	97	582%
5	96	480%
10	95	950%
6	94	564%
5	93	465%
2	92	184%
1	90	90%
44		4204%

average availability, $\bar{x} = \dfrac{4204}{44} = 95.5\%$

The historical data indicate the average ($p=.5$) availability during the first year of service was 95.5 percent. If a contractor buys a dozer representative of the above sample, he could assume that on the average ($p=.5$) his dozer will be available 95.5 percent or more of the time during its first year of service.

By way of caution, these data tell only what has occurred in the past. The data might not repeat themselves if significant changes have occurred in any of the many variables affecting equipment availability, such as design changes, production line variables (material or labor shortages causing substitution of material or skills), newer and more efficient automation, on-the-job maintenance, or severity of working conditions.

RELIABILITY, STANDARD DEVIATION

Assume for the moment the contractor feels confident that the data above adequately represent his equipment. He would like to know how reliable this average estimate is. According to the theory of statistics, if the events are normally distributed and the number of events recorded adequately represents the population under study, then a range of σ values may be determined. This is a range on both sides of the average within which 68 percent of the events will probably occur providing the events are normally distributed. Other multiples of σ will delineate other bands on both sides of the average to encompass other probable percentages of all events. The equation for the standard deviation is given as

$$\text{standard deviation,} \quad \sigma = \pm\sqrt{\frac{\Sigma(x-\bar{x})^2}{n-1}}$$

Using the data in the previous example, the standard deviation may be found as in the following tabulation.

NUMBERS OF DOZERS REPORTING (m)	PERCENT AVAILABILITY	$(x-\bar{x})$ DEVIATION FROM AVERAGE $\bar{x}=95.5$	$(x-\bar{x})^2 m$
1	100	4.5	20.25
3	99	3.5	36.75
5	98	2.5	31.25
6	97	1.5	13.25
5	96	.5	1.25
10	95	.5	2.50
6	94	1.5	13.50
5	93	2.5	31.25
2	92	3.5	24.50
1	90	5.5	30.25
44			$205.00 = \Sigma(x-\bar{x})^2 m$

$$\sigma = \pm\sqrt{\frac{\Sigma(x-\bar{x})^2}{n-1}} = \sqrt{\frac{205}{44-1}} = \pm 2.183 \text{ or } \pm 2.2$$

where σ is the plus or minus variation within which 68 percent of the events will occur, assuming the events are normally distributed.

Approximately 68 percent of the occurrences will be within ±2.2 percent of the average. Thus there is a 68 percent probability that the dozers will be available between 95.5±2.2 percent of the time or between 97.7 and 93.3 percent of the time. This same statistical theory predicts that 90 percent of the occurrences will lie within 1.64σ or ±1.64×2.2 percent of the average 95.5 percent. There is therefore a 90 percent probability that the availability will be between 91.9 and 99.1 percent. The information on average availability may be used as follows.

Example 13.2

The dozer above is used to push load a fleet of five scrapers. If the dozer is out of service, production stops. Production is usually at 1,000 yd³/hr with costs at $900 per hour and income at $1.20 per cubic yard. With the dozer out of service, the costs continue at $900 per hour but production stops. A standby unit may be rented at $35 per hour. Should the unit be rented? Assume for the moment the standby unit is available 100 percent of the time.

SOLUTION
Assume 95.5 percent availability of the dozer.

Without standby,

income $0.955 \times \$1.20/\text{yd}^3 \times 1,000 \text{ yd}^3/\text{hr}$ $= \quad \$1,146/\text{hr}$

cost $= \quad -\ 900/\text{hr}$

net $= \quad \$\quad 246/\text{hr}$

With standby,

income $\$1.20/\text{yd}^3 \times 1,000 \text{ yd}^3/\text{hr}$ $= \quad \$1,200/\text{hr}$

cost $\$900/\text{hr} + \$35/\text{hr}$ $= \quad -\ 935/\text{hr}$

net $= \quad \$\quad 265/\text{hr}$

The results indicate that a standby unit would be profitable at any cost less than $1,200 - $1,146 = $54 per hour. Since this one is available at $35 per hour it is profitable to have on hand. (This assumes the standby is 100 percent available.)

MULTIPLE PROBABILITIES

More realistically the standby unit in the previous example is not available 100 percent of the time. How to account for multiple probabilities is illustrated in the next example.

Example 13.3

The standby tractor in the previous example is an older unit and has about 30 percent downtime itself. If the regular tractor has an availability of 95.5 percent (is down 4.5 percent of the time), what are the probabilities of them both being down simultaneously?

SOLUTION
The probability of any two probable events occurring simultaneously is $p_1 \times p_2$ (Axiom 4). So the probability of them being down simultaneously is

$.30\% \times .045 = .0135 = \underline{\underline{1.35\%}}$

GRADIENT PROBABILITIES

Reliability of equipment characteristically decreases with increasing years of service. A method of accounting for gradient changes in probability is shown in the following example.

Example 13.4

Assume a contractor needs a tractor to haul a set of compaction rollers around the fill areas on a sequence of highway jobs over the next five-year period. The availability of the tractor during the first year is expected to be 95 percent but drop about 4 percent each year thereafter. The contractor is considering putting a standby tractor on the job since whenever the compaction stops, the whole earthmoving operation stops. When the compactor is in operation, production averages 200 yd^3/hr and the bid price (income) averages $1.20 per cubic yard. Expenses average $100 per hour whether the compactor is running or not. The standby tractor would cost $80,000, could be sold in five years for $40,000, and is estimated to be available 75 percent of the time without change throughout the five-year period. Find the rate of return on the proposed investment in the standby tractor (1,000 hr/yr use).

SOLUTION

Consider the probable loss of income without the standby tractor versus the probable loss of income if the standby is present. The expenses and net income are actually not involved in this particular part of the problem but would be needed to consider overall profitability. Gross maximum possible income from earthmoving is

$$\$1.20/yd^3 \times 200 \ yd^3/hr \times 1,000 \ hr/yr = \$240,000/yr$$

Probable loss of income due to downtime with *no* standby unit is

YEAR	DOWNTIME	COST OF DOWNTIME % × MAX. INCOME
1	5%	0.05 × $240,000 = $12,000
2	9%	0.09 = 21,600
3	13%	0.13 = 31,200
4	17%	0.17 = 40,800
5	21%	0.21 = 50,400

Probable loss of income due to downtime when standby is *available* is (the probability of two occurrences happening simultaneously is $p_1 \times p_2$)

YEAR	DOWNTIME TRACTOR	STANDBY
1	.05 × .25 × $240,000 =	$3,000
2	.09 × .25 =	$5,400
3	.13 × .25 =	$7,800
4	.17 × .25 =	$10,200
5	.21 × .25 =	$12,600

The amount saved by purchasing the standby unit is the difference between the last two sets of calculations.

YEAR	LOSS WITHOUT STANDBY	LOSS WITH STANDBY	SAVED BY STANDBY
1	−$12,000	−$ 3,000	+$ 9,000
2	− 21,600	− 5,400	+ 16,200
3	− 31,200	− 7,800	+ 23,400
4	− 40,800	− 10,200	+ 30,600
5	− 50,400	− 12,600	+ 37,800

Therefore the amount saved may be credited as income produced by this investment and the rate of return calculated. Notice that the amount "Saved by Standby" in this case has an arithmetic gradient of $7,200 per year.

Try i=20 percent

PW of *income − cost*

base annual $P_1 = \$9,000/\text{yr} \ (P/A, 20\%, 5)$ $\qquad = \qquad \$26,920$
$\qquad\qquad\qquad\qquad\qquad 2.991$

annual gradient $P_2 = \$7,200/\text{yr} \ (P/G, 20\%, 5)$ $\qquad = \qquad 35,322$
$\qquad\qquad\qquad\qquad\qquad 4.906$

resale in 5 yr $P_3 = \$40,000 \ (P/F, 20\%, 5)$ $\qquad = \qquad 16,076$
$\qquad\qquad\qquad\qquad\qquad 0.4019$

purchase price $P_4 = \$80,000$ $\qquad = \qquad - \ 80,000$

$\qquad\qquad\qquad\qquad\qquad\qquad NPW @ 20\% \qquad = \qquad -\$ \ 1,682$

Try i=15 percent

PW of *income − cost*

base $P_1 = \$9,000/\text{yr} \ (P/A, 15\%, 5)$ $\qquad = \qquad \$30,168$
$\qquad\qquad\qquad\qquad 3.352$

gradient $P_2 = \$7,200/\text{yr} \ (P/G, 15\%, 5)$ $\qquad = \qquad 41,580$
$\qquad\qquad\qquad\qquad\qquad 5.775$

resale $P_3 = \$40,000 \ (P/F, 15\%, 5)$ $\qquad = \qquad 19,888$
$\qquad\qquad\qquad\qquad 0.4972$

purchase $P_4 = \$80,000$ $\qquad = \qquad - \ 80,000$

$\qquad\qquad\qquad\qquad\qquad NPW @ 15\% \qquad = \qquad +\$11,636$

Interpolating,

$$15\% + \frac{11,636}{11,636 + 1,682} \times 5\% = 19.37\%$$

Under the conditions stated, the rate of return on the investment in the standby tractor is 19.4%.

PROBABILITIES INVOLVED IN SELECTING THE BETTER ALTERNATIVE

Probabilities can also be used to select the least cost alternative from among a number of possible choices. For instance, in designing culverts, an engineer may estimate the cost of several alternative culvert sizes and types. The probability of exceeding the design capacity is calculated and an estimate made of the value of damage occurring in the event capacity is exceeded. These variables may be related as follows to determine the design with the lowest net cost.

Example 13.5

A culvert is being designed to pass under a highway. The cost of repairs in case the highway is overtopped is estimated at $20,000. If i is estimated at 6 percent and $n=20$ for each culvert, which of the following is the most economical?

TYPE OF CULVERT	COST NEW ($)	PROBABILITY (p) OF OVERTOPPING IN ANY ONE YEAR
24 in. RCP	20,000	.20
2×4 ft box culvert	40,000	.10
Two 2×4 ft twin box	60,000	.05

SOLUTION
The probability of overtopping the 24-in. RCP is given as .20 in any one year, or an annual equivalent of $4,000 *expected* damage. ($20,000×0.20= $4,000/yr.) If the annual equivalent damage is calculated for each type of culvert and added to other annual costs, the lowest cost culvert may be determined as follows.

TYPE CULVERT	24-IN. RCP	2×4 FT	TWO 2×4 FT
Annual equivalent to cost new=			
$P(A/P,6\%,20)$	$1,743	$3,487	$5,230
0.08718			
Annual cost of overtopping			
$p×$20,000	$4,000	$2,000	$1,000
Total	$5,743	$5,487	$6,230

The single 2×4 ft is the most economical under these circumstances.

GRADIENT IN LOSSES

In many situations the value of the loss will change depending on the year in which it occurs. For instance, if the value of the property subject to flooding increases every year due to increased development, improvement, or inflation, the loss incurred if flooding occurs in the tenth year will be greater than in the first. Either an arithmetic or geometric gradient may be used to transform gradient losses into either present worth or annual equivalents. For example, assume the probability of loss in any one year is 10 percent or .10. Assume further that if the loss occurs the first year the damage will amount to $1,000. If it occurs the second year the damage is estimated as $1,500, and the third year $2,000. These damages may be tabulated as follows, assuming i as 7 percent.

YEAR, n (COL. 1)	AMOUNT OF LOSS, L, IF OCCURRING THIS YEAR (COL. 2)	PROBABILITY, p, OF LOSS THIS YEAR (COL. 3)	LOSS LIABILITY ACCOUNTED FOR THIS YEAR (COL. 4)	PW OF LOSS LIABILITY $Lp(P/F,7\%,n)$ (COL. 5)
1	$1,000	.1	$100	$ 93.46
2	1,500	.1	150	131.02
3	2,000	.1	200	163.26
				$387.74

The sum of the present worths of the loss liability in column 5 also may be expressed as

$$P = \Sigma \text{ col. } 5 = Lp(P/A, i, n) + \Delta Lp(P/G, i, n)$$

or

$$P = \$1,000 \times .1 \, (P/A, 7\%, 3) + \$500 \times .1 \, (P/G, 7\%, 3) = 262.44 + 125.30$$
$$\quad\quad\quad 2.6244 \quad\quad\quad\quad\quad\quad 2.5061$$

$$= \$387.74$$

If the cost of losses is desired in terms of the annual equivalent, dollars per year for each of n years, the equation may be written as $A = pA + pG(A/G, i, n)$.

Comparisons Involving Gradients and Present Worth

Comparisons of either PW or AW are equally valid for determining the lowest cost alternative. In addition a gradient factor may be needed since areas where development is occurring will probably suffer greater damage in future years. The next example illustrates both of these features.

Example 13.6

A culvert is needed in a growing neighborhood. If the culvert is overtopped, the damages initially are estimated at $40,000 per year. The cost and probability of overtopping of each of three alternatives are given as:

CULVERT	COST OF CULVERT	PROBABILITY OF OVERTOPPING IN 1 YR
A	$ 45,000	.10
B	20,000	.20
C	100,000	.05

The life of the structure is estimated as 20 years and $i=6$ percent.

1. Which is the most economical selection?
2. A new estimate is made indicating the damages from overtopping will be $40,000 if overtopping occurs the first year, but will increase $10,000 per year thereafter due to the rapid development of the area. Which is now the best choice? (Find PW of each alternative.)

SOLUTION 1

STRUCTURE	PROBA-BILITY OF DAMAGE p	ESTIMATED DAMAGES IF OVER TOPPED	EXPECTED ANNUAL DAMAGE, A	PW OF DAMAGE, $A(P/A,$ $6\%,20)$	COST OF CULVERT	PW OF TOTAL COST	
A	.10	40,000	4,000	45,880	45,000	90,880	lowest total cost
B	.20	40,000	8,000	91,760	20,000	111,760	
C	.05	40,000	2,000	22,940	100,000	122,940	

SOLUTION 2
If damage estimate increases $10,000 every year, then

gradient, $P_1 = G(P/G,6\%,20) = p \times 10,000 \times 87.230 = p \times 872,300$
87.230

base cost of damages, $P_2 = A(P/A,6\%,20) = p \times 40,000 \times 11.47 = p \times 458,800$

STRUCTURE	p	P_1 $=G(P/G,6\%,20)$ $=p\times872,300$	P_2 $=A(P/A,6\%,20)$ $=p\times458,800$	P_3 COST OF CULVERT	PW OF TOTAL COST p	
A	.10	87,230	45,880	45,000	178,110	
B	.20	174,460	91,760	20,000	286,220	lowest total cost
C	.05	43,615	22,940	100,000	166,555	

EXPECTED VALUE USES INVOLVING JOINT OCCURRENCE PROBABILITIES

Often probabilities are calculated for a series of alternatives which successively reduce the probability of damage. For example, hydrology studies may report incremental flood levels from heavy rains in terms of probability, such as one in one hundred probability ($p=.01$) of reaching 20 ft above a certain norm, with other probabilities of reaching other levels.

Such a hydrologic tabulation may appear as follows.

FLOOD HEIGHT ABOVE DATUM LEVEL (FT)	YEARLY PROBABILITY OF EXCEEDING HEIGHT
0	1.00
2.0	.15
4.0	.07
6.0	.02
8.0	.005
10.0	.0001

At first glance this table may appear to violate Axiom 2, in that the vertical sum of probabilities exceeds 1.00. However, further examination shows that the events are not *mutually exclusive*. A flood height exceeding 4 ft also exceeds 2 ft. The probabilities contain joint occurrences (not mutually exclusive) and Axiom 2 can only show that if the yearly probability of exceeding a flood height of 2.0 ft is .15, the yearly probability of *not* exceeding 2.0 ft is $1-.15$ or .85. Also, the probability of having a flood height of between 2.0 and 4.0 ft would be $.15-.07=.08$. The following example illustrates the analysis of this type of problem.

Example 13.7

An apartment complex is subject to periodic flooding. A consultant's study shows that the complex can be protected from successive increments of flood water levels but each higher level of protection costs more, as shown in the table.

ALTERNATIVE	FIRST COST ($)	ANNUAL MAINTENANCE COST ($)	PROVIDES PROTECTION AGAINST FLOOD OF DEPTH (FT)	PROBABILITY OF FLOOD EXCEEDING EXPECTED DEPTH IN ANY ONE-YEAR PERIOD	ESTIMATED LOSS WHEN GIVEN DEPTH IS EXCEEDED ($)
A	none	none	0.5	1.0	50,000
B	50,000	1,000	1.0	.2	800,000
C	100,000	1,500	1.5	.1	1,000,000
D	300,000	2,000	2.0	.05	1,400,000
E	600,000	3,000	3.0	.02	1,900,000
F	900,000	4,000	4.0	.01	2,000,000

SOLUTION

The equivalent expected costs of flood damages are calculated for each alternative and, using net worth analysis, the alternative with the lowest net equivalent cost is selected. Beginning with the highest cost alternative, the annual equivalent expected costs (EC) of flood damages for alternative F is simply

$$EC(F) = 2,000,000(.01)$$

$$= \$20,000$$

The EC of alternative E is slightly more involved since the probability of a flood over 3.0 ft is $P = .02$, but the probability of a flood over 3.0 ft also includes a flood of over 4.0 ft (joint occurrence). Therefore, the probability of a flood occurring which has a depth *between* 3.0 and 4.0 ft must be determined as $p(3-4\ \text{ft}) = .02 - .01 = .01$. Then the EC of alternative E is

$$EC(E) = 1,900,000(.02 - .01) + 2,000,000(.01)$$

$$= 19,000 + 20,000 = \underline{\$39,000}$$

In a similar manner

$$EC(D) = 1,400,000(.05 - .02) + 39,000$$

$$= 42,000 + 39,000 = \underline{\$81,000}$$

$$EC(C) = 1,000,000(.1 - .05) + 81,000$$

$$= 50,000 + 81,000 = \underline{\$131,000}$$

$$EC(B) = 800,000(.2 - .1) + 131,000$$

$$= 80,000 + 131,000 = \underline{\$211,000}$$

$$EC(A) = 50,000(1.0 - .2) + 211,000$$

$$= 40,000 + 211,000 = \underline{\$251,000}$$

Calculation of total costs on the basis of equivalent annual costs yields

ALTERNATIVE	ANNUALIZED FIRST COST ($) $P(A/P, 10\%, 20)$ 0.1175	ANNUAL MAINTENANCE COSTS ($)	ANNUAL EXPECTED FLOOD COSTS ($)	TOTAL ANNUAL COSTS ($)	
A	0	0	251,000	251,000	
B	5,875	1,000	211,000	217,875	
C	11,750	1,500	131,000	144,250	
D	35,250	2,000	81,000	118,250	lowest
E	70,500	3,000	39,000	112,500	cost
F	105,750	4,000	20,000	129,750	alternative

Based on expected values of damages the lowest cost alternative is E.

INSURANCE COSTS

People buy insurance not to avoid loss but to pool the risk of loss. Problems involving insurance costs necessarily must consider the cost of the risk of loss. This is basically a probability problem situation and is susceptible to solution as illustrated in the following example.

Example 13.8

A new civic center is proposed with an estimated insurable value of $3,600,000. You are asked to report on the feasibility of an automatic sprinkler system costing $220,000 (which will not add to the insurable value of the building). This will reduce the annual fire insurance premium cost from 1.30% to 0.50%. Half of the fire insurance premium cost is used to pay the actual insured fire loss, while the other half pays overhead, sales, and claims investigation costs and profit. The records indicate that total losses from a destructive fire will be about two and one-half times the insurable losses. The annual cost of operation and maintenance of the sprinkler system is estimated at $1,200. Assume the life of the sprinkler system is 20 years and the civic center will be financed by bond funds which are expected to cost 6 percent. Is the sprinkler system a good investment?

SOLUTION

	WITHOUT SPRINKLER	WITH SPRINKLER
Annual insurance premium cost		
$0.013 \times \$3,600,000 =$	$-\$46,800$	
$0.005 \times \$3,600,000 =$		$-\$18,000$
Actual *insured* fire loss (estimated as one-half insurance premium cost)		
Reimbursed fire loss		
$\$46,800/2 =$	$+\$23,400$	
$\$18,000/2 =$		$+\$\ 9,000$
Actual *total* fire loss		
$2.5 \times 23,400 =$	$-\$58,500$	
$2.5 \times 9,000 =$		$-\$22,500$
Annual O & M for sprinkler		$-\$\ 1,200$
First cost, sprinkler		
$A = 220,000\,(A/P,6\%,20) =$		$-\$19,180$
$\qquad\qquad\qquad 0.08718$		
	$-\$81,900$	$-\$51,880$

Conclusion. The sprinkler is recommended under these circumstances.

REQUIREMENTS FOR A PROBABILITY PROBLEM

A wide variety of practical problems are susceptible to analysis if they can be defined in terms of "probability" or "percentage-of-occurrence" problems.

The requirements are:

1. There must be an exposure to a certain type of event involving a measurable loss or gain. (Example: exposure to washout of existing 18-in. culvert.)
2. There must be enough history of similar exposures to the event and of the resulting loss or gain so that a probability p, of future similar events occurring can be estimated (p=number of events actually occurring divided by the number of exposures to the event). (Example: history of stream indicates an 18-in. pipe is not quite adequate for a 1 in 25 year storm or $p=.04$.)
3. The equivalent periodic value of the event equals the probability of the event multiplied by the value of the event (loss or gain). (Example: if loss from washout is $5,000, then the equivalent periodic value is $A = .04 \times \$5,000 = \$200$ equivalent annual loss.)
4. If the equivalent periodic value of preventing or causing the event is less than the equivalent periodic value of the event, then the investment required to prevent or cause the event is numerically justifiable. (Example: if the equivalent annual cost of enlarging the culvert, $A = P(A/P, i, n)$ is more than $200, keep the existing 18-in. culvert. If not, enlarge it.)

Example 13.9

A contractor finds that adding an extra sack of cement per cubic yard of concrete reduced the probability of concrete cylinder test failure from 1 in 20 ($p=.05$) to 1 in 50 ($p=.02$). There is one test taken for every 20 yd^3. Each failure costs him an average of $2,000 worth of additional tests, delays, arguments, and replaced concrete. Cement costs $2 per sack. Should he add the extra cement?

SOLUTION

cost of failure $\times$ probability of failure $=$ expected cost (EC) per test

without extra cement $\$2,000 \times .05 = \$100$ EC per test

with extra cement $\quad 2,000 \times .02 = \underline{\quad 40}$

expected savings per test $= \$ 60$ per test

There are 20 yd^3 poured for each test taken, so an expected savings (ES) of $60 per test results in an ES of $3 per cubic yard. The cost of preventing the failure is one sack of cement per cubic yard or a cost of $2 per cubic yard. Thus the contractor will find it profitable to add the extra sack of cement per cubic yard under the stated circumstances.

The event does not necessarily have to involve a loss. It can just as easily be a gain as the following example illustrates.

Example 13.10

A road building contractor finds that infrared air photos will sometimes reveal previously unreported adverse drainage conditions, and other times will reveal good borrow sites, likewise previously unreported. He finds these events occur about 1 time in 20 ($p=.05$). When they do occur they result in an average benefit of about \$10,000. The cost of having each prospective job photographed is \$300. Should the contractor have infrared photos made on each job?

SOLUTION
The benefit realized is $\$10,000 \times .05 = \$500$ average per job. The cost is \$300. So under the stated conditions the photos more than pay for themselves.

In comparing "probabilities" with "percentage of occurrence," a great deal of similarity can be observed. For instance if the probability of having a machine out of service at any given time is 1 in 20, then $p=.05$ and it is averaging 5 percent downtime or 95 percent availability. The two terms, probability and percentage of occurrence, can often be used interchangeably.

PROBLEMS FOR CHAPTER 13, PROBABILITY EVALUATION

Problem Group A

A.1 An apartment complex is subject to periodic flooding from storms. Whenever the floods reach floor level, the cost of damage approximates \$100,000. A consultant has calculated the following information.

ALTERNATIVE	FIRST COST (\$)	ANNUAL MAINTENANCE COST (\$)	PROBABILITY OF FLOOD REACHING FLOOR LEVEL IN ANY YEAR
Do nothing	0	0	.05
Improved storm sewer A	15,000	500	.01
Improved storm sewer B	25,000	600	.006

Neglecting salvage value, which of the three alternatives should be selected if $i = 12$ percent and the complex is expected to last 20 years?

A.2 A 100-unit apartment is being considered for investment purposes. For an initial investment of $350,000 for land and $1,000,000 for the complex, the following probabilities and net incomes are predicted.

PROBABILITY	NET INCOME ($/YR)
.4	100,000
.4	115,000
.2	180,000

a. Assuming the apartment complex will last 20 years with no salvage value and increases in rent will offset inflation, calculate the expected rate of return on the investment. (*Hint:* How do you treat the future value of the land?)

b. What is the probability that less than the expected rate of return will be realized?

A.3 A construction company adds a markup on each job of 10 percent for profit. Past records indicate the company has, on any given job, a 50 percent probability of making their 10 percent profit; a 10 percent probability of making 15 percent profit; a 30 percent probability of making zero profit; and a 10 percent probability of *losing* 10 percent.

a. On any given job what is the expected profit?

b. The company calculates it can go to computer-based estimating, and by so doing improve its probabilities of making a profit. The new profit probabilities are estimated to be:

making 10% profit   60% probability
making 15% profit   15% probability
making zero profit   15% probability
losing 10%   10% probability

If the computer will last ten years, and an *MARR* is 15 percent, how much can the company afford to spend on the computer for a contract volume of $70,000,000 per year [contract volume = contract cost $\times$ (1 + profit markup)].

c. The company purchases a computer for $5,000,000. The new probabilities for making their profit turn out to be:

making 10% profit   55% probability
making 15% profit   15% probability
making zero profit   20% probability
losing 10%   10% probability

The actual company contract volume turned out to be $55,000,000 per year. If the computer lasted ten years, what rate of return did the company make on their investment?

A.4 To protect against a robbery you are considering the purchase of a burglar alarm. The estimated cost of a robbery would be $10,000, and you figure the probabilities of robbery are .01 in any given year. The cost of the burglar alarm is $550 and the sellers tell you your probability of robbery would be reduced to .004. Assuming the alarm would last 15 years with no salvage value is this a good investment, if $i = 8$ percent?

A.5 Insurance to guard against water damage costs $1,000 per year, payable in advance. Water damage can be expected from severe storms which occur with a frequency of once every 20 years. If water damage does occur, the cost will be $22,000. Assuming interest rates of 12 percent, should the insurance be purchased, if the business expects to stay in the location for the next seven years?

A.6 Income from student parking violation fines on a university campus generated the following incomes during the last 12 months.

MONTHS	MONTHLY REVENUES FROM FINES ($)
1–4	9,000
5–9	6,500
10–12	2,000

If the revenues continued in this sequence every year for a five-year period, and were deposited at the end of each month in an interest bearing account of 0.75 percent per month, what amount would be accumulated at EOY 5?

A.7 An enterprising student offers you $1.00 if you can toss a coin three times in succession landing on heads, provided you pay her $0.25 if you fail to throw the three heads. If this student succeeds in getting 1,000 people to gamble with her, how much money should she make (or lose)? (The probability of throwing a heads is .5.)

A.8 A culvert is to be selected for installation where an arterial street will cross a creek. If the culvert is of insufficient size so that water flows over the street, damages to the street are estimated to be $10,000. (Assume that the upstream and downstream damages are unaffected by the design alternative.) For interest at 6 percent and a 25-year life, which alternative would be most economical for the city to provide?

ALTERNATIVE	CONSTRUCTION COST ($)	PROBABILITY* THAT CAPACITY WILL BE EXCEEDED IN ANY ONE YEAR
A	30,000	.25
B	45,000	.10
C	65,000	.04
D	95,000	.02

*The set of occurrences is all possible rainfalls.

A.9 Commercial Leasing Inc. has an option on a piece of property which they believe is an ideal location for construction of an office complex. However, much of the site is subject to occasional flooding. The Local Flood Plain ordinance permits construction in the flood zone as long as the improvements do not interfere with the free passage of water. The estimated loss to the building and contents is estimated at $300,000 if the flood exceeds the floor elevation. The cost of constructing the building is increased by $35,000 for each 1 foot the building is raised.

    a. Find the most economical floor elevation if the building life is 20 years and $i = 15\%$.

b. Assume the estimated loss increases by $30,000 each year.

FLOOD ELEVATION (FT)	EXPECTANCY OF FLOOD EXCEEDING GIVEN ELEVATION
360	1.0
361	.2
362	.1
363	.05
364	.02
365	.01
366	0

A.10 Provincial Drilling Ltd. has been adding 10 percent to the estimated project cost (direct plus indirect costs) of each job for management, risk, and profit. The area in which the firm operates has an extremely complex geology; therefore, cost estimates are subject to random variation. An audit to the firm's operators indicates the following.

PERCENT PROFIT (LOSS)	PERCENTAGE OF JOBS (PROBABILITY)
10% profit	50
15% profit	10
no profit	30
(10% loss)	10

a. What is the expected profit on a given job?

b. A geological engineer calculates that additional exploration and improved bid preparation procedures will result in the following probabilities.

PERCENT PROFIT (LOSS)	PROBABILITY ($)
10% profit	60
15% profit	20
no profit	10
(10% loss)	10

If a job is expected to cost $1 million, how much can the firm expend on additional exploration and bid preparation and still make the profit that they previously achieved?

A.11 A contractor in a relatively freeze-free area is considering the question of whether or not to provide weatherproofing for his building projects. An investigation of the weather records in this area shows that during the past 20 years there have been an average of 3 weeks per year of cold weather bad enough to slow down the work. It is estimated on the average that this will reduce the value of the construction produced during those weeks by about 50 percent, with no reduction in costs. Average weekly gross value put in place during normal weather is estimated at $2,000 per week. The initial investment in weather protection equipment is estimated at $15,000. The life of the weather protection equipment is 20 years. Annual labor costs for setting up the equipment is $150. Average annual fuel cost is $50. If it is assumed that this will give complete protection against loss due to cold weather will it be desirable to purchase and operate the weather protection devices? Assume interest at 8 percent.

A.12 A friend wants you to do $100,000 worth of earthmoving now to launch a speculative real estate development. She promises $200,000 cash in a lump sum in five years when she hopes the development will be booming. What percent return on your investment of $100,000 would this amount to assuming compounded interest? If you took advantage of ten opportunities similar to this, and due to their speculative nature, five failed (you receive nothing) and five succeeded (you receive the $200,000 as promised) what percent return would you realize on the group of ten (same as if the one project has a 50% probability of success)?

A.13 You are working for a contractor with 100 employees and an annual payroll of $1,600,000. He is currently paying Workmen's Compensation Insurance premiums at the rate of 3 percent of the total payroll. His accident record shows an average of $30,000 worth of insured claims per year. The total loss to the company is estimated at about $90,000 (of which $30,000 is repaid by the insurance company) due to the extra paper work, lost time of other employees, and the general confusion and lack of productivity that always accompanies an accident of any kind. The insurance company is suggesting a safety training course for all employees under the following conditions.

1. Cost of the entire course paid by the contractor is $100 per employee.
2. The course is given on company time, so loss of services will amount to about 20 hours per employee at an average of $8 per hour.
3. The insurance company's experience with the course shows that accident claims will be cut 12 percent the first year, but will gradually return to their former level over a six-year period at the rate of 2 percent per year. They therefore recommend a similar safety course be offered at the end of every four years.
4. They offer to cut their insurance charges from the present rate by 12 percent the first year, 10 percent the second, 8 percent the third, and 6 percent the fourth.

Your boss asks you to examine the insurance company's proposal and determine whether it is a wise investment of the company resources to have one of these courses once every four years ($i = 10\%$).

A.14 A new classroom building on campus is proposed with an estimated insurable value of $10,000,000. You are asked to report on the feasibility of adding more firewalls and other fireproofing elements costing $300,000 (the insurable value of the building will also rise to $10,300,000). This will reduce the annual fire insurance premium cost from 1.00 percent of the value of the building to 0.60 percent. In either case, fire insurance premium cost is based on an estimate that half the premium will be needed to pay actual insured fire loss, while the other half pays insurance company overhead, sales and claims investigation costs, and profit. The records indicate that total losses from a destructive fire will be about two and one-half times the insurable losses. (Example: the fire causes $2.50 worth of losses of business, time, records, and intangibles for each $1 worth of insured loss reimbursed by the insurance company. The owner suffers a net loss of $-\$2.50+\$1.00=-\$1.50$ in addition to insurance premium payments.) The annual cost of maintenance of the extra fireproofing is estimated at $3,000. Assume the life of the classroom building is 50 years and that it will be built on bond funds with $i=6$ percent. Is the fireproofing a good investment?

Problem Group B

Involves gradients.

B.1 The county engineer asks your help in determining the most economical culvert design for a new county road. The costs of each alternative design are tabulated as follows.

CULVERT DESIGN	DESIGNED TO BE OVERTOPPED ONCE EVERY	COST OF CULVERT ($)	ANNUAL MAINTENANCE ($/YR)	GRADIENT IN MAINTENANCE EACH YEAR ($/YR)
A	5 yr	5,000	500	50
B	10 yr	7,000	800	75
C	25 yr	12,000	1,000	100
D	50 yr	20,000	1,500	150

If the culvert is overtopped the first year the resulting damage will approximate $20,000. Due to increasing development in the area the resulting damages will increase by about $3,000 for each succeeding year. Thus if overtopping occurs during the second year, the damages will be about $23,000, and so on. Which is the most economical design if the life of the culvert is estimated as 25 years and $i=8$ percent?

B.2 You are designing a parking lot for a shopping center and are concerned about pedestrian safety for people passing to and from their parked cars. Pedestrian walkways can be constructed and protected with reinforced planter boxes at a cost listed below. Without the protected walkways, an accident involving a pedestrian with an auto is expected to occur to one customer in 100,000. If the protected walkways are installed the probability drops to 1 in 1,000,000 customers. The parking lot contains 825 parking spaces. Each space will host an average of 3.1 cars per day, and each car will contain an average of 1.5 customers. Assume 300 shopping days per year. Each accident costs an average of $5,000 if it occurs the first year, with an increment of $500 for each subsequent year in which the accident occurs. (If the accident occurs the second year the cost is $5,500, the third year $6,000, and so on.) Use $i=10$ percent, $n=20$ years, all costs charged at EOY. Cost of reinforced concrete planter boxes and pedestrian walkways are

first cost	$200,000
life	20 yr
maintenance	$10,000/yr for the first 5 yr
increase in maintenance	$1,000/yr/yr for the last 15 yr
	(e.g., $11,000 for EOY 6, $12,000 for EOY 7, etc.)

a. Find which is the more economical solution, (i) installing the planter-box protected walkways and paying for fewer accidents, or (ii) omitting the planter-protected walkways and paying for more accidents.

b. Are there other considerations than just the economics? (If yes, list).

B.3 A contractor has a problem involving security against theft. She is building a new home office and shop compound for her own use. According to police records, in any one year she has a 1 in 20 chance of being robbed (breaking and entering the compound). She expects to have an average of about $100,000 worth of easily portable office equipment, small, tools, supplies and cash on the premises which could be targets for the thieves. Thus, she has a 1 in 20 chance ($p=.05$) of losing $100,000 in any one year. The value of this property is expected to go up about $10,000 each year as the company expands and replacement costs rise. To protect the premises the contractor has four alternatives: (a) electronic detection devices connected to police headquarters, (b) guard dogs, (c) security services (watchmen), or (d) theft insurance. The costs and benefits are as listed in the table. Find the equivalent annual costs of each alternative and tell which is most economical overall if $i=10$ percent.

	PROBABILITY OF $100,000 LOSS	CAPITAL COST ($)	LIFE (YR)	SALVAGE VALUE ($)	ANNUAL O&M ($)	ANNUAL INCREASE IN O&M ($)
(a) electronic detection devices	.02	10,000	10	2,000	1,500	200
(b) Guard dogs	.04	2,000	5	0	3,000	300
(c) Security service	.03	0	perpetual	0	8,000	1,000
(d) Theft insurance	.00	0	perpetual	0	10,000	1,000

B.4 You are called to consult on the construction of a motel on the seafront in an area where the probability of hurricane force winds in any one year is 1 in 12. A standard design is estimated to cost $512,000 but is not expected to survive a hurricane without severe damage. A hurricane-resistant design is estimated to cost $702,000. It is estimated that the standard design will suffer damage of 90 percent of the value current at the time of the hurricane plus loss of business costs amounting to 40 percent of the cost of the damage. Assume further that the hurricane-resistant design will survive 98 percent of the hurricanes intact, but will suffer 50 percent loss of value (including damage and loss of business) in the other 2 percent. The replacement value of both designs increases by $40,000 per year. Use $i=9$ percent, and an estimated life of 30 years for the motel. Which design is most economical and by how much in terms of annual costs?

B.5 An earthmoving contractor has a fleet of equipment consisting of ten scrapers, three tractors, one grader, and one compactor. The average income for each machine is now $30 per hour, but is expected to increase by $3 per hour each succeeding year. Each machine should earn income for an average of 1,500 hours per year, but currently about 14 percent of this time the machine is not working because of mechanical problems of some sort. A lubrication expert advises the contractor that the downtime could be cut from the present 14 percent down to 10 percent by using additional planned preventive mainte-nance. His plan involves purchase of a lube truck for $15,000, plus hiring another person for $10,000 per year, with the salary increasing $1,000 per year each additional year. The lube truck will cost about $2,000 per year for fuel, supplies and operating expenses, and can probably be sold in five years for $5,000. The current interest rate for the contractor is $i = 10$ percent.

a. Is this a good investment over a five year period under the stated circumstances?

b. If it turns out that the downtime is actually reduced only to 12 percent, was it a good investment?

Problem Group C

$n = \infty$.

C.1 A client is planning to build an office park and is trying to decide between two alternative locations. The basic difference between the two sites is the fire protection available. Site A will cost $165,000 for the land only, whereas site B can be obtained for $125,000. The office park after construction will be insured for $3,000,000. The annual insurance premium on site A is 0.60 percent of the insured value, while the premium on site B is 0.85 percent. Over the years, about one-half of these premiums will be paid back to your client as compensation for actual fire losses occurring at your park. The total cost of the fires to your client will closely approximate three times the compensation paid him by the insurance company. With $i = 9$ percent, which is the more economical site? $n = \infty$.

1. Goal Def.

2. Relative weight

3. " " impact.

4. Indentity cond. so

Chapter 14
Benefit/Cost Analysis

Conventional B/C

$$B/C = \frac{net \ savings \ to \ users}{owner \ cap. \ costs + owners \ net \ O\&M}$$

$$= \frac{U_n}{C_n + M_n} = \frac{B_n}{C_n + M_n}$$

Modified

$$B/C = \frac{U_n - M_n}{C_n} = \frac{B_n - M_n}{C_n}$$

COMPARING BENEFITS TO COSTS

A commonly used method of evaluating the worth of a proposed project is called the benefit/cost (B/C) ratio method. As the name implies, the method consists of comparing the annual worth* of the benefits of a proposed project to the annual worth of the costs. For example, a proposal with estimated benefits of $20,000 per year and estimated costs of $10,000 per year has a ratio of $B/C = \$20,000/\$10,000 = 2.0$.

The B/C ratio method became widely used following passage of the federal government's Flood Control Act of 1936 which stated that federal funding for improvements to waterways for flood control would be justified only "if the benefits to whomsoever they may accrue are in excess of the estimated costs." The principle has since been applied to many other types of public projects, from airports to zoos and is now an accepted measure of desirability for projects at most levels of government.

A common use of the B/C ratio method is to rank various public works proposals on the basis of their B/C ratio and select the one or ones

*Other standards of comparison may be used, such as present worth, or future worth. However, annual worth is most commonly used.

with the highest B/C ratio. However, the proper ranking of projects requires further examination, and merely ranking in order of numerical B/C ratios may lead to erroneous conclusions as the following example illustrates.

Example 14.1

Analysis of several mutually exclusive roadway alignments yields the following information ($i=7\%$).

	A	B	C
Annual benefit ($)	375,000	460,000	500,000
Annual cost ($)	150,000	200,000	250,000
B/C ratio	2.5	2.3	2.0

SOLUTION

On the basis of B/C ratio, alignment A appears the best because it has the highest B/C ratio. But, all three alignments have ratios greater than 1.0 and thus all deserve consideration. The next step after finding the $B/C>1.0$ is to analyze the *incremental B/C* ratio, beginning with the lower cost pair B−A. The incremental B/C ratio is determined by subtracting lower cost A from the next higher cost B, as follows.

B minus A

$$B/C = \frac{460,000-375,000}{200,000-150,000} = \frac{85,000}{50,000} = 1.7$$

Each additional dollar invested in alignment B over A yields a benefit of $1.70. Therefore, assuming all other aspects are equal, B is preferred over A (analogous to the incremental rate of return presented in Chapter 11). Discard A. Then compare current candidate B to the next higher cost alignment C, as follows.

C minus B

$$B/C = \frac{500,000-460,000}{250,000-200,000} = \frac{40,000}{50,000} = 0.8$$

Each additional dollar invested in alignment C over B yields a benefit of $0.80. Therefore, B is preferred over C and C should be discarded. Select B.

Comment. B is the better alternative even though it has a lower B/C ratio than A (it also has a lower rate of return). But it offers the greatest benefit for the total expenditure.

B/C RATIOS

The usual method for computing the benefit/cost ratio is to calculate all benefits and costs on an annual basis. Thus initial construction costs are converted to equivalent annual costs by application of the appropriate $(A/P, i, n)$ factor. Similarly, future salvage values and gradient costs and income are also converted to equivalent annual values. Occasionally, the sign convention may be confusing. Using conventional sign notation (benefits positive and costs negative) the B/C ratio would be a negative number. To make the B/C ratio positive, treat all increases in benefits (such as reductions in user costs) as positive when in the numerator, and all increases in costs (such as increase in maintenance cost) as positive when in the denominator. Most texts presume the reader can discern any discrepancies in sign notation and adjust for them where appropriate.

User Benefits

When new highways or other public facilities are constructed, the user usually enjoys some savings in costs (such as travel costs) when the new facility is compared to the existing one. These savings may occur due to less time required to travel a certain distance (less congestion, fewer stops or delays at traffic signals, shorter route, etc.), lower fuel consumption, less wear, fewer accidents, or other similar reasons. Savings of this type are determined by estimating the total annual cost to the users for the present facility (this cost is designated U_p) and subtracting from it the total annual cost to the same number of users for the future facility (designated U_f). Thus the net user benefit, designated U_n, is calculated as $U_n = U_p - U_f$. If the net costs to users of the future facility are less, the resulting U_n will be positive. The terms $(U_p - U_f)$ and U_n both designate *net* user benefits and are used interchangeably.

Owner's Costs

Owner's costs are divided into two categories, (1) capital costs, and (2) maintenance costs.

1. Capital costs are usually considered as the construction, acquisition, or other capital costs of a facility. If a replacement facility is being considered the question arises on how to handle the present value of the existing facility. The present value is the cash salvage value that would be received now if the existing facility were sold or demolished. If the existing facility is maintained in place, this cash value is left invested in the existing facility. Therefore the capital cost of the existing facility is the cash salvage value. The cost of the proposed facility is *not* reduced by the worth (salvage value) of the existing facility since the value of the existing facility is an owner's asset that can be freely kept or spent like cash or any other asset. To handle the sign notation, the following terms are defined.

C_f = Equivalent capital cost of proposed (future) facility, usually expressed on an annualized basis.

C_p = Equivalent capital worth of the existing facility (present salvage value), usually expressed on an annualized basis. This is the value that either can be received in cash or left invested in the existing facility.

$C_n = C_f - C_p$ = *Net* capital cost of replacing the present facility with the future facility. Note that if the present facility is greater than the cost of the proposed facility (an unusual situation) the net capital cost could be negative. Normally, using this chapter's sign convention, C_n will be positive.

2. Maintenance costs are the owner's costs for operating and maintaining the facility (O&M). If a replacement facility is being considered, its O&M costs may be more, or less, than the O&M costs for the present facility. Thus, let us define

M_f = Equivalent operating and maintenance costs of the future (proposed) facility, usually expressed as an annual (or annual equivalent) cost.

M_p = Equivalent operating and maintenance costs of the present (existing) facility, expressed in the same terms as M_f.

$M_n = M_f - M_p$ = *Net* operating and maintenance cost of the proposed facility over the present facility. M_n may be either positive or negative.

SEVERAL METHODS FOR CALCULATING B/C RATIO

One of the major problems facing the engineer in determining B/C ratios is deciding which items to include in costs and which items to include in benefits. Several variations of methods exist for determining B/C ratios. As a result, considerable engineering judgment is required both in originating B/C studies, as well as in evaluating B/C studies done by others.

The two most common approaches to computing the benefit/cost ratio are known as (1) the conventional B/C or AASHTO system,* and (2) the modified B/C method. Both of these involve a comparison of a proposed facility with an existing facility. (If there is no existing facility, use the present cost of reaching the same objective. For example, if the proposed facility is a bridge over a river, and there is now no existing method of crossing the river at this point, then the present user cost is the cost of whatever route must be taken to reach the other side.)

1. *Conventional B/C*. The benefits (usually annual) are determined for users. Thus, the benefits are defined as

$B_n = U_n$ = *Net* annual benefits (savings in costs) through safety improvement, decreased gasoline consumption, decreased tire wear, etc.

*"Road User Benefit Analysis for Highway Improvements," American Association of State Highway Officials, Washington, D.C., 1960, pp. 27–28.

Sign notation: in the numerator any increase in benefits (or reduction in costs) results in a net positive benefit. Most proposed projects are designed to reduce user's costs. Therefore the net benefits are equal to the user costs of the present facility *minus* the user costs of the proposed (future) facility, or $U_n = U_p - U_f$.

$C_n + M_n$ = Costs consist of the annual equivalent costs to the *owner* of the facility, including capital costs *and* maintenance. The numerator consists of all of the user's benefits; the denominator is the sum of *all* the owner's costs. Thus,

$$\text{Conventional } B/C = \frac{\text{net savings to users}}{\text{owner's net capital cost} + \text{owner's net operating}}$$
and
maintenance cost

$$= \frac{U_n}{C_n + M_n} = \frac{B_n}{C_n + M_n} \qquad (14.1)$$

Sign notation: in the denominator any increase in costs results in a net positive cost.

2. *Modified B/C.* This method uses the same input data but net operating and maintenance costs (M_n) are treated as negative benefits (or disbenefits) rather than as costs. Thus, they are placed in the numerator rather than in the denominator. The resulting equation is

$$\text{modified } B/C = \frac{U_n - M_n}{C_n} = \frac{B_n - M_n}{C_n} \qquad (14.2)$$

The sign convention is the same as for the conventional B/C. In the numerator, any net increase in benefit (or decrease in cost) is positive, and in the denominator any net increase in cost is positive.

Recalling Chapter 8 where the net annual worth (NAW) was shown to be

$$NAW = AW \text{ benefits} - AW \text{ costs}$$

the problem of treating maintenance costs (as a cost or as a *dis*benefit) did not arise. The sign convention automatically handled the problem because the *net* value was determined. But, when a ratio is used the results can be difficult to interpret as shown in the next example.

Example 14.2 (*Comparison of conventional and modified B/C*)

Find the conventional B/C for alternate routes A and B proposed to replace an existing route between two points. $i = 8$ percent, $n = 20$ years.

	EXISTING ROUTE	PROPOSED ROUTE A	PROPOSED ROUTE B
Construction cost ($)	0	100,000	100,000
Annual equivalent to construction cost $A = \$100,000\,(A/P,8\%,20)$	0	$ 10,190	$ 10,190
0.1019			
Estimated user's cost ($/yr)	200,000	165,000	195,000
Owner's operating and maintenance costs ($/yr)	250,000	270,000	240,000
Total annual cost ($)	450,000	445,190	445,190
Annual savings over existing route ($/yr)	0	4,810	4,810

SOLUTION

Substituting in the *conventional B/C* equation yields

Route A Compared to Existing

$$\frac{U_p - U_f}{(C_f - C_p) + (M_f - M_p)} = \frac{200,000 - 165,000}{(10,190 - 0) + (270,000 - 250,000)}$$

$$= \frac{35,000}{30,190} = \underline{\underline{1.16}}$$

Route B Compared to Existing

$$\frac{U_p - U_f}{(C_f - C_p) + (M_f - M_p)} = \frac{200,000 - 195,000}{(10,190 - 0) + (240,000 - 250,000)} = \frac{5,000}{190}$$

$$= \underline{\underline{26.32}}$$

By comparison, the modified *B/C* yields the following results.

Route A

$$\text{modified } B/C = \frac{(U_p - U_f) - (M_f - M_p)}{C_f - C_p}$$

$$= \frac{(200,000 - 165,000) - (270,000 - 250,000)}{10,190 - 0}$$

$$= \frac{15,000}{10,190} = \underline{\underline{1.47}}$$

Route B

$$\text{modified } B/C = \frac{(200,000 - 195,000) - (240,000 - 250,000)}{10,190 - 0} = \frac{15,000}{10,190}$$

$$= \underline{\underline{1.47}}$$

Thus with identical capital investment, the annual savings for route A and route B are identical at $4,810 per year, yet the conventional *B/C* ratio yields *B/C* of 1.16 for A and 26.32 for B. The modified *B/C* ratio on the other hand yields B/C ratios of 1.47 in each case. *This and similar comparisons indicate that the modified B/C method usually yields more consistent results.*

COMPARISON OF B/C RESULTS WITH RATE OF RETURN

It is interesting to compare the B/C evaluation of the alternate routes with an evaluation using the rate of return method from Chapter 10.

Example 14.3

Find the rate of return of the investment in route A or route B (same for each) from Example 14.2.

SOLUTION

$$(U_p - U_f) - (M_f - M_p) = C_f - C_p$$

For both cases

$$\$15,000 = \$100,000(A/P, i, 20)$$

$$0.150 = (A/P, i, 20)$$

By interpolation,

$$i = 14\% \qquad 0.1510 = (A/P, 15\%, 20)$$

$$i = ? \qquad 0.1500 = (A/P, i, 20)$$

$$i = 13\% \qquad 0.1424 = (A/P, 12\%, 20)$$

$$i = 13 + \frac{0.0076}{0.0086} = \underline{\underline{13.9\%}}$$

Discussion:

a. Alternatives A and B each have the same rate of return, as indicated by the modified B/C method, rather than the differences indicated by the conventional B/C method.

b. The modified method indicates that for every $1 invested there will be a benefit of $1.47, which some might interpret as a 47 percent rate of return. Careful analysis indicates the actual rate of return is only 13.9 percent.

What to Compare

B/C studies by their very nature involve a comparison of two or more alternatives. One of these alternatives should be the existing facility or method, or erroneous results may occur. It is possible for instance to compare an expensive new bridge across a bay with a more expensive new tunnel. The result could be a recommendation for construction of the bridge, simply because the existing route involving a short trip around the end of the bay was not considered in the comparison. Sometimes there is no existing facility, such as in the following instances.

1. EPA requires construction of a sewage treatment plant.
2. An interstate highway requires a river, railroad, or intersecting road crossing and no existing crossing meets the standards.
3. A new facility is required due to population growth or growing demand for service.

Where there is no existing facility to which a comparison may be made caution is urged both in the formulation of B/C calculations as well as in the examination of B/C studies by others.

User Benefits: User benefits in the B/C ratio are ordinarily supposed to represent the reduction in user's cost resulting from construction of the new facility. Where there is no existing facility it is difficult to establish a legitimate reduction in cost. Some practitioners have filled this void with a substitute alternative of their own selection to replace the missing existing facility. This immediately raises some question concerning whether the favorable B/C ratio resulting from such a comparison between two proposed facilities (versus a comparison of existing versus proposed) really represents a real benefit to the users. For an extreme but plausible example, a B/C comparison between a pedestrian overpass over a quiet residential street and a pedestrian tunnel under that street could be formulated to show a high B/C ratio in favor of the one or the other, but would likely be of little actual benefit to the pedestrian, who naturally prefers a street level route. This satire on some actual B/C studies is illustrated by the following theoretically logical example.

Example 14.4

The pedestrian overpass is estimated to cost $100,000 to build, and $40,000 per year to maintain. The pedestrian tunnel will cost $200,000 to construct and $10,000 per year to maintain. The expected life of each is 20 years, and $i=6$ percent. User costs, before and after, are zero since no one is expected to use either facility.

SOLUTION
The annual equivalent of the capital costs are calculated as shown.

$$C_{\text{overpass}}=\$100,000\,(A/P,6\%,20)=\$8,720$$
$$0.0872$$
$$C_{\text{tunnel}}=\$200,000\,(A/P,6\%,20)=\$17,440$$
$$0.0872$$

The B/C calculations may then be carried out as follows.

$$\text{Modified } B/C = \frac{(U_0 - U_t) - (M_t - M_0)}{C_t - C_0} = \frac{0 - (10{,}000 - 40{,}000)}{17{,}440 - 8{,}720}$$

$$= \underline{\underline{3.44}}$$

According to the example, the investment in the tunnel shows a B/C of 3.44, or an "implied" showing of \$3.44 for every \$1 invested.

This example illustrates what can occur when B/C studies do not include a comparison with the existing facility or method. The B/C method is an excellent tool when properly applied, but obviously considerable care and some discretion is required in the use of it.

FURTHER COMMENTS ON B/C RATIOS

The reader should recognize that under some circumstances one or the other B/C ratios may not be usable at all. Considerable caution is urged when devising a B/C study, or examining one presented by others. The following cases are cited as extreme examples. However, bear in mind that as actual problem situations approach these extremes, the results may show signs of skewing toward the limits indicated by the extremes.

1. If there is a zero saving to users, then $U_n = 0$.

$$\text{conventional } B/C = \frac{U_n}{C_n + M_n} = \frac{0}{C_n + M_n} = 0$$

$$\text{modified } B/C = \frac{U_n - M_n}{C_n} = -\frac{M_n}{C_n}$$

A project should be justifiable, if the *savings* in maintenance and operating costs more than offset the capital cost. This could not be demonstrated with the conventional B/C, but could be by the modified B/C.

2. If the project results in a savings to users (U_n is $+$), and maintenance and operating costs drop more than capital costs rise ($C_n + M_n$ is $-$), then

$$\text{conventional } B/C = \frac{U_n}{C_n + M_n} < 0$$

$$\text{modified } B/C = \frac{U_n - M_n}{C_n} = > 1$$

The conventional B/C ratio becomes a negative number, whereas the modified B/C properly accounts for the benefit of lower maintenance to the owner.

3. If there is no change in the capital investment required ($C_n=0$), then the conventional B/C can be substituted to show the relative merits of the project.

$$\text{conventional } B/C = \frac{U_n}{C_n+M_n} = \frac{U_n}{M_n}$$

$$\text{modified } B/C = \frac{U_n-M_n}{C_n} = \frac{U_n-M_n}{0} = \infty$$

The modified B/C reflects the benefits derived from capital investment only. Since in this example the added capital investment required is zero, any benefit at all from zero investment yields an infinite B/C ratio. The conventional B/C compares the user's benefits to the owner's costs. Therefore, no matter how much the owner may benefit by lower capital or operating costs, the user must benefit commensurately or the project acquires a low B/C rating under the conventional B/C system. Other considerations affecting the validity of B/C output to other variables should be examined carefully when evaluating a B/C study. These are cash flow and dollar volume.

Cash Flow

The B/C method of project comparison can be relatively insensitive to timing of cash flow if low interest rates are used. Thus the timing may require separate scrutiny apart from the B/C result itself. For instance, under certain circumstances a cost now of $100,000 resulting in a $400,000 benefit occurring 60 years from now could outrank (by means of a higher B/C) an alternative expenditure of $100,000 now that would bring a $140,000 benefit immediately. The following example illustrates how this could happen.

Example 14.5

1. Use $i=1$ percent. Alternative X, which involves $400,000 benefit occurring 60 years from now, has an annual benefit equivalent to

$$B=\$400,000\,(A/F,1\%,60)=\$4,880/\text{yr}$$
$$0.0122$$

Assume an expenditure of $100,000 required right now to produce this $400,000 future benefit. The annual equivalent cost is

$$C=\$100,000\,(A/P,1\%,60)=\$2,220/\text{yr}$$
$$0.0222$$

Therefore, the B/C may be calculated as

$$\text{alternative X: } B/C = \frac{\$4,880}{\$2,220} = 2.20$$

Assume now a competing alternative project Y involves

benefit of \$140,000 occurring immediately$=$\$140,000

cost of \$100,000 occurring immediately$=$\$100,000

$$\text{alternative Y: } B/C = \frac{\$140,000}{\$100,000}\frac{(A/P,1\%,60)}{(A/P,1\%,60)} = \underline{\underline{1.4}}$$

2. For a more realistic value of i, use $i=7$ percent.

$$\text{alternative X: } B/C = \frac{400,000(A/F,7\%,60)}{100,000(A/P,7\%,60)} = \underline{\underline{0.67}}$$

$$\text{alternative Y: } B/C = \frac{140,000(A/P,7\%,60)}{100,000(A/P,7\%,60)} = \underline{\underline{1.4}}$$

SOLUTION 1
For $i=1$ percent, alternative X is preferred.

SOLUTION 2
For $i=7$ percent, alternative Y is preferred. Alternative X has a B/C ratio of less than 1 and thus should not be selected under any circumstances.

Evaluation of Benefits and Costs

In private enterprise the question of benefits is usually simple since the benefits are typically measured in dollars of income. When considering public works, however, direct income frequently does not occur and other types of benefits both tangible and intangible must be considered very carefully. Ultimately the decision on what benefits to include is a political decision (using "political" in its best sense). It must be made by elected political decision makers usually with input from both lay citizens and technical advisors. Technical advisors for their part undoubtedly can profit from a continuing input from their peers. Technical meetings of engineers particularly should schedule regular periodic panel discussions and encourage further consideration of methods of evaluating benefits and costs (both tangible and intangible) in order that technical people may contribute more intelligently to the political decision-making process.

Returning to the phrase in the federal Flood Control Act of 1936 which reads "...benefits to whomsoever they may accrue...", the American people usually prefer the users of public works to pay for those works wherever feasible. For example, gasoline tax pays for roads, rate payers support sewer, water, and electric facilities, and so on. Exceptions to user type taxes occur when users are difficult to determine, or are too poor to pay. Public policy prefers that public funds derived from a broad tax base should not be used to benefit a few, except in severe hardship cases. Thus the phrase "... benefits to whomsoever they may accrue..." possibly may be interpreted at times to infer broader coverage than most citizens would desire. Questions continually arise, such as:

1. If a canal is constructed at government expense, is the increase in value of adjoining property a proper benefit to be counted? Arguments:
   a. Pro. The act specifies "benefits to whomsoever they may accrue" which could be interpreted as a few or many without discrimination between large landholders and others. Also, higher land values mean higher property tax payments back to the government and capital gains taxes upon sale. If the property has a higher value, isn't this evidence that it is more useful (or at least more desirable)? A nation's wealth consists of the total wealth of all its citizens, thus when value is added to a citizen's property, wealth is added to the nation.
   b. Con. Government funds should be used to benefit large groups of ordinary citizens. If some few landowners reap windfall profits from government projects let them pay special taxes or assessments on it. "My taxes are not meant to make a few people wealthy" is the feeling of most.
2. Are recreational and scenic benefits as valuable as direct cost savings to users? Arguments:
   a. Pro. People spend significant percentages of their income on recreation and aesthetics. The house, furniture, clothes, car, and other important items are usually chosen with appearance (or aesthetics) uppermost in mind. People have, therefore, demonstrated with their pocketbooks that scenic aesthetic benefits are important. Likewise recreation plays an important role to the average American and every year satisfies an increasing need as evidenced by rising annual expenditures. Benefits should be valued at the cost of traveling to and using the nearest equivalent location multiplied by the number of users.
   b. Con. Not everyone appreciates the dollar value of a tree or a fishing spot. However, a much larger number of people benefit from reduced transportation costs of consumer products. Therefore, direct savings should weigh more heavily than secondary benefits. For instance, discounting secondary benefits by 50 percent seems reasonable.
3. Should recreational and scenic locations that are destroyed due to a proposed project be counted as part of the cost? Arguments:
   a. Pro. Each asset that is destroyed is as much a part of the cost as each dollar spent. Evaluate them as fairly as possible. Lost capital assets are part of the capital costs, lost income is an annual cost. Their market value depends in part on relative scarcity, what

other similar assets remain, and how accessible these remaining similar assets are.

b. Con. Progress requires a certain amount of changes in the landscape. Nothing can be preserved forever. A practical balance must be struck between people and the environment. Property owners have a right to destroy what they have purchased if the resulting project is profitable.

Approach to Applications

Applications of B/C techniques become more complex as user costs become more difficult to determine. For instance, a recreational or public park complex may not show large direct tangible benefits. Most will agree however that the intangible benefits are very important to individuals, families, and communities. One rational approach to the determination of user benefits is to:

1. Estimate how many users will use the new facility per year.
2. Assume that all of these potential users are now using an existing comparable facility somewhere else. Obviously they are not, but if they use the new facility there is evidence of frustrated desire to use such a facility but the present cost in travel time, money, and effort to get there is just too high. The benefit results from lowering the cost to a level they can afford, by building a facility close by.
3. Find the cost, in travel time and money, for them to use the existing comparable facility and use this for present user cost.
4. The cost of using the proposed new facility to these same people in terms of decreased travel time and money is the user cost of the future facility.
5. These costs are then compared to the maintenance and capital costs to determine the B/C ratio.

Applications of the B/C method to other projects associated with community growth are more complex yet. For instance, the addition of a new shopping mall or steel mill to a community introduces some interesting questions. User benefits and costs in a community situation may sometimes be estimated as an average accruing to citizens within a reasonable radius affected by the development. A new shopping center may have the following effects on those within the service area.

Benefits
1. Gives added opportunity for comparison shopping and consequent savings to consumers.
2. Shortens the travel distance needed to go shopping for some (decreases traffic on some previously traveled routes).
3. Increases land values in the area.

4. Increases tax revenues (sales tax, income tax from employees, property tax from increased investment).
5. Increases money in circulation.

Costs
1. Increases traffic in the area immediately around the shopping center, greater congestion of existing facilities.
2. Increases cost of municipal services
   a. fire and police protection
   b. utilities services, water, sewage, electric, gas, telephone, solid waste disposal.
3. New jobs bring new people with added costs of services, schools, police, roads, fire, parks, playgrounds, and other public facilities.
4. Reduces business for existing stores.
5. Increases rain runoff, heat, dust (depends upon previous use of property).

For a new steel mill, a similar type of list could be constructed. The difficulty lies in attempting to estimate the dollar's worth of benefit or cost to each of the items listed and the number of people affected. However some attempts at analysis must be made and frequently a preliminary attempt will give rise to suggestions for a more detailed and comprehensive subsequent attempt. In any case if the community is to make wise decisions, it needs as much information as possible. Engineers have for too long been largely passive followers in these decision-making processes. Engineers by training and practice have the ability to analyze difficult and complex problems affecting the employment of both our natural and constructed environment. By virtue of that ability, engineers have an obligation to actively probe and research for answers. They should assume leadership roles. Here is an area of challenging problems affecting the enjoyment of life and, some say, the very existence of millions of people. The situation certainly warrants a great deal of creative thinking and logical analysis by engineers. Due to the press of normal professional work this creative thinking will often be largely on the engineers' own time and of their own volition, at least at the outset.

Growth Factors

In past years, estimators could assume a growing number of users as time went on. Now not only growing user numbers may be assumed, but growing unit costs as well. Growth when it occurs can be approximated by any of three methods or combinations thereof.

1. Arithmetic growth of a certain set sum each period.
2. Geometric growth of a certain percentage each period.
3. Step growth, using a certain number for each group of periods. The steps need not be equal, nor need the groups of periods be equal.

Road Improvements

Problems requiring an evaluation of the benefits accruing to the public from the expenditure of public road improvement funds frequently can be resolved by use of B/C studies. The user's benefits usually accrue to the motoring public while the maintenance and capital costs are incurred by the governmental agencies involved, which of course are financed by the public. The following example illustrates a traffic signal study.

Example 14.6

A city is considering the installation of a traffic activated signal system at a busy intersection now controlled by a single preset traffic light. Traffic counts indicate that 20,000 vehicles cross the intersection daily. A major consideration is the number of accidents occurring in the intersection. The existing signal is a single traffic light with maintenance costs averaging $300 per month. The new system has an initial cost of $6,867 with monthly maintenance costs estimated at $500. The life of the new system is assumed as five years. No salvage value is assumed for either system. Current traffic studies indicate the number of vehicles stopping at the intersection is now 8,000 per day. The new traffic activated system will stop about one half this many, or 4,000 cars per day. The average cost for a motorist to stop at this light is $0.025. Experience with similar installations indicates the new system will reduce accident costs from a current rate of $9,528 per year to an anticipated $300 per year. Find the B/C for the proposed installations (assume $i=6\%$).

SOLUTION

The owner's capital cost, O&M costs, and user costs are calculated and then inserted into the B/C equation as follows.

	ANNUAL EXISTING	ANNUAL FUTURE PROPOSED
DOT owner's cost per year (C)	0	$6,867 ($A/P$,6%,5)=$1,630
O&M costs (M)	$3,600	6,000
Facility user costs (C):		
accidents	9,528	300
Stop and start		
$0.025/vehicle×365 days/yr×vehicle=	73,000	36,500
Total user costs	$82,528/yr	$36,800/yr

$$\text{modified } B/C = \frac{(U_p - U_f) - (M_f - M_p)}{C_f - C_p}$$

$$\text{modified } B/C = \frac{(82{,}528 - 36{,}800) - (6{,}000 - 3{,}600)}{1{,}630 - 0}$$

$$= \frac{43{,}328}{1{,}630} = \underline{\underline{26.6}}$$

By way of comparison the conventional $B/C = 11.4$ for this case.

Intangible Benefits: In addition to the readily quantifiable benefits accounted for in the equation, the following intangible benefits are anticipated.

    a. Injury—reduction of pain, suffering, lost productivity, mental anguish of friends and loved ones, and so forth. Direct accident costs are accounted for in the data, but the real cost of accidents often exceed by far the costs listed.

    b. Air pollution—4,000 less stops per day will enhance air quality in the area.

    c. Driver fatigue and irritation will be reduced by fewer stops and starts.

APPLICATIONS OF *B/C* RATIOS

Where user costs and savings are easily determined, the application of the B/C method is simple and straightforward. This includes many traffic situations such as signalized intersections or alternate routes. A traffic count is made and projected ahead. An estimate is made of user cost and maintenance cost before and after. Capital cost is estimated. The conversion of capital cost to annual cost depends on finding valid n and i. The determination of the useful life, n, is discussed in detail in Chapter 21. Selection of i for public projects is a matter of some controversy. As noted in the examples of this and other chapters, the value of i will strongly influence the result. Very low values of i (promulgated by some public officials) will favor long-term investments yielding low benefits each year. Higher values of i (promulgated by most public officials) will favor shorter-term investments yielding more immediate benefits. Since most public agencies borrow money through the issuance of bonds or some other financial investment (see Chapter 17), they do experience time value of money in much the same way as private companies. Hence, realistic rates of interest, at least comparable to bond rates for tax exempt bonds, should be used. For a more complete discussion on this topic, see Eugene L. Grant et al., *Engineering Economy*, Sixth Edition, Ronald Press Company, New York, New York, pp. 447–477.

Increase in Number of Users

A problem in arithmetic frequently arises when considering user costs. More people are attracted to the new facility, hence the total user cost rises even though the cost per individual user may drop. This can produce misleading results as illustrated in the following example.

Example 14.7

A new bridge is proposed to replace an existing bridge. The data below show current and projected costs. Assume $i=6$ percent, $n=25$.

	OLD BRIDGE	NEW BRIDGE
User trips per year	2,000,000	8,000,000
User cost ($/user trip)	.50	.25
Annual user cost ($/yr)	1,000,000	2,000,000
Maintenance ($/yr)	200,000	200,000
Capital cost ($) ($P$)	0	5,000,000
Annual equivalent to capital cost $A = P(A/P,6\%,25) = \$5,000,000 \times 0.0782 =$ 0.0782		$391,000/yr

SOLUTION

The conventional B/C ratio yields the following

$$B/C = \frac{U_p - U_f}{(C_f - C_p) + (M_f - M_p)}$$

$$= \frac{2,000,000 \times \$0.50 - 8,000,000 \times \$0.25}{(\$391,000 - \$0) + (\$200,000 - \$200,000)}$$

$$= -2.56$$

The modified B/C ratio also yields the same since the maintenance costs are a net of zero.

$$B/C = \frac{(U_p - U_f) - (M_f - M_p)}{C_f - C_p} = \frac{-\$1,000,000}{\$391,000} = -2.56$$

The results in both cases are negative, not only less than 1, but less than zero. Intuitively the investment appears profitable. If a person can cross the existing old bridge at a cost of $0.50 and could cross the new one for only $0.25 there should be a demonstrated saving.

The solution to the dilemma is revealed in Figure 14.1, which shows the relationship between cost and number of user trips per year. The number of users attracted to the facility increases as the cost per trip decreases. Therefore the response is said to be "price elastic," and the increase in

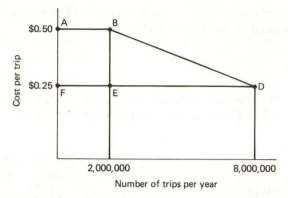

Figure 14.1   User benefits resulting from increasing number of users.

number of trips is assumed to be roughly proportional to the reduction in cost (varying according to line *BD* in Figure 14.1). Actually the price-elastic response line could be represented by a variety of configurations between points *B* and *D*, but a straight line can usually serve as a reasonable approximation. Examining Figure 14.1 it is evident that all 2,000,000 existing user trips benefit by the full $0.25 savings, and the total savings for those user trips will amount to (2,000,000×$0.25)=$500,000/year. In Figure 14.1 this saving is represented by the area of the rectangle *ABEF*. If the cost were reduced only from $0.50 to $0.49 per trip, all 2,000,000 existing trips would benefit (by $0.01 per trip), but only a few new user trips would be attracted to the facility. The user benefit would be calculated as the number of existing trips times $0.01 per trip plus the *average* number of new trips times $0.01 per trip. Each added decrement in cost attracts an added increment of trips, forming an incremental *triangle* of trips times savings per trip. The new trips added with each cost decrement are assumed to be attracted at incremental points in time because previous decrements in cost did not represent a savings to them and only with the most recent decrement did a real savings occur which then attracted that increment of trips. For Example 14.7, reduction of costs from $0.50 to $0.25 results in an added 6,000,000 trips, and the incremental benefit can be calculated as the area of the triangle *BDE*, or [($0.50−$0.25)×(8−2)×10^6]/2= $750,000/year benefit to new users plus the $500,000 per year benefit calculated previously for the existing users. Using this approach, both the conventional and the modified methods yield:

$$B/C = \frac{\$500,000 + \$750,000}{\$391,000} = \$3.20$$

EVALUATING COSTS OF LIFE AND PERSONAL INJURY

Attempting to equate life itself with a dollar value seems crass and repugnant. However, every day many lives are lost through accident or other cause. If the cause of death involved negligence, very often the negligent party is required to compensate to some extent the heirs of the deceased. What is just compensation for the loss of a life? Difficult as a judgment may be, opinions are required and decisions and awards are made based upon these opinions. And difficult as it may be, engineers are often required to rank proposed public works projects in order of priority recognizing that capital is limited and not all of the proposals can be funded. For example, if funds are available to improve the skid resistance of highway A or highway B, which one should be selected? One consideration is the savings from reduced accidents. Continuing efforts are underway to evaluate the costs of life and personal injury for use in engineering cost and feasibility studies. For example, data indicate that travel on interstate routes is much safer than travel on conventional rural or urban roads. Given the same amount of travel, one life per year is saved for every 5 mi of road raised to interstate standards. If the road is assumed to have a life of about 20 years and costs somewhere near $2,000,000 per mile, this works out to a cost of about $500,000 per life saved. In addition, a large number of nonfatal accidents and injuries are prevented and obviously there are additional costs and benefits as well. When experts are asked to evaluate the cost of a death in order to seek compensation in a court of law, they will frequently use the following approach.

1. Loss of earnings. Assume the deceased was 30 years old, was making $15,000 per year and could expect raises of $1,000 per year until retirement at age 65. The present worth may be calculated at some reasonable investment rate, say 8 percent for illustration purposes, as

$$P_1 = 15,000\,(P/A, 8\%, 35) + 1,000\,(P/G, 8\%, 35)$$
$$11.6546$$
$$P_1 = 174,800 + 116,100 = \$290,900$$

Thus the jury would be requested to award the survivors the sum of around $290,900 as compensation for loss of earnings.

2. In addition the attorneys may also ask compensation for loss of (a) love and affection, and (b) physical comfort and other named benefits now lost due to the untimely death. In addition, if the death was accompanied by or caused (c) pain and suffering, and (d) mental anguish or other trauma of a similar nature, compensation may be requested for these damages. In short, almost any type of loss that can be explained and evaluated to a jury's satisfaction (and is not ruled out of order by the judge) can be listed as a compensable loss. The jury then uses its own discretion and judgment in evaluating the claim and awarding whatever compensation, if any, is justified. A similar approach may be taken by the engineer in evaluating priority of expenditure. One unfortunate difference occurs however. The jury need not regard the ability of the defendant to pay the damages in making the award, but the engineer must always abide by the budgetary restraints of the funding agency. However, most funding agencies have some

flexibility and it is quite possible that a proper presentation of the benefits versus the cost of life and injury-saving installations may wring extra funding from even the most adamant agency.

B/C Analysis for Private Investments

Although developed for public projects where benefits are difficult to determine, this method of analysis can be used for private investments. In fact, often it is a very easy method of analysis because benefits and costs are easily quantified. Use of the B/C ratio is analogous to use of the rate of return method described in Chapters 10 and 11. The steps to be followed are exactly the same and the incremental benefits to incremental costs must be determined in order to obtain valid results. The analysis of Example 11.6 using the B/C ratio is illustrated in the following example.

Example 14.8

An engineering firm is considering five different small computers to aid the firm in running complicated and repetitive calculations. Analysis of the various machines, when applied to the firm's business, indicates the following cash flow information ($MARR = 12\%$).

	G	H	F	I	J
First cost ($) ($P$)	18,000	22,500	25,000	28,400	32,000
Annual net income ($)	3,800	4,800	5,000	6,000	6,600
Useful life (yr)	8	8	8	8	8
Present worth of benefits ($)	18,878	23,845	24,838	29,806	32,787
$P_B = A(P/A, 12\%, 8)$					
B/C	1.04	1.06	0.99	1.05	1.02

SOLUTION

Since H has the highest B/C ratio, one might be tempted to select H, but that would be wrong. At this point the only conclusion that can be made is to discard F because its B/C ratio is less than 1.0 (which means its rate of return is less than 12.0%). To solve the problem, each *incremental* B/C ratio must be determined. Thus,

H–G:

$$\frac{\Delta B}{\Delta C} = \frac{23,845 - 18,878}{22,500 - 18,000} = \frac{5,057}{4,500} = \underline{\underline{1.12}}$$

Since $B/C > 1.0$, discard lower cost alternative (G).

I–H:

$$\frac{\Delta B}{\Delta C} = \frac{29,806 - 23,845}{28,400 - 22,500} = \frac{5,981}{5,900} = \underline{\underline{1.01}}$$

Since $B/C > 1.0$, discard lower cost alternative (H).

J–I:

$$\frac{\Delta B}{\Delta C} - \frac{32,787 - 29,806}{32,000 - 28,400} - \frac{2,981}{3,600} \underline{-0.83}$$

Since $B/C < 1.0$ discard higher cost alternative (J). The remaining alternative is I. Therefore, select I. This is the same conclusion reached using incremental rate of return, net present worth, and so forth (see Example 11.2).

COST EFFECTIVENESS: UTILITY/COST METHOD OF EVALUATING INTANGIBLES

Often, when comparing alternative candidates for purchase or investment, some of the desirable characteristics of the competing purchases or investments are intangible and difficult to quantify.

For instance, when selecting a particular exterior surfacing for a building wall, how does the visual appearance of brick compare with the appearance of concrete block or cedar shakes or other alternatives? When buying a car, how should an evaluation be made of comfort, noise level, styling, prestige, convenience, and other similar characteristics?

The concept of utility value can help quantify these characteristics in order to facilitate a more direct and objective comparison. The utility value itself is a derived number, obtained by comparing the actual level of performance of the candidate system to an ideal desired level of performance or performance goal for an idealized system.

The utility/cost (U/C) format may be employed on either of two levels, defined by the matrix dimension. Either a one-dimensional or two-dimensional matrix may be used.

The one-dimensional matrix consists of a left-hand column listing a series of performance goals, together with a top row of column headings containing the alternative candidates to be evaluated against the performance goals.

The two-dimensional matrix employs a separate matrix for each candidate. The left-hand column now contains a list of desirable system features generally comparable to the features exhibited by each of the candidate systems, while the top row of column headings contains the list of performance goals.

The step-by-step process for each matrix system is detailed as follows:

One-Dimensional Matrix

1. List a series of desirable features or performance goals in a column to the left side of the page.
2. Evaluate each individual feature or goal in terms of its contribution

to the overall objective. Place a number in each row reflecting the relative percentage contribution of each feature to the objectives as a whole, such that the sum of all these numbers adds up to 100 percent.

3. List the names of each candidate under consideration for acquisition, as column headings across the top of the page.

4. Draw a matrix of boxes joining the columns and rows. Then draw a diagonal across each box from upper left to lower right.

5. In the top right triangle of each box, evaluate the performance of each candidate with respect to the goal. Use a 10 for "fulfills goals perfectly", and 0 for "not qualified".

6. In the bottom left triangle, place the product of the evaluation, determined in (5) above, multiplied by the percentage number found in (2) above.

7. Sum the figures in each column from the bottom left triangles only. This sum represents the utility value of each candidate.

8. Divide the sums in each candidate's column by the cost of that candidate and determine a utility/cost ratio. The higher values are preferred, reflecting a higher number of utility value points per dollar spent.

9. Plot the utility value (*y* axis or *ordinate*) versus the cost (*x* axis or *abscissa*) of each candidate. Begin at the origin and draw an envelope through all points closest to the *y* axis. Normally the selection is made from one of the points on the envelope.

Example 14.9

A contractor needs to purchase a supply of hard hats for all job personnel on a large construction site. Four different brands are available each with its own special features and price. Determine the preferred selection.

SOLUTION

The contractor decides on the performance goals for the hard hat and lists them in the left-hand column of Figure 14.2. Each goal is weighted for relative importance to the overall objective. Next each candidate brand of available hard hat is listed at the column head and the evaluation process begun. Each brand of hard hat is rated from 1 to 10 according to its ability to fulfill the individual performance goals and the rating number entered in the upper right triangle of each matrix rectangle. Then the ratings are multiplied by the relative importance and the product entered into the lower left triangle of each matrix rectangle.

The products are summed vertically and the resulting sum is the utility value of each candidate brand of hard hat. The utility value is divided by the cost to find the U/C ratio and a graphical plot made as shown in Figure 14.3.

Performance goals	Relative importance of this goal to overall objectives	Brand A		Brand B		Brand C		Brand D	
		Rating importance	Rating 1 to 10	Rating importance	Rating 1 to 10	Rating importance	Rating 1 to 10	Rating importance	Rating 1 to 0
Safety	50%	450	9	300	6	350	7	350	7
Appearance	15%	30	2	135	9	45	3	75	5
Weight	15%	45	3	30	2	105	7	75	5
Comfort	20%	60	3	80	4	140	7	100	5
Total utility value	100%	Σ = 585		545		640		600	
($)Cost/each		$30.00		$22.00		$27.50		$20.00	
Utility/Cost		19.5		24.77		23.27		30.00	

Figure 14.2 One-dimensional U/C matrix for Example 14.9.

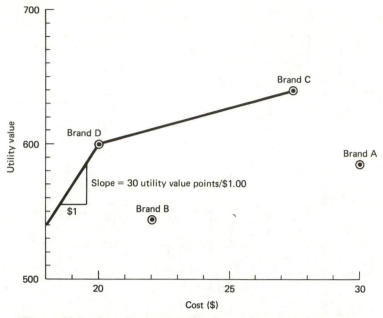

Figure 14.3 One-dimensional U/C graph for Example 14.9.

It is evident from the U/C ratio that Brand D is the better buy in terms of utility rating points per dollar expended. However, for top quality rating, an extra 40 utility points may be obtained by purchasing Brand C for an additional $7.50 over Brand D. The extra points are purchased at a rate of $40/\$7.50=5.33$ utility points per dollar, which is a far more expensive rate than the 1.5 utility points per dollar paid for Brand D. Brands B and A are obviously out of the competition since they provide fewer utility points at a higher price. Therefore, the evident choice is Brand D.

Two-Dimensional Utility/Cost Ratio

The two-dimensional matrix can be employed where a series of desirable system features contribute to achieving the performance goals, and each candidate may contain from one to all of the system features to some degree.

Example 14.10

Assume a county engineer needs a good dependable earth moving excavator that can operate with reasonable economy as a front end loader and as a trencher. The following steps are used to set up the two-dimensional utility/cost matrix.

1. Establish the primary goals and objectives of the system and enter these system goals as column headings at the top of the matrix. In

this case the system goals are:

a. Front end loading

b. Trenching

c. Economy of operation

d. Durability and dependability

2. Weight the system goals according to their importance in terms of their percent contribution to the overall objectives.

   In this example assume the county engineer's priorities are expressed as follows:

   a. Front end loading   35%

   b. Trenching   20%

   c. Economy of operation   20%

   d. Durability and dependability   25%

        Total   100%

3. Each candidate machine has certain system features which in some way contribute to the attainment of the goals listed above. These system features are listed in the left-hand column of the matrix. For the example they are listed as:

   a. Front bucket size and design

   b. Rear backhoe size and design

   c. Engine type and horsepower

   d. Quality of construction

4. Consider in turn each system goal listed at the top of each column, and in each matrix rectangle below the column head evaluate (in terms of percent) how much each system feature contributes to each system goal. Thus, for the system goal of "front end loading" the front bucket size and design will contribute heavily to achieving the

Candidate Machine A

System features	System goods	Front end loading 35%	Trenching 20%	Economy of operations 20%	Durability and dependabiltiy 25%	Total 100%
Front end bucket size and design		40	0	5	20	
Rear back hoe size and design		0	45	5	25	
Engine type and horespower		30	25	50	20	
Quality of construction		30	30	40	35	
Total		100	100	100	100	

Figure 14.4   Matrix features versus matrix goals for Machine A.

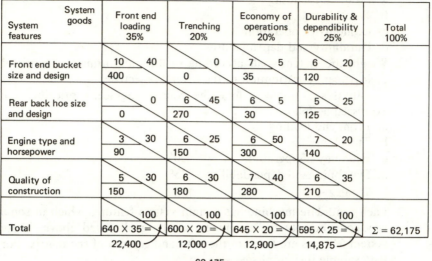

Candidate Machine A, cost = $27,000

System features	System goods	Front end loading 35%	Trenching 20%	Economy of operations 20%	Durability & dependability 25%	Total 100%
Front end bucket size and design		10 \ 40 — 400	0 — 0	7 \ 5 — 35	6 \ 20 — 120	
Rear back hoe size and design		0 — 0	6 \ 45 — 270	6 \ 5 — 30	5 \ 25 — 125	
Engine type and horsepower		3 \ 30 — 90	6 \ 25 — 150	6 \ 50 — 300	7 \ 20 — 140	
Quality of construction		5 \ 30 — 150	6 \ 30 — 180	7 \ 40 — 280	6 \ 35 — 210	
Total		100 640 × 35 = 22,400	100 600 × 20 = 12,000	100 645 × 20 = 12,900	100 595 × 25 = 14,875	$\Sigma = 62{,}175$

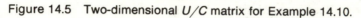

$$U/C = \frac{62{,}175}{\$27{,}000} = 2.30$$

Figure 14.5 Two-dimensional U/C matrix for Example 14.10.

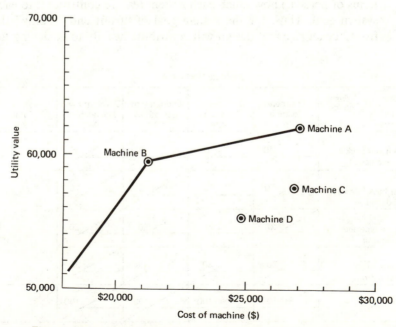

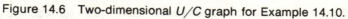

Figure 14.6 Two-dimensional U/C graph for Example 14.10.

300

goal, whereas rear backhoe size and design will have very little bearing on front end loading. The matrix now appears as shown in Figure 14.4.

5. Each candidate machine is rated (evaluated) on a basis of 1 to 10 on each of the system features using one full matrix for each candidate. Thus, if candidate machine A has an exceptionally good front end bucket size and design, it rates a 10 in the upper left triangle of the appropriate matrix rectangles, as shown in Figure 14.5.

6. The ratings found in (5) above are multiplied times the percentage found in (4) and entered into the lower left trapezoid of each matrix rectangle.

7. In each column, the products entered into the trapezoids are summed vertically and entered into the trapezoid at the bottom of the column.

8. The sums found in (7) above are multiplied by the percentage amounts found at the top of the column, the product is entered in the lower right triangle at the bottom of each column, and summed to the right-hand column for the total utility value score for this candidate machine.

9. The utility value score is divided by the cost of the machine to determine utility/cost ratio, and the utility value versus the cost are plotted on a graph with other competing candidate machines for greater visual clarity, as shown in Figure 14.6.

For Example 14.10, Candidate Machine B is the preferred selection based on the U/C ratio.

SUMMARY

This chapter presented a useful method of analysis of public works projects, called the benefit/cost analysis, in which the ratio of benefits to costs are determined on an equivalent basis. If the ratio is greater than 1.0, benefits exceed costs, and the project deserves further consideration.

In determining the ratio, problems can arise in defining benefits and costs. While benefits are widely understood to mean cost savings, accident reductions, intangible service, and so forth, to the users of the facility, and costs are widely understood to mean the cost to construct the facility, a problem arises in considering the costs of operating and maintaining the facility. Such costs can be considered as either increases in the costs (denominator) or as reductions in benefits (disbenefits) in the numerator. Thus, there are two forms of the B/C ratio.

$$\text{conventional } B/C = \frac{B_n}{C_n + M_n} \tag{14.1}$$

$$\text{modified } B/C = \frac{B_n - M_n}{C_n} \tag{14.2}$$

Use of these two formulas can result in different conclusions and the reader is cautioned to closely examine the arguments presented in the chapter concerning which formula to use. Limitations of this method are also presented, which are very important to its judicious use.

For additional reading see C. W. Smith, *Engineering Economy*, Third Edition, Iowa University Press, Iowa City, Iowa, 1979, chap. 11.

PROBLEMS FOR CHAPTER 14, BENEFIT/COST ANALYSIS

Problem Group A

Given data for alternatively continuing an existing facility or constructing a new one, formulate the conventional and/or modified B/C ratios.

A.1 A group of citizens are petitioning to have a certain gravel road paved. The county engineer works up the following estimates of cost data. $i=6$ percent, $n=15$ years.

	Existing gravel	Proposed pavement
Annual maintenance	$2,500/mi	$1,000/mi
Annual gradient increase in maintenance	$200/mi/yr	$10/mi/yr
Road user costs, annual	$10,000/mi	$8,000/mi
Annual gradient increase in road user costs	$2,000/mi/yr	$1,000/mi/yr
New capital investment required	0	$20,000/mi

In addition, there will be some safety benefits that have not been quantified.

Find the B/C by both conventional and modified methods and comment briefly, in one or two sentences, on your recommendation for which to use. Assume there is another project competing for these funds and it shows a $B/C=2.5$. Comment on which project should have priority and why. Would you recommend further study? Why or why not?

A.2 Same as Problem A.1 except $i=8$ percent and $n=25$ years.

A.3 Compare these two new alternate routes at an interest rate of 6 percent.

	EXISTING LIFE	EXISTING COST	ALTERNATE A LIFE	ALTERNATE A COST	ALTERNATE B LIFE	ALTERNATE B COST
R/W	∞	0	∞	$400,000	∞	$1,000,000
Pavement	5 yr	40,000	10 yr	100,000	20 yr	200,000
Structure	10 yr	50,000	20 yr	100,000	40 yr	500,000
Maintenance		20,000/yr		10,000/yr		50,000/yr
User costs		1,000,000/yr		970,000/yr		900,000/yr

NOTE: The annual cost of a capital asset with an infinite life is simply the cost of the interest payments.

A.4 Same as Problem A.3 except that $i=7$ percent.

A.5 A bottleneck exists on University Avenue west of 34th Street where a two-lane bridge constricts the flow of traffic. Using an $i = 8$ percent for road bond funds, assume the following data and find the modified B/C.

traffic count, year around average	20,000 vehicles/day
annual gradient in traffic count	1,000 vehicles/day/yr (second year 21,000 vehicles/day, third year 22,000 vehicles/day, etc.)
traffic mix and costs	commercial traffic, 20% @ $10.00/hr Noncommercial, 80% @ $2.00/hr

	EXISTING	PROPOSED
Cost to DOT for maintenance	$2,000/yr	$4,000/yr
Annual maintenance gradient	200/yr/yr	400/yr/yr
Travel time to pass bottleneck area	6 min	4 min
Capital cost	0	$2,000,000
Estimated life	25 yr	25 yr

A.6 A harbor dredging and enlarging project is proposed for a coastal city. If the harbor is dredged and enlarged, larger container-type ships will be able to dock at the city, resulting in savings of approximately $0.50 per ton on shipping charges. Current tonnage through the port is 1,000,000 tons per year, increasing at the rate of 50,000 tons per year. The cost of the dredging is estimated at $5,000,000. Maintenance of the current harbor now runs about $600,000 per year, increasing about $50,000 per year each year. The life of the project is estimated at 20 years. If the harbor is dredged, the maintenance will cost $800,000 per year and increase by $40,000 per year each year.
   a. Find the modified B/C ratio, assuming that bond funds are available for the project at 6 percent.
   b. Find the rate of return on the $5,000,000 investment. (Omit the information concerning bond funds at 6 percent, and find i when the PW of costs $= PW$ of savings. *Hint*: For a first approximation, try $i = 10\%$.)

A.7 Same as Problem A.6 except that current tonnage is increasing at the rate of 5 percent per year.

A.8 A new regional park facility is planned and you are asked to do (a) a benefit/cost study, and (b) a rate of return analysis. The data are:

Land: Purchase price $1,000,000. Life, perpetual. This land was bringing in a tax return of $10,000/yr which will now be lost to the community.

Improvements: Swimming pool, tennis courts, ball fields, picnic area, pavillion. Construction cost $5,000,000. Life, 40 yr. O&M costs are $400,000/yr, increasing $10,000/yr each year.

Use: It is estimated that an average of 1,000 people/day, 365 days/yr will use this facility at the outset. This figure should grow by about 60 people/day/yr (1,060/day in the second year, 1,120/day in the third year, etc.). If this proposed facility were not constructed, these people would have to travel an average of ten extra miles round trip to find a comparable park. The extra time and travel costs are estimated at $2.00/person. Therefore, the new park is assumed to save each user $2.00/visit.

a. Find the modified benefit/cost ratio if funds are available at 6 percent.

b. Find the rate of return on the capital investment of the taxpayers' funds. First, try 10 percent.

A.9 Same as Problem A.8 except the average of 1,000 people per day should grow by about 6 percent per year.

A.10 Assume that a new direct route from Sarasota to Melbourne is proposed with two alternate alignments. The data on each alternative are given in the table below. The anticipated life of the road is 25 years, with a value at the end of that time estimated at 50 percent of the cost new. Use the modified B/C to determine which is the best alternative. Use $i=7$ percent.

	EXISTING ROUTE	ALIGNMENT A	ALIGNMENT B
Cost of R/W and construction ($)	0	300,000,000	500,000,000
Annual maintenance ($)	2,000,000	8,000,000	5,000,000
User costs per vehicle trip ($)	58.52	42.37	39.16

Anticipated traffic on either of the new alignments is estimated at 4,000,000 vehicles per year the first year, increasing by 500,000 vehicles per day until reaching capacity at 10,000,000 vehicles per day. Traffic on the existing route is currently 1,200,000 vehicles per day and growing by 100,000 vehicles per day.

A.11 A new road is under consideration for connecting two municipalities in the Smoky Mountains by a more direct route. The existing route is 25 mi long and has an annual maintenance cost of $4,800 per mile. Every ten years major maintenance and resurfacing costs an additional $1,000,000, and one is due at the present time. The new direct route will be just 15 mi long and will cost $15,000,000. Annual maintenance is estimated at $3,200 per mile, and major maintenance and resurfacing every ten years is expected to cost $520,000.

The average speed on the existing route is 45 mph, whereas the new route should sustain 55 mph average. The average annual traffic is estimated at 800,000 vehicles per year of which 20 percent are commercial. Traffic is expected to increase at about 40,000 vehicles per year until reaching capacity at 2,000,000 vehicles per year. The cost of time for commercial traffic is estimated at $12.00 per hour and for noncommercial traffic at $3.00 per hour. Time charges do not include vehicle operating costs which are estimated on the existing route at $0.30 per mile for commercial traffic and $0.12 per mile for noncommercial. Operating costs on the new route should be 10 percent less per mile. The anticipated life of both proposals is 40 years, with zero salvage value for both. Using $i=7$ percent, compare the B/C using the modified and the conventional methods.

A.12 A water supply flood control project is proposed for a certain river valley in central Florida. There is a question as to the location of the dam, and the number of alternate sites has been narrowed down to two, site A and site B. Estimates of the costs associated with each of these sites are listed below. The funds to construct the project are available from bonds bearing interest at 6 percent. The anticipated life of the project at either site is 40 years, and no salvage value is expected. Use the modified B/C to reach a recommendation on which alternative is best.

	EXISTING SITUATION	SITE A	SITE B
Cost of construction	0	$3,000,000	$8,000,000
Annual O&M costs	0	56,000	94,000
Annual cost of flood damages	$620,000	290,000	120,000
Annual loss due to land lost to reservoir	0	20,000	46,000

A.13 Same as Problem A.12 except $i=7$ percent and $n=60$ years.

A.14 The federal government is planning a hydroelectric project for a river basin. The project will provide electric power, flood control, irrigation, and recreation benefits in the amounts estimated as follows.

Initial cost	$-$ $24,000,000
Power sales	$+$ $900,000/yr increasing by $50,000/yr
Flood control savings	$+$ $350,000/yr increasing by $20,000/yr
Irrigation benefits	$+$ $250,000/yr increasing by $12,000/yr
Recreation benefits	$+$ $100,000/yr increasing by $5,000/yr
Operating and maintenance costs	$-$ $200,000/yr increasing by $10,000/yr
Loss of production from lost land	$-$ $100,000/yr increasing by $5,000/yr

The dam would be built with money borrowed at 6 percent, and the life of the project is estimated as 50 years. The salvage value at the end of 50 years is estimated as $1,000,000. Find the B/C by the (a) conventional and (b) modified methods.

A.15 A university is considering building 12 additional tennis courts at a cost of $40,000 *each*. Assuming their life is 25 years, annual maintenance for each court will be $100 the first year, increasing by $20 each year thereafter, and there will be no salvage value, is this project justifiable if an average of two students use each court five hours per day for 210 days each year and a "benefit" of $1.50 is assigned to each student-hour of use? Use the *modified benefit/cost ratio analysis* and a rate of return of 6 percent.

A.16 Same as Problem A.15 except the life is 15 years and $i=8$ percent.

Problem Group B

The existing facility is no longer adequate, so two or more proposed facilities are compared.

B.1 A small city is experiencing rapid growth and has taken on a number of new city employees over the past several years in new service areas such as traffic, planning, recreation, and so on.

The one operator at the city hall manual switchboard can no longer handle all the calls and the telephone company offers the following alternatives for service for the next ten years (use $i=6\%$ and, for simplicity, use $n=10$ yr rather than 120 months).

1. Add a second operator. Each operator gets $450 per month ($5,400/yr), including fringe benefits. This salary is expected to increase by $50 per month each year for each operator ($600/yr). A second manual switchboard would cost $5,000 plus $100 per month ($1,200/yr) service charges for each switchboard. The existing switchboard service could be sold for $1,000 if not continued in service.
2. Install an automatic system for $40,000 plus $600 per month ($7,200/yr) on a ten-year service contract.

a. Find the modified B/C. (*Hint*: Consider alternative 1 as the existing system and alternative 2 as the proposed. All operating expenses are either M_p or M_f.)
b. Does the conventional B/C equation give valid results? Why?
c. Can the rate of return method be used to compare these alternatives? Why?

Problem Group C

Formulate a B/C study of your own.

C.1 Make up a hypothetical B/C study of one of the proposed new installations listed below or an equivalent of your own selection.

Assume you are a consultant to the city council. They want to know what the benefits and costs to the community will be if the proposed installation is constructed in or near your city.

Set up a hypothetical situation of your own devising. Assume any reasonable values. The main value of this assignment will come from applying your own creative imagination to determining methods of solving this very common but difficult problem. Assuming a reasonable amount of time and financing, spell out in detail how you propose to determine the value of each item in the B/C study. Find the resulting B/C.

Try to keep your solution process as simple as possible. The more complex the process gets, the less it will be understood and used.

shopping center	civic auditorium
water works	university activities center
sewage treatment plant	bond issue to build new schools
steel mill	municipal electric generating plant

Problem Group D

B/C method for company investments.

D.1 A firm is considering three alternatives as part of a production program. They are

	A	B	C
Installed cost ($)	10,000	20,000	15,000
Uniform annual benefit ($)	1,265	1,890	1,530
Useful life (yr)	10	20	30

Assuming a minimum attractive rate of return of 6 percent, which alternative, if any, would you select, based on B/C ratio method?

D.2 Two possible pollution control devices are under consideration. If a firm's minimum attractive rate of return is 10 percent, which alternative should be selected, using the B/C method?

	A	B
Purchase price ($)	720,000	1,250,000
Annual maintenance		
cost ($)	82,000	23,000
Useful life (yr)	10	10
Salvage value ($)	0	0

D.3 Estimates for alternate plans in the design of a waste water facility are

	PLAN R	PLAN S
First cost ($)	50,000	90,000
Life (yr)	20	30
Salvage value ($)	10,000	0
Annual costs ($)	11,000	5,600

For an $i=12$ percent, which plan should be selected, using the B/C method?

D.4 Three alternative proposals are available for investment. Using the incremental B/C method, determine which alternative should be selected, assuming $i=20$ percent and $n=15$ years for all alternatives.

	T	U	V
Initial investment ($)	12,000	24,000	16,000
Annual net income ($)	2,300	5,500	3,600
Salvage value ($)	0	0	0

Part III
PROJECT INVESTMENT EVALUATIONS

Chapter 15
Taxes

KEY EXPRESSION IN THIS CHAPTER

IRS= The Internal Revenue Service, the agency of the federal government
responsible for collecting the income tax.

Most thinking persons appreciate such taxpayer-government sponsored
services as schools and colleges, national defense, law enforcement, fire
protection, construction and operation of roads, parks, and so on. The
government charges us for these services through various kinds of taxes,
including property, sales, and income taxes. For the investor these taxes are
a necessary part of the cost of doing business and must be accounted for
when estimating costs versus income on any investment.

SOURCES OF TAX REVENUE

Traditionally the greater part of the revenues needed to support the various
levels of government has come from three major sources.

LEVEL OF GOVERNMENT	MAJOR SOURCE OF REVENUE
Local (county, city, schools)	Property ad valorem tax
State	Sales tax
Federal	Income tax

In addition to these three prominent tax sources, legislators have devised a number of additional means (excise taxes, license fees, franchise taxes, inheritance taxes, etc.) for raising the revenue needed to fund the government. The main criteria for evaluating a potential tax source is generally:

1. The cost of collection should be low. (For example, the cost of collecting income tax is less than 2% of revenue received, whereas the cost of collecting a $0.10 toll at a bridge could easily exceed one-half of the revenues.)
2. The amount of tax collected should relate to the ability to pay. (For example, the ad valorem property tax relates to the value of the property owned, sales tax relates to purchasing power, and the graduated income tax is proportional to income.)

Millage

mill rate (levy) set by legislative body
county assessor assigns worth

Property is usually taxed on the basis of millage. One mill equals one tenth of one cent (1 mill = 0.1 cent), and the tax rate expressed in millage signifies the number of mills the tax collector expects to collect on each dollar of assessed property value. For instance, if the property is valued by the tax assessor at $30,000 (or, "the assessed valuation is $30,000") and the millage rate is 25 mills, then the tax owed will be

$$\$30,000 \times (25 \text{ mills}/\$1 \text{ valuation}) = 750,000 \text{ mills}$$

Since 1,000 mills = $1, the tax bill is $750.

The county property appraiser (tax assessor) sets a value on each piece of property in the county, while the local governing board sets the millage rate annually. The county tax collector then sends out bills and collects taxes.

Property Taxes

Most local governments (county, municipality, school board) depend on property taxes as an important source of revenue. Property taxes are usually assessed in proportion to the value of the property and thus are termed "ad valorem" taxes. Basically there are two types of property for tax purposes.

1. *Real property* consists of relatively unmovable property such as land, buildings, and fixed attachments thereto. It is subject to annual ad valorem tax usually by local government only (city, county, schools).
2. *Personal property* consists of any kind of readily movable property. It is subdivided into (a) tangible, and (b) intangible.
   a. *Tangible personal property* is any readily movable property that has inherent value (rather than being a piece of paper representing value), for example, a bulldozer or other movable construction equipment, a car, clothing, jewelry, and appliances that are

not fixed to a building. A kitchen range that is free-standing would usually be classified as personal property, whereas a built-in range is real property. Personal property held for income producing purposes is usually taxed by local government, but personal property held for private use only often is exempted or ignored by taxing authorities.

b. *Intangible property* consists largely of paper that represents a right or title to the value inherent in some other object, property, or enterprise, for example, money, bonds, stocks, mortgages, notes, accounts receivable, and other certificates of indebtedness or promises to pay. Intangible property is often subject to a one-time or annual ad valorem state or local tax, but usually at a very low rate. (For example, Florida tax rate varies from 0.1 mill to 1 mill on annual levies, and is 2 mills on nonrecurring assessments such as on mortgages.)

Sales Taxes

The largest single source of income for state government in about two-thirds of the states is the sales tax. This usually consists of a fixed percentage of tax on retail sales, collected by the retailer at the time of sale to the customer, and forwarded periodically to the state treasury. The amount typically ranges from 3 to 5 percent. Some cities and counties are empowered by their respective states to levy an additional sales tax, and have done so.

Income Tax

The federal government derives most of its revenue from income taxes, as illustrated in Figure 15.1.

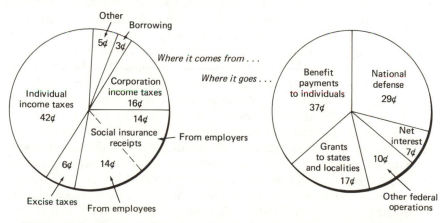

Figure 15.1 Federal revenue and expenditures. (SOURCE: U.S. Treasury Department)

criterin
1. collection costs low
2. Ability to pay

314 PROJECT INVESTMENT EVALUATIONS

FEDERAL INCOME TAX LAW

Federal tax law originates in the U.S. House of Representatives, which is a body of 435 individuals each with his or her own ideas of what fair and proper tax law should be. The law begins as a proposal or bill in a committee where members with somewhat different viewpoints reach as equitable a compromise as possible. The bill then may go through other committees, and finally to the floor. At each point some additional input, revisions, and compromises may occur until finally the bill is in a form satisfactory to a majority of the members of both houses (including the Senate with another 100 members) at which point it is voted upon, passes, and upon approval by the president becomes law. Obviously it cannot be expected that the resulting document will be an epitome of clarity, nor be without some ambiguity. After the bill becomes law, it is the responsibility of the Internal Revenue Service (IRS) to interpret the law and administer it. In doing this they have additional input from the justice department, and from the courts. Where taxpayers disagree with the interpretation given by the IRS, the matter can be taken to the courts, where a binding decision is made either upholding or overturning the IRS interpretation. The IRS of course takes note of the findings of the courts in any future interpretations of the tax law. Since tax law does become rather involved (there are over 10,000 pages of current federal regulations regarding income taxes), it is recommended that specialists be consulted when important investment decisions with tax ramifications are involved.

Partners with the Federal Government

In a real sense the federal government is a partner in every taxpaying business. When the business prospers the government shares in that prosperity, since often over 50% of the profit is paid to the government in taxes. Conversely, when business loses a dollar, the government loses over $0.50 of potential income. Current (1982) income tax rates for most corporations now are at 46 percent on all corporate net income over $100,000. (Some corporations with less than 15 stockholders may qualify as Subchapter S Corporations and be taxed once as individuals rather than as corporations.) Thus a large corporation pays an income tax of approximately 46 percent on net income. After income taxes and all other expenses for the year are met, the remaining profit may be invested in new plant and equipment or distributed to the stockholders as dividends. Many firms invest about one-half the profit in expansion and distribute the remaining one-half in dividends. When the dividends are received by the stockholders they may be taxed a second time as personal income. (As a minor concession to this double taxation, the first $100 of dividend income per individual is currently income tax free.) If the individual stockholder is in the 40 percent tax bracket, the corporate net income dollar, already reduced to $0.54 by the corporate tax of 46 percent, is further reduced by a personal income tax of (40%×$0.54=) $0.216. Thus the stockholder-owner of the corporation gets to keep about ($1−$0.46−$0.216=) $0.32 out of every corporate dollar. The federal government by comparison receives $0.68. Thus it is obviously in the interest of government itself, as well as the national interest, for Congress to pass laws aimed at creating a climate of prosperity for American business and industry.

Investment Tax Credit

Tax law is one very powerful tool that modern governments use to help stabilize and stimulate the national economy. Experience teaches that gains in national productivity are necessary for sustained long-term increases in standards of living. One key to increasing productivity is to continually update and modernize productive plant and equipment. Many governments in progressive nations around the world now provide tax incentives to business to encourage continual modernization of productive capacity. With this concept in mind the U.S. Congress passed an investment tax credit provision in 1962 allowing a percentage of the first cost of certain qualifying categories of investment in plant and equipment as a direct deduction from the amount of federal income tax owed. For instance the current investment tax credit allowance of 10 percent on a qualifying investment costing $100,000 results in a *reduction* of taxes owed by $10,000, but only in one year. The amount of tax credit allowed changes from time to time, depending upon whether the government feels the economy needs stimulation or restraint. For instance in 1966 the investment credit was suspended entirely, and then restored in 1967, all according to the government's interpretation of national needs.

Capital Gains

When an asset is sold, part of the money received usually represents the original purchase price, plus any other expenses involved in the costs of purchase and ownership. If the selling price is higher than these costs, the balance represents a profit, and if the selling price is lower, then the difference represents a loss. Profits on qualified assets held longer than one year may be considered as long-term capital gains and 60 percent of the gain may be excluded from taxable income. For example, assume an investor, in the 30 percent tax bracket, purchases land for $10,000. Two years later he sells the land for $12,000, for a capital gain of $2,000. Only 40 percent of the gain is taxed at the investor's ordinary income tax level of 30 percent.

Thus,

selling price	=$12,000
purchase price	= 10,000
capital gain	=$ 2,000

40% of capital gain reported as ordinary income

$$0.4 \times \$2,000 \qquad\qquad = \$800$$

income tax due @ 30%, $0.3 \times \$800 = \$240$ tax

If the profit results from the sale of an asset held for less than one year, the profit is a short-term capital gain and is taxed at the same rate as ordinary income.

Exceptions to the capital gains treatment include assets held as stock for regular business. Investors in real estate sometimes find that profit from sale of real estate held more than one year is ruled taxable as ordinary income, if the IRS has reason to regard them as dealers, and the land as their stock in trade. This may occur where an investor makes a regular habit of buying and selling real estate, or buys large tracts and subdivides for resale.

For the convenience of the student, the federal income tax is usually shown in textbooks as a percentage of net pretax income, where net income is the income left over after subtraction of allowable expenses and deductions. Allowable expenses usually include all normal business expenses such as interest payments, payroll, other taxes, and depreciation allowances.

In actual practice taxpayers usually pay a fixed amount based on their income strata plus a percentage of any net income overlapping into the next higher strata. Thus the percentage used in textbook examples and problems represents the marginal tax bracket, or the percent tax on each *additional* dollar earned in the highest strata attained by the taxpayer in question. For instance, using the slightly simplified textbook approach, a taxpayer in the 30 percent tax bracket with a gross income of $50,000 and allowable expenses and deductions of $30,000 would pay a tax of

$$(\$50,000 - \$30,000) \times 0.30 = \$6,000 \text{ income tax}$$

The tax bracket percentage rises by steps as the amount of net income increases, so that taxpayers with $30,000 net incomes normally find themselves in a higher percentage tax bracket than those with $20,000 net incomes.

Similarly, corporations pay income taxes on a graduated scale. The Internal Revenue Service (IRS) schedule for corporations is

| TAX RATE | | TAXABLE INCOME |
1982	1983	
16%	15%	First $25,000
19%	18%	Second 25,000
30%	30%	Third 25,000
40%	40%	Fourth 25,000
46%	46%	All over $100,000

In addition to the income tax paid to the federal government, about 44 out of 50 states and over 40 cities levy income taxes also, although the rate is usually much lower than the federal income tax rate.

Personal Income Taxes

Taxable income is defined as earned income minus adjustments and minus allowable deductions. The adjustments and allowable deductions depend primarily upon the type of entity (person or corporation).

Key Expressions Used in This Section

AGI = Adjusted gross income
AD = Allowable deductions
PE = Personal exemptions
TI = Taxable income
EI = Earned income
AJ = Adjustments (deductions from earned income)
ITC = Investment tax credit

Earned income (EI) of individuals includes all wages, salaries, tips, and anything else of value (money, goods, or services) received from an employer or through self-employment for services performed. Subtracted from this income are a number of items, termed adjustments (AJ), which include such items as moving expenses in connection with employment, employee business expenses not reimbursed by an employer, and payments to a retirement plan. After subtraction the amount remaining is termed adjusted gross income (AGI). From the AGI is subtracted personal exemptions of $1,000 per person ($PE$) plus either the standard deduction (automatically figured into Internal Revenue Service (IRS) Tax Tables A through D in their *Instructions for Preparing Form 1040*) or itemized allowable deductions (AD) (includes excess medical expenses, certain state and local tax expenses, interest expenses, contributions, normal business expenses,

1980 Tax Rate Schedules

If you cannot use one of the Tax Tables, figure your tax on the amount on Schedule TC, Part I, line 3, by using the appropriate Tax Rate Schedule on this page. Enter the tax on Schedule TC, Part I, line 4.

Note: Your zero bracket amount has been built into these Tax Rate Schedules.

SCHEDULE X—Single Taxpayers

Use this schedule if you checked Filing Status Box 1 on Form 1040—

If the amount on Schedule TC, Part I, line 3, is:

Not over $2,300....... 0

Enter on Schedule TC, Part I, line 4:

Over—	But not over—		of the amount over—
$2,300	$3,400	14%	$2,300
$3,400	$4,400	$154+16%	$3,400
$4,400	$6,500	$314+18%	$4,400
$6,500	$8,500	$692+19%	$6,500
$8,500	$10,800	$1,072+21%	$8,500
$10,800	$12,900	$1,555+24%	$10,800
$12,900	$15,000	$2,059+26%	$12,900
$15,000	$18,200	$2,605+30%	$15,000
$18,200	$23,500	$3,565+34%	$18,200
$23,500	$28,800	$5,367+39%	$23,500
$28,800	$34,100	$7,434+44%	$28,800
$34,100	$41,500	$9,766+49%	$34,100
$41,500	$55,300	$13,392+55%	$41,500
$55,300	$81,800	$20,982+63%	$55,300
$81,800	$108,300	$37,677+68%	$81,800
$108,300	....	$55,697+70%	$108,300

SCHEDULE Y—Married Taxpayers and Qualifying Widows and Widowers

Married Filing Joint Returns and Qualifying Widows and Widowers

Use this schedule if you checked Filing Status Box 2 or 5 on Form 1040—

If the amount on Schedule TC, Part I, line 3, is:

Not over $3,400....... 0

Enter on Schedule TC, Part I, line 4:

Over—	But not over—		of the amount over—
$3,400	$5,500	14%	$3,400
$5,500	$7,600	$294+16%	$5,500
$7,600	$11,900	$630+18%	$7,600
$11,900	$16,000	$1,404+21%	$11,900
$16,000	$20,200	$2,265+24%	$16,000
$20,200	$24,600	$3,273+28%	$20,200
$24,600	$29,900	$4,505+32%	$24,600
$29,900	$35,200	$6,201+37%	$29,900
$35,200	$45,800	$8,162+43%	$35,200
$45,800	$60,000	$12,720+49%	$45,800
$60,000	$85,600	$19,678+54%	$60,000
$85,600	$109,400	$33,502+59%	$85,600
$109,400	$162,400	$47,544+64%	$109,400
$162,400	$215,400	$81,464+68%	$162,400
$215,400	....	$117,504+70%	$215,400

Married Filing Separate Returns

Use this schedule if you checked Filing Status Box 3 on Form 1040—

If the amount on Schedule TC, Part I, line 3, is:

Not over $1,700....... 0

Enter on Schedule TC, Part I, line 4:

Over—	But not over—		of the amount over—
$1,700	$2,750	14%	$1,700
$2,750	$3,800	$147.00+16%	$2,750
$3,800	$5,950	$315.00+18%	$3,800
$5,950	$8,000	$702.00+21%	$5,950
$8,000	$10,100	$1,132.50+24%	$8,000
$10,100	$12,300	$1,636.50+28%	$10,100
$12,300	$14,950	$2,252.50+32%	$12,300
$14,950	$17,600	$3,100.50+37%	$14,950
$17,600	$22,900	$4,081.00+43%	$17,600
$22,900	$30,000	$6,360.00+49%	$22,900
$30,000	$42,800	$9,839.00+54%	$30,000
$42,800	$54,700	$16,751.00+59%	$42,800
$54,700	$81,200	$23,772.00+64%	$54,700
$81,200	$107,700	$40,732.00+68%	$81,200
$107,700	....	$58,752.00+70%	$107,700

SCHEDULE Z—Heads of Household (including certain married persons who live apart (and abandoned spouses)—see page 6 of the Instructions)

Use this schedule if you checked Filing Status Box 4 on Form 1040—

If the amount on Schedule TC, Part I, line 3, is:

Not over $2,300....... 0

Enter on Schedule TC, Part I, line 4:

Over—	But not over—		of the amount over—
$2,300	$4,400	14%	$2,300
$4,400	$6,500	$294+16%	$4,400
$6,500	$8,700	$630+18%	$6,500
$8,700	$11,800	$1,026+22%	$8,700
$11,800	$15,000	$1,708+24%	$11,800
$15,000	$18,200	$2,476+26%	$15,000
$18,200	$23,500	$3,308+31%	$18,200
$23,500	$28,800	$4,951+36%	$23,500
$28,800	$34,100	$6,859+42%	$28,800
$34,100	$44,700	$9,085+46%	$34,100
$44,700	$60,600	$13,961+54%	$44,700
$60,600	$81,800	$22,547+59%	$60,600
$81,800	$108,300	$35,055+63%	$81,800
$108,300	$161,300	$51,750+68%	$108,300
$161,300	....	$87,790+70%	$161,300

Figure 15.2 Tax rate schedules from *1980 Instructions for Preparing Form 1040.* (SOURCE: U.S. Treasury Department, Internal Revenue Service)

etc.). The remainder is termed taxable income (*TI*) and was taxed in 1980 in accordance with Figure 15.2. Thus, the taxable income for individuals is:

$$TI - AGI - PE - AD \tag{15.1}$$

where

$$AGI = EI - AJ \tag{15.2}$$

The following example illustrates the use of the table.

Example 15.1

A married couple with two children filing a joint income earned $42,754 from wages and salaries. In addition the husband received honorariums of $2,500 from making after-dinner speeches. During the year they changed employment and spent $1,532 in moving expenses of which the employer reimbursed them $1,000. Itemized allowable deductions for the year totaled $6,532. How much federal income tax do they owe?

SOLUTION

$$
\begin{array}{lll}
EI(42,754 + 2,500) & = & \$45,254 \\
AJ(1,532 - 1,000) & = - & 532 \\
AGI & = & 44,722 \\
\text{exemptions } (4 \times 1,000) & = - & 4,000 \\
AD & = - & 6,532 \\
TI & = & \$34,190
\end{array}
$$

From Schedule Y (Figure 15.2)

$$\text{tax} = 6,201 + 0.37(34,190 - 29,900)$$

$$= 6,201 + 1,587 = \underline{\underline{\$ \ 7,788}}$$

Corporate Income Taxes

Corporations, like individuals, must pay income tax on profits they earn, on graduated scale. Their taxable income (*TI*) is determined by taking total earned income (*EI*) and reducing it by allowable deductions:

$$TI = EI - AD \tag{15.3}$$

Allowable deductions include a much greater range of expenses than allowed to individuals, consisting of such items as all normal costs incurred in doing business, depreciation on capital items (Chapter 16), depletion allowance on consumption of natural resources, research and development expenditures, advertisement costs, and interest costs on debt (to name only a few).

Effective Tax Rate

Since the tax rate of both individuals and corporations increases on a graduated scale, the question arises as to what tax rate should be used when analyzing a proposed investment. There are two basic approaches to this question.

• *Marginal or Incremental Tax Rate.* One approach is to treat the return from the proposed investment as income added on top of present income and thus subject to tax at the highest marginal or incremental rate applicable. For example, if a single taxpayer already has a *TI* of $29,000, then every $1 of added income from the new investment will be taxed at the 44 percent rate (Figure 15.2), unless the profit from the investment increases the *TI* to more than $34,100 and then the tax rate goes even higher.

• *Average Tax Rate.* The other approach is to use an income tax rate that is equal to the average tax rate for the total *TI* including the added new income. For example, the single taxpayer earning $29,000 (*TI*) would pay income taxes of $7,434+0.44 times his incremental income over $28,800. If he anticipates an additional *TI* of $1,000 from the proposed investment then his average tax rate is calculated as

present *TI*	$29,000
proposed *TI*	+ 1,000
new total *TI*	$30,000
next lower IRS tax step (Figure 15.2)	− 28,800
increment of *TI* above step	1,200
incremental tax rate (Figure 15.2)	× 0.44
incremental tax	528
IRS base schedule tax (Figure 15.2)	+ 7,434
total income tax	= $ 7,960

$$\text{average tax rate} = \frac{\text{total tax}}{\text{taxable income}} = \frac{7,960}{30,000} = \underline{\underline{27\%}}$$

EFFECT OF TAXES ON CASH FLOW

When evaluating the feasibility of any proposal subject to taxes, the taxes may be treated as just another expense of doing business.

Example 15.2

Assume an investor has an opportunity to invest $10,000 for five years in a proposal estimated to yield the net cash flow *before* taxes, illustrated in Figure 15.3.

Since the $10,000 principal is returned intact at EOY 5, the *before*-tax rate of return is obviously

$1,000/$10,000 = 10\% \ ROR$

However, if the investor is required to pay taxes on the income, and she is in the 40 percent tax bracket, she will pay the government $0.40 of each $1 of income or interest earned by this investment. The $10,000 invested at EOY 0 is *not* taxed upon withdrawal from the investment at EOY 5, since this is old investment capital returned and not new earned income or interest. The after-tax cash flow diagram now appears as in Figure 15.4.

The after-tax rate of return now is

(annual income − income tax)/(intact principal) = *ROR*

or,

($1,000 − 0.40 × $1,000)/$10,000 = 6\% *ROR* after taxes

Notice that the numerator (annual income − income tax) can be expressed more simply as [income × (1 − tax rate)], or in this case [$1,000 × (1 − 0.4)], or $1,000 × 0.6.

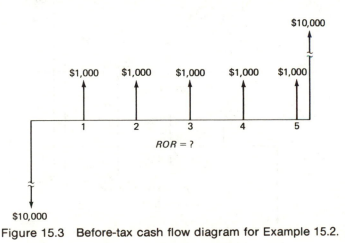

Figure 15.3   Before-tax cash flow diagram for Example 15.2.

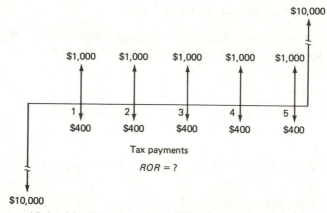

Figure 15.4   After-tax cash flow diagram for Example 15.2.

For the more common cases where the original investment is *not* returned intact, the after-tax rate of return is calculated as in previous rate of return problems with but one simple innovation; that is, the taxes are now calculated and added on as a new cost item. An example involving taxes follows.

Example 15.3

A client proposes a project with estimated net cash flows illustrated in Figure 15.5, and asks you to calculate the after-tax rate of return. He is in the 40 percent tax bracket (of every $1 of net income, $0.40 is paid in taxes). For this investment assume that the $2,000 profit from resale is *not* eligible for taxation at the lower capital gains rate discussed later.

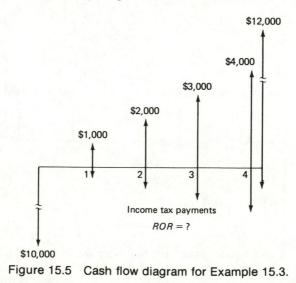

Figure 15.5   Cash flow diagram for Example 15.3.

SOLUTION

All the annual net incomes are taxable at 40 percent, so the client retains $0.60 out of every $1 of income after taxes. At EOY 4, $12,000 of principal is returned where only $10,000 was invested. Therefore only the extra $2,000 (from $12,000−$10,000=$2,000) is taxable at 40 percent, and the net after-tax amount received at EOY 4 is $10,000+$2,000×0.6=$11,200. Thus the after-tax net cash flow diagram will appear as in Figure 15.6, with after-tax incomes equal to 0.6 times before-tax incomes.

The after-tax rate of return is determined as in previous chapters by formulating present worth equations and substituting trial values of i until the value of i is found at which the total present worth equals zero.

		15%	20%
$P_1=-\$10,000$	$=-\$10,000$	$-\$10,000$
$P_2=+(1,000-400)(P/G,i,5)(F/P,i,1)=+$	3,984.89	$+$ 3,532.46
$P_3=+(12,000-800)(P/F,i,4)$	$=+$ 6,403.71	$+$ 5,401.31
	$+$ 388.60	$-$ 1,066.23

$$i=15\%+5\%\times\frac{388.60}{388.60+1,066.23} \qquad = \qquad 16.3\% \text{ after tax } ROR$$

After-Tax Revenues

For every $1 of taxable income received, the government wants a certain percentage by way of taxes. If the taxpayer is in the 30 percent tax bracket, he gets to keep $0.70 out of every $1 of income taxable at that rate and pays the other $0.30 in taxes. Thus when entering such revenues into time-value

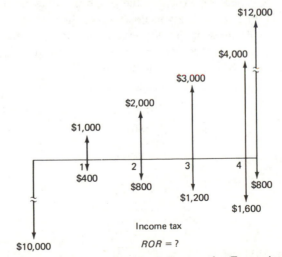

Figure 15.6   After-tax cash flow diagram for Example 15.3.

calculations, simply multiply the pretax revenue by (1 − tax rate) in order to obtain the after-tax revenue.

Example 15.4

Find the after-tax present worth of taxable revenues of $10,000 per year for ten years with $i = 10$ percent. The taxpayer is in the 30 percent tax bracket.

SOLUTION

Find the after-tax cash flow as

after-tax cash flow = taxable income × (1 − tax rate)

after-tax cash flow = $10,000/yr × (1 − 0.3)

after-tax cash flow = $10,000/yr × 0.7

then P = $10,000 × 0.7 ($P/A$, 10%, 10) = $43,010
            6.1446

After-Tax Expenses

The government recognizes legitimate business expenses and does not tax revenue dollars that are used to pay these expenses. For instance, if a consulting engineer receives gross revenues of $50,000 per year, but has expenses of $20,000 per year, then the government only taxes the $30,000 net income after expenses.

For every $1 spent on expenses the engineer saves the tax rate times that $1. Thus every dollar of expense only costs the consultant $1 × (1 − tax rate). If the consultant is in the 30 percent tax bracket and takes several clients out to a business lunch and spends $30 as a business expense, it only costs him $30 × 0.7 = $21. Here is how it works.

	BEFORE LUNCH	AFTER $30 LUNCH
Gross income	$50,000	$50,000
Expenses	− 20,000	− 20,030
Net income before taxes	30,000	29,970
Tax bracket	× 0.3	× 0.3
Income tax	9,000	8,991
After-tax income		
Before lunch	$21,000	
After lunch	− 20,979	$20,979
After-tax cost of $30 lunch	$ 21	

Thus the $30 lunch actually reduced the consultant's after-tax income by only $21, and it is evident that the government shares both in our income as well as our business expenses required to earn that income.

SUMMARY

Income tax and other tax effects on investments are significant and should be considered carefully during feasibility studies. Thus, the engineer should keep abreast of current trends in federal, state, and local tax laws and their affect on economic analysis of engineering projects. In a real sense the federal government is a silent partner, sharing in a company's profits and easing the burden of some expenses. Entities subject to income taxes are persons and corporations. Taxes are assessed as a percentage of taxable income on a graduated basis; the higher the taxable income the higher the tax rate. Taxes reduce the income received from investments and thus significantly affect cash flow. On a profitable investment the after-tax rate of return will be less than the before-tax rate of return.

There are three special provisions of current tax law which significantly influence the cash flow. They are (1) depreciation (covered in the next chapter), (2) investment tax credit, and (3) capital gain.

The specific tax rates and examples contained in this text should be used as guidelines *only*, as the laws as well as the interpretations thereof are subject to change.

PROBLEMS FOR CHAPTER 15, TAXES

Problem Group A

Find taxes or tax rate.

A.1 John and Mary have one child and earned $24,762 during the year. They had allowable deductions of $5,680 and use the tax schedule shown in this chapter.
   a. How much taxes do they owe?
   b. What tax bracket are they in?
   c. What is their average tax rate?

A.2 A corporation estimates its total income this year will be $1,750,000 of which $1,432,566 will be allowable deductions and are subject to the corporate tax schedule shown in this chapter.
   a. How much taxes will they owe?
   b. What is their average tax rate?

A.3 The corporation in Problem A.2 is considering a major capital investment of $175,000, which qualifies for an *ITC* of 10 percent. If they purchase this investment how much will their taxes be reduced?

A.4 A major corporation (more than $100,000 per year in profits) purchased some common stock in another company four years ago for $2,758,500. They now have an opportunity to sell the stock for $5,752,400. If they sell how much income tax will they pay on the capital gain?

Problem Group B

Find the before-tax and after-tax rates of return.

B.1 In Problem A.4 what are the before-tax and after-tax rates of return over the four-year holding period on the sale of the stock, and assuming no dividends have been paid?

B.2 An investor buys a corporate bond for $1,000. The bond pays $100 cash interest per year for ten years. At EOY 10 the bond matures and the $1,000 investment is refunded. The investor pays an annual income tax of 25 percent on the interest income and an annual intangible tax of 0.1 percent on the $1,000 face value of the bond. What is his after-tax rate of return?

B.3 Assume you purchased a lot five years ago for $2,000 and now have an offer to sell for $10,000. You have been paying annual real estate taxes at 30 mills on assessed value of $1,500. Assume real estate taxes are deductible from ordinary income at the end of the year in which paid. Assume you are in the 42 percent tax bracket. If you sell, you will pay 42 percent income tax on 40 percent of the capital gain (sale price less cost). What is the rate of return after taxes?

B.4 A client of yours (who is in the 42 percent tax bracket) purchased 80 acres for $80,000 five years ago. Her expenses have averaged $2,000 per year and have been listed as tax deductible items in the year in which they occurred. Therefore, they have not added to the book value (tax basis) of the land and the actual cost to the client was [expense × (1 − tax rate)]. Your client now has two offers from people who would like to buy the land. Offer A is for the entire 80 acres for $150,000. The $70,000 profit would be taxable as capital gains, therefore she pays a 42 percent tax on 40 percent of the capital gain.

Offer B involves subdividing the 80 acres into 5-acre lots and selling for $3,000 per acre. Your client is advised that subdividing and selling will classify her as a dealer and the profit is taxable as ordinary income. Find the after-tax rate of return for each offer.

B.5 You are in the 42 percent tax bracket and have $10,000 to invest. Tax free municipal bonds maturing in 20 years are available with 6 percent coupons (coupons paying 6% of the face value annually). One such bond with a face value of $10,000 can be purchased for $9,200.

An alternative $10,000 corporate bond (income taxable) matures in 20 years with 7 percent coupons. It can be purchased for $9,000. The annual coupon income is taxable at ordinary income rates and the difference between the $9,000 purchase price and $10,000 redemption value is taxable as capital gain at the time of redemption. Which is the better investment if both bonds are of equal quality or risk? Assume 40 percent of capital gains is taxed at the ordinary income tax rate.

B.6 Your client is in the 35 percent tax bracket and is considering an investment in a proposed project whose anticipated cash flow is tabulated below.

EOY	EXPENDITURE	INCOME
0 purchase land	100,000	
1 construction	1,000,000	
2 operating expenses and income	200,000	350,000
3 operating expenses and income	220,000	375,000
4 operating expenses and income	240,000	400,000
5 operating expenses and income	260,000	425,000
6 operating expenses and income	280,000	450,000
6 resale		1,500,000

Assume there is no allowable depreciation taken on this project. The capital investment consists of the expenditures for "purchase land" and "construction," and the difference between these expenditures and the "resale" income is taxed as capital gain. The client asks you to determine if the proposed project meets his *MARR* objective of 15 percent after taxes.

B.7 You are offered the opportunity of investing in a new engineering office supply business. For an investment of $10,000 you are told you can expect to receive $15,000 at the end of four years.

a. What before-tax rate of return will you receive on this investment?

b. Assuming the entire profit on the investment will be treated as ordinary income and you expect to have a taxable income of $38,000 four years from now (not counting the $5,000 profit you will receive that year), what after-tax rate of return will you receive on the investment?

B.8 As an alternative to receiving $15,000 in one lump sum payment as stated in Problem B.7, you are offered $3,500 for each of four years (for a total of $14,000 over the four years). Each year, if you count $3,500 as payment on the principal and $1,000 as earned income, and your taxable yearly income (without this additional $1,000) is expected to be constant at $38,000, what after-tax rate of return will you receive on this investment?

B.9 Compare the two alternative proposals in Problems B.7 and B.8 and discuss intangible factors that could affect your decision as to which alternative you would prefer.

Chapter 16
Depreciation

KEY EXPRESSIONS IN THIS CHAPTER

P = Purchase price of depreciable asset = book value at time zero.

N = Depreciable life of the asset.

F = Estimated salvage value of asset at time N.

m = Age of asset at time of calculation.

D_m = Annual depreciation amount taken for year m.

R_m = Depreciation rate for year m, expressed as a decimal value.

R = Depreciation rate for declining balance depreciation method.

BV_m = Book value of asset at a particular age, m, ($BV_0 = P$)

SL = Straight line method of depreciation.

SOY = Sum of years. Sum of the ordinal digits for years 1 through N.

$SOYD$ = Sum-of-the-years digits method of depreciation accounting.

DB = Declining balance method of depreciation accounting.

DDB = Double declining balance method of depreciation accounting.

DEPRECIATION: A LOSS IN VALUE

Depreciation is the loss in value of a property such as a machine, building, vehicle, or other investment over a period of time, caused by one or more of the following.

1. Unrepaired wear.
2. Deterioration.
3. Obsolescence.
4. Reduced demand.

1. Wear accumulates as a function of hours of use, severity of use, and level of preventive maintenance. Each hour of wear on moving parts of an engine, for instance, brings them closer to the point of requiring repair or salvage. From time to time, decisions are made whether or not to invest in repairs, and to what extent. The investment in repair is economically justifiable only if the value added by the repair exceeds the cost of the repair. When the engine reaches the point in its life where the cost of a needed repair exceeds the value added to the engine, it is typically sold for whatever it will bring, or junked.
2. Deterioration, the gradual decay, corrosion, or erosion of the property, occurs as a function of time and severity of exposure conditions. It is similar to wear depreciation in many ways, except it occurs whether or not the property has moving parts in actual use. Deterioration can usually be controlled by good maintenance. Many structures that are hundreds of years old (in a few instances, one or two thousand years old) are still in usable condition, usually due to a consistent level of good maintenance.
3. Obsolescence depreciation is the reduction in value and marketability due to competition from newer and/or more productive models. Obsolescence can be subdivided into two types: (a) technological and (b) style and taste.
   a. Technological obsolescence can be measured in terms of productivity. Over the short term, technological obsolescence has typically occurred at a fairly constant rate. Earthmoving equipment, for instance, has typically increased in productivity at the rate of about 5 percent per year over the last 20 years, according to the manufacturers. In theory however, sudden drastic breakthroughs are possible in any area. For instance, in the construction industry, invention of a new cheap building material, light, strong, and easily workable could make billions of dollars worth of existing manufacturing facilities obsolete with commensurate rapid depreciation as a result. As a practical matter technological progress has historically come gradually at a fairly constant rate, with a few notable exceptions.

  b. Style obsolescence depreciation occurs as a function of customers' taste. This is much less predictable, although just as real in terms of lost value. As styles change, both men and women typically refuse to purchase clothing, cars, houses, and so forth, that do not incorporate features considered "up-to-date." The "split level" house was "in" during the 1940s along with tail-finned cars, but in the 80s consumers prefer newer styling. Styles and tastes are to some extent influenced by large expenditures of funds for advertising and PR, although a notable exception was the failure of a large advertising budget to influence the public's taste in the case of the Edsel.

4. Reduced demand normally results in a decreased value (increased depreciation) of an asset in accordance with the marketplace "law" of supply and demand. Reduced demand results from a number of factors, such as a general business recession, change in development patterns, consumer habits, or even changes in weather patterns. The onset of a prolonged dry weather cycle, for example, may reduce the demand for drainage facilities and enhance the demand for irrigation, resulting in demand-induced changes in depreciation. The effect of the combined types of depreciation is shown graphically in Figure 16.1. The typical pattern is a gradually decreasing value, interrupted by jogs indicating decisions to invest in repairs, overhaul, or renovations. For each such investment the value added is presumed to exceed the cost. Finally, there comes a time when the cost exceeds the value added and the item is junked, sold, or abandoned.

MEASURING DEPRECIATION

Since depreciation is the loss in value over a period of time it may be traced by simply graphing the market value as a function of time. In determining

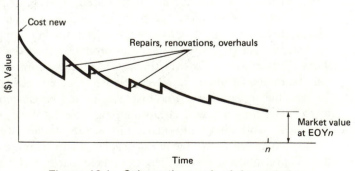

Figure 16.1 Schematic graph of depreciation.

market value, several variables are usually encountered. For instance, in the automobile market there is one market price for retail and one for wholesale, with other variables affected by mileage, condition, location, and so forth.

As an example, a graph of values of a typical automobile depreciating with time appears in Figure 16.2.

At times a vehicle or structure may *appreciate* rather than depreciate. When the October 1973 oil embargo was imposed, many makes and models of fuel-saving small used cars rose several hundred dollars in value. Buildings typically appreciate in value over the prime years of their lives due to the rising cost of construction of their newer competitors. Therefore, the actual depreciation (or appreciation) is the actual change in market value. This can and frequently does follow an erratic course. To simplify the bookkeeping involved, several methods have been devised based upon the premise that values of all depreciable properties will decline from the new purchase price to salvage sooner or later.

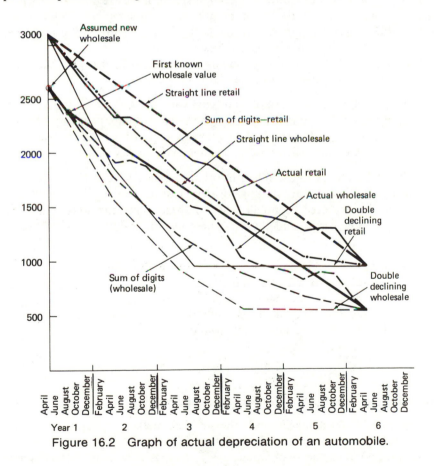

Figure 16.2 Graph of actual depreciation of an automobile.

DEPRECIATION ACCOUNTING

Depreciation accounting is the systematic division of the depreciable value of a capital investment into annual allocations over a period of years. There are two basic reasons for needing depreciation accounting.

1. To provide owners and managers with an estimate of the current value of their capital investment. Depreciation accounting for this purpose should approximate actual market values.
2. To account for depreciation in a manner that yields the maximum possible tax benefits. Depreciation accounting for this purpose may not exceed strict legal guidelines but need *not* approximate market values.

Since the two purposes are so divergent, usually two different methods of depreciation accounting are employed simultaneously on the same capital item. This double accounting is quite legal and ethical and is done openly. While the government requires strict adherence to its tax regulations regarding depreciation accounting, it does not expect the methods allowed or required by tax law are going to provide the type of depreciation information required to efficiently manage a business.

Estimates of three important items are required in order to calculate depreciation by any depreciation method. They are

1. Estimate the purchase price or cost when new.
2. Estimate the economic life (time between purchase new and disposal at resale or salvage value), or recovery period for tax purposes.
3. Estimate the resale or salvage value (zero for tax purposes).

Based on these three items of information, accountants have devised a number of methods of accounting for depreciation at a regular predictable rate over the expected economic life. Of these many methods four have emerged as noteworthy and the first three have found wide acceptance in professional practice.

1. Straight line method.
2. Sum-of-the-years digits method.
3. Declining balance method.
4. Sinking fund method (this method is not widely used and is discussed only for general background).

Straight Line Method of Depreciation Accounting

Straight line (SL) depreciation is the simplest method to apply and is the most widely used method of depreciation. The annual depreciation, D_m, is constant and thus the book value, BV_m, decreases by a uniform amount each year. The equation for *SL* depreciation is

$$\text{depreciation rate, } R_m = \frac{1}{N} \tag{16.1}$$

$$\text{annual depreciation, } D_m = R_m(P-F) = \frac{P-F}{N} \tag{16.2}$$

$$\text{book value, } BV_m = P - mD_m \tag{16.3}$$

An example and graph of straight line depreciation follows.

Example 16.1

Given the data below, find the annual depreciation and plot a graph showing the depreciation each year.

dragline, purchase price, $P = \$80,000$
  resale value after 7 yr, $F = \$24,000$

SOLUTION

purchase price, P $\qquad\qquad\qquad\qquad\qquad = \$80,000$

future resale or salvage value, F $\qquad\qquad\qquad \doteq \;\; 24,000$

value to depreciate in $N = 7$ equal installments, $P - F$ $\quad = \$56,000$

depreciation taken each year, $D_m = (P-F)/N = \$56,000/7 = \$\;8,000$

A graph of the values is shown in Figure 16.3.

The Sum-of-Years Digits Method

This is an accelerated depreciation (fast write-off) method, which is a term applied to any method that permits depreciation at a rate faster than straight line. This method calculates depreciation for each year as the total original depreciable value times a certain fraction. The fraction is found as

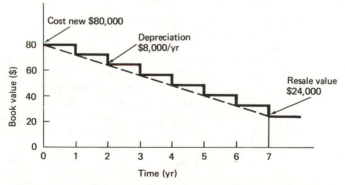

Figure 16.3 Straight line depreciation for Example 16.1.

follows: the denominator of the fraction is the sum of the digits including 1 through the last year of the life of the investment. The numerator is the number that represents the years of life remaining for the investment.

Equation for Sum-of-Years Digits Method

The sum of the ordinal digits for each of the years 1 through N is

1. $SOY = \dfrac{N(N+1)}{2}$ (16.4)

2. The annual depreciation D_m, for the mth year (at any age, m) is

$$D_m = (P-F) \frac{N-m+1}{SOY}$$ (16.5)

3. The book value at end of year m is

$$BV_m = P - (P-F) \left[\frac{m(N-m/2+0.5)}{SOY} \right]$$ (16.6)

An example of sum-of-years digits depreciation is shown below.

Example 16.2

Using the same $80,000 dragline as in Example 16.1, find and plot the allowable depreciation using the $SOYD$ method.

SOLUTION
A machine with a seven-year life uses a denominator of $1+2+3+4+5+6+7=28$, or $SOY = N(N+1)/2 = 7(7+1)/2 = 28$. The depreciation allowed for the first year is $\frac{7}{28}$ of the total depreciation, the allowance for the second year is $\frac{6}{28}$ of the total, and so on to $\frac{1}{28}$ for the seventh year. Using the sum-of-the-years digits method,

dragline, purchase price, P = $80,000

resale value after $N=7$ yr, F = $\underline{24,000}$

depreciable value, $(P-F)$ = $56,000

YEAR, m	DEPRECIATION ALLOWED FOR THIS YEAR, D_m	BOOK VALUE, BV_m AT EOY m
1	$\frac{7}{28} \times \$56,000 = \$14,000$	$80,000 - 14,000 = 66,000$
2	$\frac{6}{28} \times 56,000 = 12,000$	$66,000 - 12,000 = 54,000$
3	$\frac{5}{28} \times 56,000 = 10,000$	$54,000 - 10,000 = 44,000$
4	$\frac{4}{28} \times 56,000 = 8,000$	$44,000 - 8,000 = 36,000$
5	$\frac{3}{28} \times 56,000 = 6,000$	$36,000 - 6,000 = 30,000$
6	$\frac{2}{28} \times 56,000 = 4,000$	$30,000 - 4,000 = 26,000$
7	$\frac{1}{28} \times 56,000 = 2,000$	$26,000 - 2,000 = 24,000$
	$\overline{\$56,000}$	

A graph of values from Example 16.2 is shown in Figure 16.4.

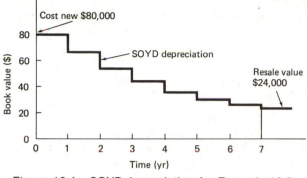

Figure 16.4 *SOYD* depreciation for Example 16.3.

To find the book value in any year, either use the equation

$$BV_m = P - (P-F)\left[\frac{m(N-m/2+0.5)}{SOY}\right]$$

or

1. Find the unused depreciation by:
   a. sum the fractions for the remaining years
   b. multiply this sum times the total depreciable value ($56,000 in this case).
2. Add this unused depreciation to the resale (salvage) value ($24,000 in this example).

Example 16.3

Find the book value at the end of the fourth year for Example 16.2.

SOLUTION

The fractions for the remaining three years of the seven-year life total $\frac{3}{28} + \frac{2}{28} + \frac{1}{28} = \frac{6}{28}$. Multiplying times the depreciable value yields $\frac{6}{28} \times \$56,000$ = $12,000. The book value at the end of the fourth year then equals the unused depreciation plus the salvage value, or $12,000 + $24,000 = $36,000 book value at end of fourth year.

The equation, of course, yields the same value.

$$BV_4 = 80,000 - (56,000)\frac{4(7 - \frac{4}{2} + 0.5)}{28} = \$36,000,\text{ book value at EOY 4}$$

Declining Balance Methods

The declining balance methods are accelerated depreciation methods that provide for a larger share of the cost of depreciation to be written off in the early years, less in the later years. This system more nearly approximates the actual decline in the market value for many types of property, especially mechanical equipment. Using this system the annual depreciation is taken as 2.0 (for the 200% rate called double declining balance method or *DDB*), or 1.5 (for the 150% rate), or 1.25 (for 125% rate), times the current book value of the property divided by the total years of economic life. For example, if the economic life is seven years, then if it were not for accelerated depreciation, $\frac{1}{7}$ of the value could logically be deducted each year. The *DDB* method doubles this deduction to $\frac{2}{7}$ of the *current value each year*. The machine is depreciated down to the resale value and then no more depreciation is taken. Book value, BV_m, is *not permitted* to fall below salvage value, *F*.

Equations for Declining Balance Methods

The abbreviation R = depreciation rate for declining balance depreciation.

1. The depreciation rate, R, is the depreciation multiple divided by the estimated life, n.

$$\text{for double declining balance depreciation,} \quad R = 2/n$$
$$1.75 \text{ declining balance depreciation,} \quad R = 1.75/n$$
$$1.5 \text{ declining balance depreciation,} \quad R = 1.5/n$$

2. The depreciation, D_m, for any given year, m, and any given depreciation rate, R, is:

$$D_m = RP(1-R)^{m-1} \quad \text{or} \quad D_m = (BV_{m-1})R \tag{16.7}$$

3. The book value for any year BV_m is:

$$BV_m = P(1-R)^m \quad \text{provided} \quad BV_m \geqslant F \tag{16.8}$$

4. The age, m, at which book value, BV_m, will decline to any future value, F, is:

$$m = \frac{\ln(F/P)}{\ln(1-R)} \tag{16.9}$$

Note that the depreciation amount D_m is determined by the book value only, and is *not* influenced by the salvage value F (excepting that $BV_m - D_{m-1} \geqslant F$). Therefore, the book value during later years may follow any of the following three tendencies:

1. If F is zero or very low, then BV_m may never reach F.
2. BV may intersect F before N (BV is not permitted to be less than F).
3. BV may intersect F at N (very rare occurrence).

Graphically these three situations are given in Figure 16.5. Since the book value must never be less than salvage value in the declining balance method the latter values of BV_m are usually force fit to equal F, as depicted in Figure 16.4.

An example of double declining balance depreciation, is shown below using the same $80,000 dragline as in Example 16.3.

Example 16.4

Find the depreciation allowed each year, using double (200%) declining balance method, (DDB),

depreciation rate, $R = \frac{2}{7}$

dragline purchase price, P $= \$80,000$

resale value, F, after $n = 7$ yr $= \underline{\ 24{,}000\ }$

depreciable value, $P - F$ $= \$56,000$

SOLUTION

YEAR, m	BOOK VALUE$\times DDB$ FACTOR $(BV_m \times R)$		DEPRECIATION, D_m ALLOWED FOR THIS YEAR
1	$\$80{,}000 \times \frac{2}{7}$ $-22{,}857$	$=$	$22,857
2	$\overline{57{,}143 \times \frac{2}{7}}$ $-16{,}327$	$=$	16,327
3	$\overline{40{,}816 \times \frac{2}{7}}$ $-11{,}662$	$=$	11,662
4	$\overline{29{,}154 \times \frac{2}{7}}$ $-8{,}330$	$=$	(8,330) N/A, use $5,154
	$(20{,}824)$		This is less than the resale value of $24,000 so only $(29{,}154 - 24{,}000 = 5{,}154)$ is allowed for depreciation this year
5	$24,000		0
6	24,000		0
7	24,000		0
			$\overline{\$56{,}000}$ total depreciation using double (200%) declining balance method of depreciation

During the fourth year the calculated depreciation of $\frac{2}{7} \times \$29{,}154 = \$8{,}330$ would, if used, reduce the book value to $20,824, which is less than the

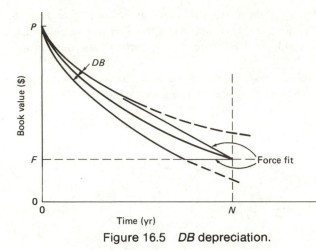

Figure 16.5 *DB* depreciation.

salvage value of $24,000. In this case the fourth year's depreciation is reduced to $5,154 in order to force the book value to equal the salvage value, *F*.

A graph of the values obtained by the double declining balance depreciation method are shown in Figure 16.6.

Sinking Fund Method

This method first sets up a savings account (sinking fund) and calculates the amount of equal annual payments required to go into the fund so that accumulated principal plus interest will equal the accumulated depreciation on the asset by the end of the asset's economic life.

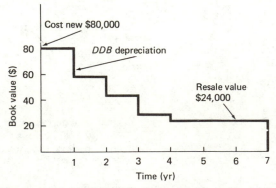

Figure 16.6 *DDB* depreciation for Example 16.4.

Example 16.5

Find the depreciation allowed for each year, using the sinking fund method. Assume $i = 10$ percent.

dragline purchase price $= \$80,000$

resale value after 7 yr $= \underline{24,000}$

depreciable value $ = \$56,000$

SOLUTION

Theoretical annual deposit into the sinking fund

$$A = \$56,000\,(A/F,\ 10\%,\ 7) = \$5,902.70$$
$$0.105406$$

Then the accumulated depreciation is simply the sinking fund balance of equal annual deposits plus accumulated interest at the end of each year. The accumulated depreciation at the end of each year for the $\$80,000$ dragline with $\$24,000$ resale value after seven years is tabulated below.

END OF YEAR	SINKING FUND BALANCE OR ACCUMULATED DEPRECIATION
1	$\$5,902.70(F/A,\ 10\%,\ 1) = \$\quad 5,903$
2	$5,902.70(F/A,\ 10\%,\ 2) = \quad 12,903$
3	$5,902.70(F/A,\ 10\%,\ 3) = \quad 19,588$
4	$5,902.70(F/A,\ 10\%,\ 4) = \quad 27,395$
5	$5,902.70(F/A,\ 10\%,\ 5) = \quad 36,037$
6	$5,902.70(F/A,\ 10\%,\ 6) = \quad 45,543$
7	$5,902.70(F/A,\ 10\%,\ 7) = \quad 56,000$

The book value at the end of year m is

$$BV_m = P - (P - F)(A/F, i, n)(F/A, i, m) \tag{16.10}$$

or book value after four years $BV_4 = \$80,000 - \$27,395 = \$52,605$.

The amount of depreciation allowed in any one year can be determined by either of two methods.

1. Simply subtract the accumulated depreciation at the end of the prior year from the depreciation at the end of the current year. For instance, the depreciation for the seventh year is simply $(\$56,000 - \$45,543)$ or $\$10,457$.

2. The depreciation allowed for any year m may be determined as the future value of one lump sum deposit of A made $(m-1)$ years previously. For instance the depreciation for the seventh year equals $\$5,902.70$ $(F/P, 10\%, 6) = \$10,457$. The depreciation allowed for each individual year

may be totalled to equal the full allowable depreciation for the seven-year period, as follows.

END OF YEAR	FUTURE VALUE OF INDIVIDUAL DEPOSITS ACCUMULATES TO FULL ALLOWABLE DEPRECIATION	
1	$5,902.70(F/P, 10\%, 0) = \$$	5,903
2	$5,902.70(F/P, 10\%, 1) =$	6,493
3	$5,902.70(F/P, 10\%, 2) =$	7,142
4	$5,902.70(F/P, 10\%, 3) =$	7,856
5	$5,902.70(F/P, 10\%, 4) =$	8,642
6	$5,902.70(F/P, 10\%, 5) =$	9,507
7	$5,902.70(F/P, 10\%, 6) =$	10,457
	Accumulated depreciation	$56,000

The sinking fund is rarely used because (1) the company usually has more immediate needs for accumulated depreciation reserves than to place them in an interest bearing account, (2) the method is cumbersome and does not lend itself to uniform yearly charges, and (3) it provides the poorest income tax benefits since the greatest depreciation allowance usually comes in later years.

COMPARISON OF DEPRECIATION ACCOUNTING METHODS

A comparison of the four methods for the dragline example is shown in Figure 16.7. Examination of the graph reveals:

1. The straight line method takes an *equal fraction* of the *depreciable value* each year. Thus, when trying to set up a rental rate for a dragline or other depreciable equipment, the *SL* method is the simplest since a constant rate for depreciation charges can easily be established.
2. The sum of years method takes a *changing fraction* of the *depreciable value*.
3. The declining balance methods take an *equal fraction* of the *current book value*.
4. Both the *SOY* and *DB* methods accelerate the depreciation during the early years at the expense of later years. This can have a significant influence on the income tax consequences of an investment.
5. The *DB* method does not allow book values to fall below the salvage value. The IRS permits one change in method in later years to transition to F at EOY N.

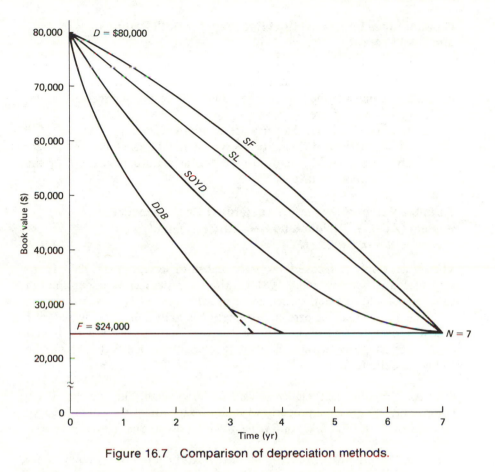

Figure 16.7 Comparison of depreciation methods.

CAPITAL COST RECOVERY AND INVESTMENT TAX CREDIT PROVISIONS OF THE ECONOMIC RECOVERY TAX ACT OF 1981

Nomenclature

Recovery property is any depreciable, tangible property (personal or real) used in business or held for the production of income.

Recovery period is the current nomenclature for the depreciation period of recovery property for tax purposes. The recovery period is the time period over which capital cost is recovered (depreciated). The recovery period is usually shorter than the useful life of the capital investment in order to provide a quicker write-off and thus encourage capital investment, stimulate the economy, and improve productivity.

General Rules for Capital Cost Recovery for Both Personal and Real Property

1. The recovery property may typically be depreciated down to zero value. Salvage value need not be accounted for in depreciation, but any gain over the depreciated value will be taxed either as capital gains or recapture.
2. Both new and used property are treated the same.
3. Either accelerated capital recovery system (*ACRS*) tables or straight line recovery (*SL*) may be used to calculate depreciation, at the discretion of the taxpayer.

Tangible Personal Property: Accelerated Cost Recovery System (*ACRS*), Straight Line Recovery (*SL*), and Investment Tax Credit (*ITC*)

These rules apply to recovery property placed in service after 1980. (Property placed in service before 1981 continues to depreciate under the old rules unless sold to a new owner-taxpayer after 1980.)

The cost of tangible personal property generally can be recovered over a cost-recovery period of 3, 5, 10 or 15 years. Cost-recovery periods and *ITC*'s are assigned to each class of recoverable property based on the following definitions.

• *3-Year Property.* This class consists of autos, light-duty trucks, R&D equipment, and personal property with an average life of 4 years or less. An investment tax credit (*ITC*) of 6 percent will apply to this class of property.

• *5-Year Property.* Most other equipment, except long-lived public utility property, is included in the 5-year category. The 10 percent *ITC* applies to qualified 5-year property.

• *10-Year Property.* This class includes public utility property with an average life greater than 18 but not greater than 25 years; railroad tank cars; manufactured homes; and real property with an average life of 12.5 years or less (e.g., theme-park structures). The 10 percent *ITC* applies to qualified 10-year property.

• *15-year Property.* This class consists of public utility property with an average life exceeding 25 years. The 10 percent *ITC* applies to qualified 15-year property.

ACRS TABLES FOR TANGIBLE PERSONAL PROPERTY

The recovery deductions (from taxable income) for tangible personal property may be accelerated through the use of accelerated recovery tables (rather than using the traditional methods of *SOYD*, *DDB*, etc.). Table 16.1

Table 16.1 1981 TAX ACT: *ACRS* STANDARD RECOVERY PERCENTAGES

RECOVERY YEAR	PROPERTY ACQUIRED DURING 1981–1984				PROPERTY ACQUIRED IN 1985				PROPERTY ACQUIRED AFTER 1985			
	3-YEAR	5-YEAR	10-YEAR	15-YEAR PUBLIC UTILITY	3-YEAR	5-YEAR	10-YEAR	15-YEAR PUBLIC UTILITY	3-YEAR	5-YEAR	10-YEAR	15-YEAR PUBLIC UTILITY
1	25%	15%	8%	5%	29%	18%	9%	6%	33%	20%	10%	7%
2	38	22	14	10	47	33	19	12	45	32	13	12
3	37	21	12	9	24	25	16	12	22	24	15	12
4		21	10	8		16	14	11		16	14	11
5		21	10	7		8	12	10		8	12	10
6			10	7			10	9			10	9
7			9	6			8	8			8	8
8			9	6			6	7			6	7
9			9	6			4	6			4	6
10			9	6			2	5			2	5
11				6				4				4
12				6				4				3
13				6				3				3
14				6				2				2
15				6				1				1

343

shows the designated annual recovery percentages based on the original cost of the property. The recovery percentage for the first year and last year is based on a "half-year convention," which allows one-half year's recovery deduction in the year of acquisition but no recovery deduction in the year of disposition.

SL DEPRECIATION FOR TANGIBLE PERSONAL PROPERTY

If *SL* recovery is selected for depreciation of personal property, the same one-half-year convention applies, allowing one-half-year's recovery deduction in the year of acquisition but no recovery deduction in the year of disposition.

RECAPTURE OF DEPRECIATION ON TANGIBLE PERSONAL PROPERTY

When personal property is disposed of, regardless of whether depreciated by *ACRS* or *SL*, any gain in excess of depreciated book value is recaptured by taxing the gain as ordinary income (rather than capital gain), but only to the extent of any prior depreciation. Thus if the personal property is sold for more than the original purchase price, the gain from depreciated book value to original purchase price is recaptured as ordinary income, while the gain from the original purchase price to resale value is treated as capital gain (see Figure 16.8).

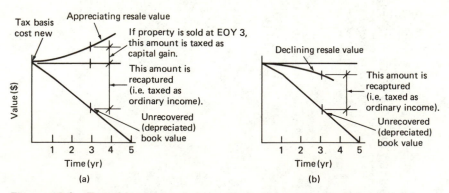

Figure 16.8 Tangible personal property: treatment for capital gain and recapture if sold at EOY 3: (a) Appreciating resale value; (b) Declining resale value.

RECAPTURE OF INVESTMENT TAX CREDIT (*ITC*) FOR TANGIBLE PERSONAL PROPERTY

The *ITC* is subject to recapture if tangible personal property is sold at an early date, as follows:

Ownership Year in Which Disposed	Percentage of ITC to Be Recaptured	
	Three-Year Property	*Other Property*
1	100%	100%
2	67	80
3	33	60
4	0	40
5	0	20

The *ITC* may be applied to used personal property costing up to $125,000 in 1982 and $150,000 in 1985 and thereafter. In case of non-recourse loans on leveraged investments, where the "at risk" amount is less than the value of the property, the *ITC* is limited to the amount for which the taxpayer is personally liable (bears the risk of loss).

Real Property: *ACRS* and *SL*

GENERAL RULES

Depreciable real property generally is assigned a 15-year recovery period using either *ACRS* or *SL*. The *ACRS* recovery schedule, prescribed by regulations, is based on the 175 percent declining balance method (200 percent declining balance method for low-income housing), with an automatic change to the straight-line method to maximize the deduction. Recovery in the year of acquisition as well as in the year of disposal is prorated according to the actual months of service during each of these years. If desired, the rate of recovery may be slowed considerably by the election to use the straight-line recovery method and extending the recovery period.

NONRESIDENTIAL REAL ESTATE

If *ACRS* depreciation is used, then on the disposition of nonresidential real property acquired after 1980, *all* gain is treated as ordinary income to the extent of *all* prior depreciation. The *full* gain from *ACRS*-depreciated book value to original purchase price is taxed at the ordinary rate (unlike the rule for property acquired before 1981, where the gain from *SL* to net selling price is treated as capital gain). If the purchase price is lower than the selling price, the gain from original purchase price to net selling price is taxed as capital gain.

At the taxpayer's discretion, the straight line (*SL*) depreciation method may be employed for nonresidential real estate using a life of 15, 35, or 45 years. The *SL* method gives lower depreciation deductions each year, but the full gain on sale is treated as capital gain (unchanged from the rule for property acquired before 1981, where the full difference between *SL*-depreciated book value and net selling price is treated as capital gain), as seen in Figures 16.9 and 16.10.

RESIDENTIAL RENTAL REAL ESTATE

For residential rental property, either *ACRS* or *SL* may be used. If the *ACRS* is used, the gain between *ACRS* and *SL* depreciated book value is recaptured at ordinary rates, whereas the gain between *SL* and net selling price is treated as capital gain. Thus *ACRS* is usually most advantageous for residential rental property in spite of the limited recapture (see Figure 16.11).

CARRYOVER PERIOD

The carryover period for net operating losses, investment credits, work incentive credits, and job credits is extended to 15 years.

Recapture

The term "recapture" refers to the process of taxing income that was not previously taxed due to offsetting deductions (such as deductions for accelerated capital recovery, investment tax credit, etc.). For instance, when residential rental property is sold, the Economic Recovery Tax Act of 1981 provides for complete recapture of all previous tax deductions due to

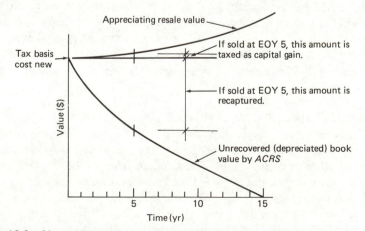

Figure 16.9 Nonresidential real estate recovered by *ACRS*—treatment for capital gains and recapture for real property acquired after 1980.

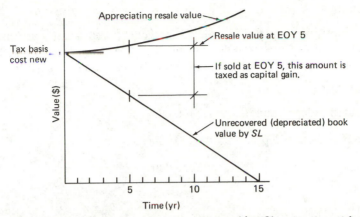

Figure 16.10 Nonresidential real estate recovered by *SL* —treatment for capital gains and recapture for real property acquired after 1980.

accelerated capital recovery in excess of straight line (*SL*) depreciation (with some minor exceptions). This means that if a taxpayer (1) owns residential rental property, (2) has claimed accelerated capital recovery, and (3) sells the property for *more* than the depreciated book value, then the gain between the accelerated depreciation book value and the *SL* book value is taxable as ordinary income (recapture). Any additional gain between *SL* book value and the selling price is taxable as capital gains, providing the taxpayer has owned the property for one year or more. Usually only about 40 percent of the capital gain is taxable at ordinary income tax rates.

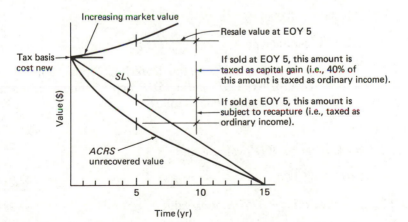

Figure 16.11 Residential real estate—treatment for capital gains and recapture for real property acquired after 1980.

Example 16.6

For example, a taxpayer in the 45 percent tax bracket purchased an apartment house for $550,000 and resold it five years later for $825,000. Find the amount of capital gain and recapture tax, assuming the following:

purchase price

land (not depreciable)   =$100,000

improvements (depreciable)= 450,000

total   =$550,000

SOLUTION

The depreciation schedule for the depreciable improvements, with a 15-yr recovery period and zero salvage value, is as follows:

	1.75DB		STRAIGHT LINE	
YEAR	ALLOWABLE DEPRECIATION	BOOK VALUE EOY	ALLOWABLE DEPRECIATION	BOOK VALUE EOY
0		$450,000		$450,000
1	52,500	397,500	30,000	420,000
2	46,375	351,125	30,000	390,000
3	40,965	310,160	30,000	360,000
4	36,185	273,975	30,000	330,000
5	31,964	242,011	30,000	300,000

selling price   =$825,000

SL book value $-$ SL depreciated value $+$ land

  =$300,000+100,000   = 400,000

gain taxable as capital gain   =$425,000

If the $1.75DB$ method is used to calculate the depreciated book value for tax purposes the difference between SL and $1.75DB$ depreciation is taxed as recapture at ordinary income tax rates, as follows.

selling price   =$825,000

$1.75DB$ book value$=1.75DB$ value$+$land

  =$242,011+100,000   = 342,011

total gain (selling price$-1.75DB$ book value)   = 482,989

capital gain (selling price$-SL$ book value)   = 425,000

recapture taxed as ordinary income (SL book value$-1.75DB$)=$ 57,989

Since the investor is in the 45 percent tax bracket the tax is

40% of capital gain at ordinary rate, $0.40 \times \$425,000 \times 0.45 = \$\ 76,500$

recapture at ordinary rate, $0.45 \times \$57,989$ $= \underline{\quad 26,095}$

total tax liability due to sale $= \$102,595$

These relationships are illustrated in Figure 16.12.

DEPRECIATION VERSUS AMORTIZATION

Depreciation should not be confused with amortization. A common example of amortization occurs when a debt P is amortized (paid out) by n end-of-period payments each of amount A, where i is the rate at which interest periodically accrues on the unpaid balance, and $A = P(A/P, i, n)$. The differences between depreciation and amortization are outlined as follows.

1. Depreciation does *not* involve cash flow (except at times of purchase and resale) whereas amortization *does* involve periodic cash flow payments of principal and interest.
2. Depreciation does *not* involve interest and simply provides a time table for orderly accounting for the loss in property value from purchase price to resale value. Amortization does involve periodic

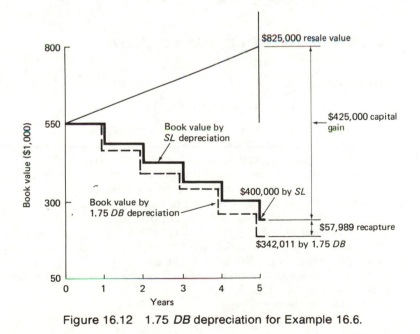

Figure 16.12   1.75 *DB* depreciation for Example 16.6.

payments of interest accruing at rate i on the unpaid balance, in addition to payments on principal.

3. The depreciation allocated for each year is a noncash flow expense which is usually tax deductible for the allocated year. Thus while depreciation is not a cash flow item it often does affect after-tax cash flow. For amortization, only the interest portion (not the principal) of the periodic payment is tax deductible.

SUMMARY

Depreciation accounting is a systematic allocation of the loss in value of a capital asset over its useful life. Depreciation as reflected by the book value does not necessarily track market value. There are several methods of depreciation accounting currently being used. They include straight line, sum-of-years digits, declining balance, and sinking fund. The straight line method is the simplest and most often used and allocates the depreciable cost of the asset uniformly over its life. Sum-of-years digits and declining balance methods accelerate the rate of depreciation during the early life of the asset. These methods usually yield the most advantageous income tax effects, but are slightly more complicated than the straight line method. The sinking fund method retains its place as a standard method but is seldom encountered in professional practice.

Depreciation can have a significant effect on after-tax rates of return for alternatives involving investment in depreciable assets and is an important aspect of engineering feasibility studies.

PROBLEMS FOR CHAPTER 16, DEPRECIATION

Problem Group A

Before-income-tax application of depreciation methods.

A.1 A friend purchased an asphalt plant 6 years ago for $250,000. She expects to sell it when it is 12 years old for $70,000. She has an offer to sell now and asks you to find the current depreciated book value (tax basis) for the plant. You explain that there are three common depreciation methods for finding book value. Find what the current book value is by all three methods.

A.2 Assume you are looking for a building in which to locate your professional office, and one is available for $50,000. The appraisal is divided into $10,000 for the land and $40,000 for the building. The building is new and can be depreciated over a 20-year period at 1.50 declining balance method. Salvage value after 20 years is $20,000 for the building.

a. In your first year of ownership, how much can you deduct for depreciation?
b. How much in the second year?
c. How much in the twentieth year?

A.3 The property tax assessor is willing to accept the book value calculated by double declining balance method as a basis for his assessments if you will provide the calculations. His property tax is collected at the end of each year. The tax rate is 30 mills (3% tax) on the book value of the property at the beginning of the year.

   a. Compared to depreciation by the straight line method, how much cash flow can be saved on property purchased for $100,000 and resold in five years for $40,000?

   b. What is the annual equivalent of this amount if $i = 8$ percent? (*Hint*: First find *PW* of each year's savings.)

Problem Group B

Find after-tax *PW* of savings or *ROR* of *DDB* versus *SL*.

B.1 A friend of yours has just purchased a pickup for use in his business. He expects to keep it for four years and then sell it. He would like some suggestions on which depreciation method to use for tax purposes, since he is in the 37 percent tax bracket. Assume that $i = 12$ percent and show how much (in terms of present worth of tax payments over the four-year period) can be saved by using the double declining balance method as compared to the straight line method. Assume the pickup cost $8,000 new and sells at the end of four years for $3,000.

B.2 An automatic block making machine is available for $50,000. Your best estimates indicate that it will be worth $10,000 when you expect to dispose of it at the end of five years. It is capable of producing 100,000 blocks per year at a net profit (not including cost of the machine) before taxes of $0.10 per block. Use 40 percent tax rate.

   a. Find the cash flow after taxes for the machine using (i) straight line depreciation and (ii) double declining balance depreciation.

   b. Find the rate of return on the investment in the machine using (i) *SL* depreciation and (ii) *DDB* depreciation.

B.3 A contractor is considering the purchase of a new concrete batch plant, and has offers from two vendors for plants that are essentially identical except for the salvage values. The data on each alternate are listed below. Both are available fully financed with no down payment and 10 percent interest on the unpaid balance.

	BIG BATCH	SUNCRETE
Cost new	$240,000	$240,000
Life	22 yr	22 yr
Salvage value	80,000	zero
Net cash flow before taxes	52,000/yr	52,000/yr
Depreciation method	Straight line	Straight line

The contractor is in the 54 percent tax bracket. Find the difference between the alternates in the net cash flow *after* taxes.

B.4 You are consultant to a contracting firm with an annual before-tax income averaging $500,000 per year subject to income tax of 46 percent on all income over $100,000. During the course of your consulting, you find the concrete plant can be made to yield extra revenue over the next ten years by any one of the following mutually exclusive alternatives. Funds to finance these alternative proposals are available out of working capital. Working capital is currently earning the company 10 percent after taxes. Find which of the following alternatives has the highest after-tax present worth at $i = 10$ percent. Assume all taxes are paid at the end of the year.

a. An extra $22,000 of before-tax revenue may be earned by investing $100,000 in new equipment. The new equipment is eligible for a 10 percent investment tax credit (10% of $100,000 is deducted from the amount of tax due for the first year only). In addition, the $100,000 may be depreciated for tax purposes over a period of ten years at $8,000 per year. At the end of ten years, the equipment should have a $20,000 salvage value.

b. Increase before-tax revenue by $20,000 per year for the ten-year period by rearranging the flow of materials. This requires disassembly of the several large steel structures, but no capital investment. The $100,000 cost of doing the work will be spent and charged off in one lump sum at the end of the first year.

c. An extra $18,000 per year of pretax revenue for the ten-year period may be added by hiring an additional plant technician at a cost of $10,000 the first year, increasing $1,000 per year thereafter.

B.5 Your client is interested in building and operating a small shopping center of 100,000 ft^2 and asks you to calculate the rental rate needed per square foot in order to make a 15 percent rate of return after taxes on all funds invested. Your client is in the 50 percent tax bracket for ordinary income. The future costs and income for the project are estimated as follows.

EOY	
0	Purchase land. $100,000.
1	Finish construction of $2,000,000 worth of buildings and other improvements depreciable over a 40-yr life by straight line depreciation method, with zero salvage value at the end of 40 yr. Begin rental and operation of the center, but rental income and operating costs are not credited or debited until EOY 2.
2	First year's operating and maintenance costs charged at EOY 2 = 100,000/yr. First year's income deposited after income taxes are paid at a tax rate of 50% × net income (where net income = gross income − O&M − depreciation).
3–6	Same as EOY 2. (O&M costs continue at $100,000/yr.)
6	Sell the shopping center for original cost plus appreciation estimated at 4%/yr compounded. The difference between the depreciated book value and the sale price is taxable as capital gains.

a. Draw a cash flow diagram indicating the time and amount of each cash flow occurrence.

b. Find the rental rate in terms of $/ft^2/yr to obtain 15% return after taxes on all funds invested.

Comment: For convenience make equation time zero coincide with EOY 1.

B.6 Your engineering office is considering the purchase of a new computer for $100,000. You expect to sell it at EOY 5 for $40,000.

a. Fill in the depreciation schedule below for the years indicated.

Straight line depreciation

Year	book value at beginning of the year	$value	×multiplier	=	depreciation taken this year
1	_____	_____ × _____		=	_____
2	_____	_____ × _____		=	_____

Double declining balance depreciation

| 1 | _____ | _____ × _____ | | = | _____ |
| 2 | _____ | _____ × _____ | | = | _____ |

b. The computer would be 100 percent financed and will be paid off in five equal annual installment payments with interest at 10 percent on the unpaid balance. The first payment is due at EOY 1. Fill in the amortization schedule for the years indicated.

Amortization schedule

Year	amount of equal annual payment	interest payment for this year	principal payment for this year	balance owed after EOY payment
0				_____
1	_____	_____	_____	_____
2	_____	_____	_____	_____

c. Assume your engineering firm now earns a net cash flow income of $60,000 per year after all salaries and expenses, but before considering payments of principal and interest on the computer, and before accounting for computer depreciation and before paying income tax. Your firm is in the 40 percent tax bracket and uses *DDB* depreciation. (i) How much income tax would your firm owe for the second year's operation at EOY 2? (ii) What is the net cash flow income after taxes for the second year?

Problem Group C

Find the after-tax incremental rate of return.

C.1 Your firm is replacing its fleet of ten executive sedans (all the same model) and has narrowed the replacement selection down to a field of four different models with life-cycle cost characteristics listed below. The cars will be driven 15,000 mi per year and sold at the end of three years. All four models are acceptable to the firm's policy makers and they ask you to recommend a choice of one out of the four based on *after-tax* incremental rate of return.

The tax status of the firm is as follows:

ordinary income tax bracket 40%	
investment tax credit	10% of the initial cost of the sedans are credited one time only at EOY 0
depreciation method	straight line for the estimated life listed below
Assume the following tax regulations regarding income tax	deduct depreciation and O&M costs from income before calculating tax
	deduct the 10% investment tax credit from the actual tax due

TYPE MODEL	COST NEW	RESALE VALUE @ EOY 3	O&M COSTS ($/MI)	GRADIENT O&M COSTS ANNUAL INCREASE ($/MI/YR)
A. Century	$9,000	$4,800	0.175	0.012
B. Cordova	8,500	4,600	0.185	0.015
C. Monarch	9,700	4,900	0.160	0.014
D. Starfire	8,800	4,300	0.172	0.013

a. Find the incremental rate of return *after taxes* between the alternative with the lowest initial cost, and the one with the next lowest initial cost. (*Hint:* Try between 15% and 25%.)
b. Assume the incremental rates of return for the other alternatives are as listed below. Draw an arrow diagram showing the four alternatives and the direction of each arrow between the alternatives.

	D	A	C
B		18.8	11.1
D		17.3	9.3
A			4.5

c. On the arrow diagram, show by dashed arrows the method of reaching the better alternative of the four if the *MARR* is 15 percent.

C.2 A contractor needs to buy a new excavator and currently has four different models under consideration. The life-cycle costs of each model are tabulated below. In addition, the following data are supplied concerning income taxes for the contractor.

ordinary income tax bracket 40%
investment tax credit 10% of the initial cost of the excavator,
       credited one time only, at EOY 0
depreciation method straight line with n as listed
       for estimated life below

TYPE	A	B	C	D
Initial cost	$120,000	$140,000	$100,000	$160,000
Salvage value	50,000	60,000	40,000	80,000
Estimated life (yr)	10	10	10	10
Productivity (yd^3/hr)	200	210	190	220
Productivity decline gradient (productivity drop in yd^3/hr each year)	10	10	10	5
O & M	80,000/yr	75,000/yr	90,000/yr	72,000/yr
O & M gradient	8,000/yr/yr	9,000/yr/yr	6,000/yr/yr	7,500/yr/yr

The income for each cubic yard of earth moved is $0.10/yd^3, and the excavator will work 2,000 hours per year.

a. Find the incremental rate of return *after taxes* between the alternative with the lowest initial cost, and the one with the next lowest initial cost. (*Hint:* Try between 15% and 25%.)

b. Assume the incremental rates of return for the other alternatives are as listed below. Draw on arrow diagram showing the four alternatives and the direction of each arrow between the alternatives.

	A	B	D
C		14.6	22.4
A		10.5	23.0
B			29.7

c. Show by dashed arrows the method of reaching the better alternative of the four if the *MARR* is 15 percent.

Chapter 17
Bond Financing for Public Works and Corporate Investment

FINANCING PUBLIC WORKS

Like the governments of many other nations around the world, our government consists of a three-level structure: (1) federal, concerned with national and international affairs, (2) state, concerned with improving the conditions within its jurisdiction, and (3) local, concerned with the problems on the county, city, and public-school level. According to the U.S. Constitution, residual power not otherwise delegated to the federal or local governments is supposed to remain with the individual states.

The budgets for each of these levels of government usually are subdivided into an operating budget and a capital budget. The operating budget includes all expenditures necessary for maintaining the day-to-day operating and maintenance functions of each level. Salaries and wages are usually the largest single item in an operating budget, which also includes supplies and other consumables with a usable life expectancy of less than one year.

To fund their operations budget the three levels of government in the United States have primarily relied upon income from the following sources.

LEVEL OF GOVERNMENT	TRADITIONAL SOURCE OF REVENUE
Federal	Income tax
State	Sales tax
Local, county, city, school	Real estate and personal property tax

Over the years, however, as the need for additional sources of revenue became more urgent, the various government levels sought tax revenues wherever they could be found, and some mixing of tax revenue sources resulted. Currently, the federal government is still obtaining about 60 percent of its revenue from income taxes (with corporations paying one-quarter of the total and individuals paying the remaining three-fourths). An additional 27 percent of the federal revenue comes from Social Security taxes, and 6 percent from excise (sales) taxes. State and local governments still obtain over 75 percent of their revenue from sales and property taxes, although 44 states and over 45 cities now assess income taxes against their citizen taxpayers.

Expenditures for items with lives of greater than one year are usually defined as capital expenditures and belong in the capital budget. (As a matter of practice operating budgets often contain a few capital items such as typewriters, desks, etc.) Thus funding for new buildings, lands, and other large projects with a long life expectancy is done through the capital budget.

On the state and local level, often the income from regular tax revenues is totally consumed by the day-to-day operating expenses of government, leaving little for needed capital expenditures. Thus, when large capital expenditures are needed, funds to finance the project usually must be borrowed. (Approximately 60 percent of state and local construction is financed by borrowed money secured by bond issues.) Long-term borrowing for state and local government is usually done through bond issues. A bond is simply a long-term public debt of an issuer that has been marketed in a convenient denomination, backed with substantial assets, bearing interest at an agreed rate, with the principal (face value or redemption value) to be repaid on an agreed maturity date. The bond market for public works at all levels is quite large. In one recent year alone, more than $200 billion was borrowed by issuing bonds from all levels of government in this country as compared to only $40 billion borrowed by privately owned corporations issuing bonds. Purchasers of bonds include a wide variety of individual and corporate entities, with varying investment objectives. Among the purchasers of a large percentage of bonds are organizations and firms with long-term capital to invest such as commercial banks, insurance companies, and pension funds.

BONDS

Bonds are normally issued in multiples of $1,000 and have a stated nominal interest rate, payments period, and maturity date (or life). An example might be a $5,000 bond bearing 8 percent interest payable semiannually and maturing in 20 years. This means the bond issuer (governmental entity) agrees to pay the bond purchaser (investor) $0.04 \times \$5,000 = \$200$ every six months for 20 years (40 payments). Thus, the actual interest payment is 4 percent semiannually while the nominal annual interest is 8 percent. With

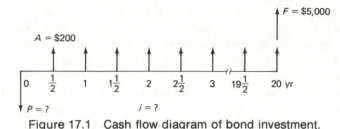

Figure 17.1   Cash flow diagram of bond investment.

the last payment the investor receives the $5,000 face value of the bond. The cash flow to the purchaser is shown in Figure 17.1.

Notice the purchase price of the bond, P, does not show a dollar amount. If the purchaser pays $5,000 for the bond she will earn a nominal annual interest rate of 8 percent and an effective interest rate of 8.16 percent ($i_e = (1+0.08/2)^2 - 1 = 0.0816$). But she may not be willing to pay $5,000 for the bond. She may offer to purchase the bond for $4,500. If the seller agrees, what interest rate will be earned? *Note that the payments are not changed.* The bond, issued as a $5,000 bond, contains coupons which may be redeemed through certain banks on their redemption date for $200. At maturity, the purchaser surrenders her last coupon and the bond, for which she receives $5,200. To determine the rate of return, simply set up a net worth equation and solve for the unknown i. (The number of compounding periods must relate correctly to the interest per period, of course.)

$$NPW = -\$4,500 + 200(P/A, i, 40) + 5,000(P/F, i, 40)$$

By trial and error

$$i = 4.6\%$$

$$i_n = 2 \times 4.6 = 9.2\%$$

and

$$i_e = (1.046^2 - 1)/100 = 9.4\%$$

Here the interest earned by the purchaser (and paid by the issuer) is greater than the interest rate shown on the bond itself. If the bond sells for more than $5,000 the resulting earned interest would be less than the rate shown on the bond.

While bonds can be bought and sold between individuals if desired, most transactions go through a broker. Some brokers specialize in bonds only, while other brokers handle stocks, bonds, and other financial paper. Because bonds are backed by the good faith (and sometimes the assets) of the governmental entity-issuer, they usually are readily marketable and are bought and sold in large quantities every business day. Thus, during the life of a bond (time until it matures or is called) it may be owned by numerous owners. Prices fluctuate each day, depending upon a number of variables such as demand, money supply, rating of the bond, public confidence in the governmental entity-issuer, and the anticipated inflation rate.

TAX-FREE MUNICIPAL BONDS

To stimulate investment in bonds the federal government provides that the interest earned from most state and local bonds are not subject to federal income taxes. Therefore, taxpaying investors will lend money at a lower interest rate for this tax exempt type government bond and still receive competitive after-tax rates of return. Tax exempt bonds are frequently referred to as "municipal bonds" or "municipals" even though they may be issued by a state, county, school board, or special governmental authority (e.g., Florida Turnpike Authority) as well as by a municipality. Interestingly the income from most federal bonds (such as Savings Bonds) is not income tax exempt.

There are several different types of bonds currently available depending upon the assets backing the bonds.

1. General obligation bonds (GO bonds). These bonds are backed by the full faith and credit of the city, county, or state issuing them. Before issuing GO bonds a public referendum is required so that the citizens may vote on whether or not they wish to pledge all of the community real estate assets to repay the bonds if called upon. Usually they are also asked to pledge a small increase in real estate taxes to repay the bonds over a set number of years. Since GO bonds do have such excellent backing, investors will frequently buy them at $\frac{1}{2}$ percent to 1 percent lower interest rates than other types of municipal bonds.

2. Revenue bonds (sometimes called revenue certificates). These bonds are backed by some special source of revenue, such as cigarette tax, sales tax, or receipts from the facility being financed (dormitories, stadiums, parking garage, electric power plant, etc.). These do not require a public referendum but simply action by the governing authority, such as the city commission, county commission, school board, and so forth. These bonds are frequently used to borrow funds to construct a facility whose receipts will then be used to pay off the bonds. Examples are Florida Turnpike Bonds, Chesapeake Bay Bridge and Tunnel, Gainesville Electric Water and Sewer and many, many others. If the revenue from the pledged source is inadequate, the bond holders usually have lien rights on the revenues (sometimes on the properties). From time to time an isolated few bonds do default temporarily. The Chesapeake Bay Bridge and Tunnel, and the West Virginia Turnpike are well-known examples of bonds that have been behind on payments for some time. The bond holders will undoubtedly be paid off some day, but meanwhile the

bonds are reselling for considerably less than their face value. (Chesapeake Bay Bridge and Tunnel bonds sold as low as 23, or 23% of their face value.)

3. Special assessment bonds. These bonds, as the name implies, represent money borrowed for a project of special benefit to a few property owners (such as a sewer line or a residential street paving project that benefits the fronting property owners). The principal and interest is paid from special annual assessments against the improved property for a specified number of years.

4. Industrial revenue bonds. Local governments are permitted to finance and/or build facilities to lure new industries or businesses into their areas. The objective of course is to provide more jobs, growth, and so forth, and generally stimulate the local economy. The money to finance these projects can be obtained by selling industrial revenue bonds. Typically the income from rental or lease of the facility is the only backing for the bonds. If a large local industrial tenant shuts down or moves out or fails to pay rent, the bonds could be in trouble. Also the tax exemptions on these bonds have been limited in recent years. For these reasons industrial revenue bonds typically are required to yield higher interest rates before investors will purchase them.

5. Housing authority bonds. Many local governments have housing authorities to plan, construct, and manage housing for low-income families (e.g., Gainesville Housing Authority). To assist them, the Housing Assistance Administration, an agency of the federal government, is authorized to guarantee annual contributions toward repayments of bonds issued to finance the housing. Thus these bonds indirectly have federal government backing.

BOND RATINGS

Evaluating and rating of all types of bonds are a necessary part of the bond market. Most fixed income securities, such as bonds, are regularly evaluated and rated in terms of their overall quality, that is, the perceived ability of the entity to pay the obligation involved in insuring the bond. Three private rating organizations provide this service: Standard & Poor's Corporation, Moody's Investor Service, and Fitch Investor Service. They each use a combination of letters to indicate *relative* quality of bonds. The highest rating is AAA (Standard & Poor's) or Aaa (Moody's) which represents issuers with a strong capacity to pay principal and interest. A slightly lower quality rating would be AA (Aa) and the ratings descend to D (Standard & Poor's) for issues in default with principal and/or interest payments in arrears. While differences do occasionally appear, the ratings services generally report similar ratings and are heavily relied upon by investors.

MECHANICS OF FUNDING PUBLIC WORKS THROUGH
A MUNICIPAL BOND ISSUE

Steps through which a municipal bond issue for public works or other investment by a local government must go before the money for the project is in hand will vary somewhat with the agency and the type of bond issue, but will follow generally the steps listed below for a typical municipality. As two typical cases, consider (1) a revenue bond issue for a parking facility, and (2) a general obligation bond issue for a swimming pool, recreation center, and park facility for a city.

1. *Revenue bond.*
   a. The city manager and staff study the city's needs. They draw up a list of needed facilities of top priority. They recommend financing these facilities by borrowing money on revenue bonds.
   b. Bond consultant is brought in. He studies the city's financial position and renders an opinion and recommendations on how much can be borrowed, how well the bonds should sell, and what bond rating and interest rate may be expected.
   c. The city commission passes a motion authorizing the city attorney to draw up a bond covenant (often in consultation with a bond counsel) and seek validation by the method required in that particular state.
   d. The bond covenant is drawn, and approved in resolution form by the city commission. The bond covenant, as the name implies, is a list of promises made to the potential bond buyers. The promises include safeguards of the investment, such as pledges of escrow, ratio of income to principal and interest (P&I) payments, retaining competent consultants, and so forth.
   e. The bond issue is then submitted for validation. This procedure varies from state to state. In North Carolina and New Jersey there are state commissions to which municipal bond issues must be submitted for approval. In Texas, bond issues must be submitted to the attorney general's office for approval. Other states have other procedures. In Florida all bond issues must be validated by the circuit court after a properly advertised hearing. Appeals by aggrieved parties may be taken to the Florida Supreme Court. In accordance with these requirements, for the example at hand, the city attorney requests the circuit court to hold a hearing on the proposed bond issue and if no reasonable arguments to the contrary are presented, the court approves (validates) the bond issue as a legal exercise of the city's powers. The validation proceeding protects the bond buyer by preventing a subsequent suit by irate taxpayers or others seeking to challenge the city's

action. City officials are legally barred from acting outside of their constituted authority. Should a plaintiff prove in a successful suit that the bond money was obtained for purposes or by procedures not specifically authorized by law, the bonds could subsequently become worthless. After successful validation proceedings, any would-be plaintiff is informed that the question of the legality of the bonds has already been tried and upheld by the court.

f. The bond consultant tries to aim for a lull or dip in interest rates on the bond market to maximize the amount of money the city will receive and announces the sale of the bonds. Prospectuses are sent out, and presentation date and sale date are arranged.

g. Prospective buyers and bond raters are invited to a meeting to interview top city officials. An informal presentation and sales pitch are given in support of the bonds. Involved are growth rate, stability, progressive outlook and the rosy future for the area.

h. The rating services (Moody's and Standard & Poor's) rate the bonds (for a fee). A city may expend considerable effort to convince the rating services that a high rating is warranted, since a high rating results in a lower interest rate.

i. On the sale date wholesale buyers (bond brokers) arrive with sealed bids in terms of a net interest rate. These are opened at the specified time and the bids read aloud. A few days later, after checking on qualifications, the qualified bidder with *lowest net interest rate* pays the city the present worth price and in return gets the bonds. The qualified bidder usually is a broker or brokerage group who then retails the bonds at the going market rate to individual investors, banks, or other institutions.

j. The bonds are then serviced by a designated fiscal agent, usually a bank that contracts this service for a fee. Servicing includes accepting regularly scheduled deposits from the city for payment of interest coupons and redemption of the bonds as they mature.

2. *General obligation* (GO) bonds differ only in that a referendum is held between (d) and (e). A referendum is not just a routine step in which the approval of the voters automatically occurs. Ordinarily, between one-half to one-quarter of all GO bond referenda fail. A number of good articles have been written on planning successful campaigns to promote voter approval of GO bonds. (Refer to Bradshaw and Mercer, "The Anatomy of a Successful Capital Improvements Bond Campaign," *Nation's Cities*, Aug. 1972, p. 27, and *Guide Books, How to Pass a Water Bond Referendum*, Cast Iron Pipe Research Association, 1211 West 22 St., Oak Brook, Illinois 60521.)

FEASIBILITY OF PROJECTS FINANCED BY REVENUE BONDS

Since a bond is essentially an IOU promising to pay interest and principal on the agreed upon dates, potential buyers of the bonds are quite interested in learning as much as possible about the anticipated costs and income (and consequent ability to make P & I payments when due) of the project that the bond funds are to finance. For this reason the sponsors of the project will usually have available a feasibility study for the project. These feasibility studies are typically done by a consulting engineer during the initial planning stages of the project and are often the basis upon which the sponsors make a firm decision on whether or not to proceed with the project. A part of the study usually details how much money must be borrowed by means of bonds and provides a schedule of repayments of the bond funds. This repayment schedule shows how much revenue is needed for payment of interest as well as principal needed to repay the bonds that mature each year. In order to adequately protect their investment, bond buyers prefer that either or both of two standard provisions be included in the bond covenant. The first provision is that a reserve fund be set aside, usually equal to one year's payment of principal and interest. The second provision is that the user's rates be set so that the anticipated income designated for P & I payments be equal to between 1.25 and 1.5 times the actual amount needed for P & I payments. The actual feasibility study and accompanying background information frequently exceeds 30 or more pages in length, including full-color pictures and attractive binding. A simplified example of the project cost analysis involved in a typical report follows.

Example 17.1 *(Portion of a simplified feasibility study for an authority owned parking garage financed by revenue bonds)*

Assume that a downtown parking authority (DPA), which is an agency of the city authorized to issue revenue bonds, proposes to construct a 200-stall parking garage in a specified downtown location. They engage a consulting engineer (CE) to do a feasibility study and determination of hourly parking charges. The CE presents the following information based on calculations and current field surveys.

Interest: The project is to be financed by revenue bonds. The P & I on the bonds will be repaid by the income revenue collected from parking fees. The interest rate on the bonds of this quality is currently estimated to be 7 percent. (The actual interest rate will not be known until the bonds are let out for bid.)

Construction cost: The construction cost for the project is estimated at $3,000 per parking space for a total of $3,000 × 200 spaces = $600,000.

Cost of land: Estimated cost of the land required for the structure is $200,000.

Life of project: The project is expected to be in service for 30 years after the date of completion.

Salvage value: After 30 years of service the parking structure is estimated to have a market value of $400,000, either for resale, refinancing, or salvage. However, any estimate of actual market value that far into the future is questionable at best, and not considered as high-quality security for the bondholders. Therefore the anticipated income from salvage is not considered as income available for repayment of the bonds. Whatever value remains at that time will simply be an additional asset of the DPA, accruing over the life of the structure.

Time of construction: It is expected to take two full years between the time the bonds are sold and the time the parking garage is open for business. The bond funds not needed for current progress payments to the contractor or other similar costs will be in interest-bearing accounts during this time and will earn income for the project. However, this income will be allocated for contingencies and will not be counted in these calculations.

First income: One full year's income is expected to accrue during the first year of operation. That portion not allocated to O & M will be credited at the end of year (EOY) 3.

Expected patronage: Since the project is located in a downtown area that is expected to suffer a continuing shortage of parking spaces, the level of occupancy should continue to be relatively high and constant over the life of the structure. Current surveys indicate that similar parking garages in this area are experiencing 90 percent occupancy for nine hours per day, five days per week for 52 weeks per year.

Operating and maintenance costs (O & M): These costs are expected to average $400 per space ($80,000/yr) during the first year of operation and increase at a rate of 5 percent per year thereafter. O & M costs will commence when the project opens at EOY 2, and will be paid out of working capital which in turn is constantly replenished from current income.

Working capital: To finance the day-to-day operation of the parking structure, the DPA needs working capital of $15,000 (approximately two months of O & M costs). This is actually a revolving fund from which current expenses are paid and income deposited. It will be needed for the entire life of the structure. The DPA decides to amortize this fund as part of capital costs, so that when the bonds are paid off the working capital fund will still be available for the continued operation of the structure if needed. When no longer needed, it becomes an asset of DPA accrued over the life of the project.

Parking fee rate increases: Parking fees may be increased each year by an amount sufficient to cover the increasing O & M costs.

Bond P & I: The bonds are scheduled for sale at EOY 0, as funds are required at this time to buy land and begin construction. During the two-year construction period, interest only will be paid, with one interest payment due at EOY 1, and another at EOY 2. Extra funds must be borrowed to finance these interest payments, since the $800,000 previously designated is needed to buy land and construct the building. The first payment of both interest and principal is scheduled for EOY 3. The payment of principal in this case means that a certain number of bonds will be scheduled to mature at that date. In succeeding years from EOY 3 through EOY 32, bonds will be scheduled to mature in accordance with the amount of income available for payment of principal as determined by the modified amortization schedule.

Bond covenants to protect the bondholders: To enhance the marketability of the bonds (and hopefully induce investors to buy at a lower interest rate) the DPA covenants to:

1. Set up a reserve fund equal to the first year's payment of P&I.
2. Set the fees for parking at a level sufficient to cover P&I payments by a factor of 1.25.

How much to borrow? The project needs the funds listed below on the dates noted. Since bonds are usually sold in $1,000 denominations the total amount borrowed and the maturity schedule for the bonds will be rounded to reflect this.

1. $800,000 for land and construction of the parking structure, needed at time zero.
2. Working capital fund of $15,000, needed to begin parking operations at EOY 2.
3. Interest payments during the two-year construction period at 7 percent on whatever funds are borrowed. Payments of interest are due at EOY 1 and EOY 2. The amount needed is as follows. Let B represent the amount borrowed, including the $800,000 for land and building, and the $15,000 working capital fund, as well as the two annual payments of 7 percent interest. Then,

   $$B = \$815,000 + 2 \times 0.07B$$

   Solving for B yields,

   $$B = \$947,674$$

   the total amount needed to cover the $815,000 plus the two interest payments.
4. Reserve fund equal to one year's payment of P&I. The reserve fund may be treated as a self-sustaining trust fund whose assets are reinvested elsewhere at 7 percent if possible. The interest income from the investment is used to set up the reserve fund. These bonds would be scheduled to mature at the end of the 30-year service life of the parking structure, at which time the reserve fund is liquidated and the bonds paid off. If the reinvestment rate is more or less than 7 percent, an annual charge or credit is added to account for the difference.

A cash flow diagram is shown in Figure 17.2 to illustrate the incomes and expenditures involved.

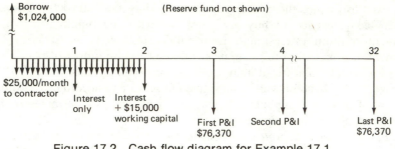

Figure 17.2 Cash flow diagram for Example 17.1.

Repayments: The bonds may be repaid in equal annual installments or by any other reasonable means mutually convenient to the lenders and borrowers. If the DPA were anticipating higher income in future years they could schedule more bonds for maturity in later years. For this example equal annual payments are used. Annual repayments will be made on the $947,674 amount calculated above. In addition a future lump sum payment will be scheduled at the end of 30 years to liquidate the reserve fund, using the proceeds from that fund. Thus, the equal annual payments required are

$$A = \$947,674(A/P,7\%,30) = \$76,370$$

The amount required in the reserve fund is equal to one annual payment of P&I, so the total amount to be borrowed is

$ 947,674

  76,370

$1,024,044, rounded to the nearest $1,000 to $1,024,000 by reducing the reserve fund by $44 to $76,326

These funds will be spent as follows.

EOY

0	$	76,326	establish reserve fund trust, and reinvest at 7%
		200,000	purchase land
0–2		600,000	progress payments to contractor during construction
1		66,337	interest on $947,674 @ 7% during construction
2		66,337	interest on $947,674 @ 7% during construction
2		15,000	working capital revolving fund for operations
		$1,024,000	

The repayment schedule for the bonds may be determined by using an amortization schedule, with minor modifications. The bonds are usually sold in denominations of $1,000, and therefore must be scheduled to mature in whole multiples of $1,000. A normal amortization schedule may be modified to take this into account as shown in Table 17.1.

Income, charges for parking: The amount charged for parking must meet two criteria:

1. The amount charged must be sufficient to cover O&M costs, as well as P&I costs to repay the bonds.
2. The amount charged must be competitive with other parking structures in the area.

The amount required to cover costs is as follows. The amount of space-hours available per year is

200 spaces×9 hr/day×5 days/week

$$\times 52 \text{ weeks/yr} \times 0.9 \text{ occupancy} \quad = \quad 421,000 \text{ space-hr/yr}$$

hourly charge for P&I $76,370/421,000 = $0.181/hr

25% for bond covenant coverage of 1.25 = 0.045/hr

hourly charge for O&M $80,000/421,000 = 0.190/hr

total hourly charge required $0.416/hr

During the second year the O&M costs are expected to rise by 5 percent so the charges would be raised by $0.05 \times 0.190 = \$0.010/hr$. Other increases are expected to follow in like manner throughout the 30-year life of the project.

Allocating the hourly charges: According to the above results the hourly charges during the first year should average $0.416 per hour. In order to reach this average it may be desirable (or more equitable) to charge different amounts to different classes of customers. For instance higher occupancy is encouraged by charging all-day parkers less per hour than transient cars parked for a shorter period of time. If studies indicate that 30 percent of the cars are parked all day, and 70 percent are transient, then a tentative schedule of rates could be determined using the following equation where H represents the hourly rate for all-day parkers and T represents the hourly rate for transients.

$$0.3 \times H\$/hr + 0.7 \times T\$/hr = \$0.416/hr$$

Since there are two unknowns in the equation, the value of one will have to be assigned. For this case suppose that a reasonable daily rate for all-day parkers (H) is $2.00 per day. The hourly rate for transient parkers

Table 17.1 MODIFIED 30-YR AMORTIZATION SCHEDULE: EOY*

EOY*	BONDS OUTSTANDING†	INTEREST PAYMENT I @ 7%	CALCULATED ANNUAL PAYMENT, A, OF P&I	INCOME AVAILABLE FOR PAYMENT OF $P (A - I)$	USED FOR PAYMENT OF P (VALUE OF BONDS MATURING THIS YEAR)	RESIDUE OF P & I LEFT OVER THIS YEAR	CUMULATIVE RESIDUE NOT USED FOR P
3	$947,674	$66,337	$76,370	$10,033	$10,000	+ $ 33	$ 33
4	937,674	65,637	76,370	10,733	10,000	+ 733	769
5	927,674	64,937	76,370	11,433	12,000	− 567	202
6	915,674	64,097	76,370	12,273	12,000	+ 273	475
7	903,674						

*End of year counting from time bonds were issued at beginning of construction period.
†Bonds outstanding except self-sustaining reserve fund, to be liquidated at the end of 30 years with proceeds from the fund plus income that year. ($1,024,000 − 76,326 = $947,647.)

may then be derived as

$$0.3 \times \$2.00/9 \text{ hr} + 0.7T = \$0.416$$

$$T = (-0.3 \times \tfrac{2}{9} + 0.416)/0.7 = \$0.499/\text{hr}$$

If desired this could be broken down further into different charges per hour for one hour, two hours, one-half day, and so forth, providing that the volume of each type of customer can be estimated with reasonable accuracy. The resulting charges are customarily rounded upward to the nearest convenient nickel, dime, or other convenient module of our monetary system. Thus, for the above example, if all the projections prove accurate, the following is one of a number of alternative rates sufficient to amortize the structure. In addition, adequate safeguards are provided for the bond-holders and a significant equity accrues for the DPA.

All-day parkers—$2.00/day
Hourly parkers—$0.50/hr

The consultant must evaluate whether or not these rates are competitive and comment on his findings.

CORPORATE BONDS FOR CAPITAL PROJECTS

Private corporations do a major part of their capital financing through issuance of corporate bonds. These are not tax exempt and frequently not as secure as many municipals, and therefore require higher interest rates to attract investors' money. There are two principal types.

1. Mortgage bonds, backed by some guaranteed right, usually a mortgage to the property being financed.
2. Debenture, backed by the full faith and credit of the corporation, but not by a specific mortgage.

INTEREST RATE

The interest rate on bonds is influenced not only by risk and supply and demand for loan funds, but also by inflation. Conservative investors seem to traditionally anticipate about 3 percent return on a good liquid (can be easily sold or liquidated, "turned into cash," should the desire or need occur) investment after inflation and taxes. On municipal bonds, since there are usually no taxes and the security is typically sound, inflation is a prime consideration. Bonds are typically long-term investments and the investors who buy them are more concerned about long-term inflation.

Until recent years even though there were short peaks in inflation in excess of 10 percent, the average long-term inflation rate was averaging less than 6 percent. Investors were satisfied with yields of 6 to 8 percent on tax exempt bonds of high quality. More recently, however, inflation has reached

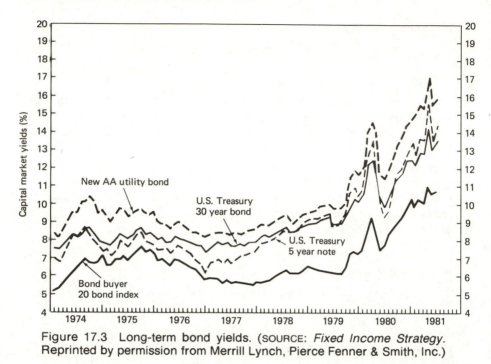

Figure 17.3 Long-term bond yields. (SOURCE: *Fixed Income Strategy.* Reprinted by permission from Merrill Lynch, Pierce Fenner & Smith, Inc.)

unprecedented levels (peaks of up to 20%) and investors have been reluctant to purchase bonds with traditionally lower yields. Some government entities have been forced to either cancel prospective bond issues until more favorable rates could be obtained, or pay uncharacteristically high interest rates for their borrowed money. Buyers of lower-quality bonds expect a higher rate of return (interest) to compensate for the additional risk. (If a bond has a risk probability of 1 chance in 50 of defaulting, then an extra 2% return is needed to compensate for the risk.) A graph of bond interest rates is shown in Figure 17.3.

RESALE VALUE

As discussed earlier there is an active market in trading bonds of all types which is separate from the issuing agency or corporation. While there are "registered" issues, the more common type of bond is the "bearer bond" in which the issuer must redeem the periodic coupons, for the value stated, to the bearer, and must pay the cash value of the surrendered bond at maturity. Once the issuer makes the initial sale and receives its funds, it no longer enters into any of the subsequent sales transactions, nor records ownership changes of bearer bonds.

When a bond is resold before maturity the buyer may offer more or less than the face value, as presumed earlier. The following example illustrates the calculations involved.

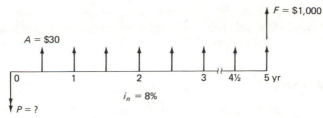

Figure 17.4   Cash flow diagram for Example 17.2.

Example 17.2

A $1,000 bond paying 6 percent semiannually ($30 semiannually) has five years to go until maturity. The current market for these bonds is 8 percent. That is, investors are willing to buy the bond if they can receive a nominal interest rate of 8 percent on their investment. How much will the bond be worth?

SOLUTION

The cash flow diagram for this problem is shown in Figure 17.4. Find the present worth of the remaining ten coupons representing $30 income every six months plus redemption value of $1,000 after five years, at 8 percent. Since the payments are semiannual, $n = 5 \times 2 = 10$ and $i = 8/2 = 4.0$ percent.

PW of 10 semi-annual payments $= P_1 = A(P/A, i, n) = \$30\,(P/A, 4\%, 10) = \$243$
8.1109

PW of the $1,000 redemption value in 5 yr $= P_2 = F(P/F, i, n) = \$1,000\,(P/F, 4\%, 10) = \underline{\;676\;}$
0.6756

total $P_1 + P_2 = \underline{\underline{\$919}}$

The bond should sell for $919 for an 8 percent nominal interest rate yield to maturity ($i_e = 8.16\%$).

TWO YIELDS

Bonds with long-term maturities are frequently listed with two different figures for yields. The first is (a) "current yield" and the second (b) "yield to maturity." The current yield is simply the amount of coupon interest received each year divided by the *current* price of the bond. To determine the current yield for Example 17.2 assume a 6 percent bond has two semiannual coupons for $30 payable each year, and the bond is currently selling for $919; the current yield is determined simply as $60/\$919 = 6.53$ percent.

The "yield to maturity" is simply the rate of return including the redemption value (face value) of the bond upon maturity, which is shown in Example 17.2 as $i = 8\%$ or $i_l = 8.16\%$.

THE BOND MARKET

Daily reports on activities of selected portions of the bond market are available in most daily newspapers. The reports are confined largely to trading on the New York Stock Exchange market and to corporate issues rather than municipals. A typical excerpt appears in Figure 17.5 (the format for presentation of bond market data will vary somewhat with each reporting agency). Interpretation of the data is not difficult, as illustrated by the following example. Step 1 of course is to find a particular bond issue of interest. For instance, find the first AT&T bond listed. AT&T is the abbreviation used for the American Telephone and Telegraph Company, whose activities and finances are detailed in reference journals available at most libraries and stock brokers' offices. An example illustrating the meaning of each item in a typical bond listing follows.

Example 17.3

Find the *first* American Telephone and Telegraph Company bond listed in Figure 17.5 and explain the listing.

SOLUTION

AT&T 8.80 s 05 AAA 14 114 $64\frac{5}{8} - \frac{1}{8}$

Item 1: Abbreviations of the company name (AT&T)

Item 2: The next number and letter following the name is 8.80 s. This tells the interest coupon rate carried by this particular bond and the payment period. That is, assuming the face amount of the bond is $1,000, the bond coupons can be cashed in for 8.80 percent of $1,000 or $88.00 each year. The "s" means serial payments. If the bond has semiannual coupons there would be two coupons due each year, dated six months apart, at $44.00 each.

AT&T	8.80s05	AAA	14.	114	64 5/8	− 1/8
AT&T	8 5/8s07	AAA	14.	193	62 3/4	− 1/4
AT&T	7 3/4s82	AAA	8.2	35	95	− 5/8
AT&T	7 1/8s03	AAA	13.	335	53 5/8	− 1/4
AT&T	7s2001	AAA	13.	212	53 7/8	− 1/4
AT&T	4 3/8s85	AAA	5.7	55	76 3/8	+ 1/2
AT&T	3 7/8s90	AAA	7.2	66	54	......
AT&T	3 1/4s84	AAA	4.2	26	78	− 1/2

Figure 17.5 Bond quotes. (SOURCE: UPI, St. Petersburg Times, August 29, 1975)

Item 3: The next number, 05, gives the last two digits of the maturity date for the bond, the year 2005. In that year the bond will be redeemed by AT&T for the face amount, usually $1,000.

Item 4: The next letters, AAA, indicate the bond rating according to the rating service that specializes in evaluating the security of bonds. Unrated bonds are not necessarily inferior, they just may not have applied for or received a rating.

Item 5: The next numbers indicate the current yield is 14 percent instead of the coupon rate of 8.80 percent. This results from the bond selling at below the $1,000 face value.

Item 6: The number of bonds sold on the day reported is shown in the next column, in this case 114 bonds were sold.

Item 7: The market price of the last bond sold for the day (closing market price) is listed in the next column as a percent face value. In this case the listed price is $64\frac{5}{8}$. If the bond has a face value of $1,000 (as many bonds do) then the last bond sold before market closing sold for $64\frac{5}{8}$ percent of $1,000 or $646.25.

Item 8: The change from the last closing price is listed in the next column as $-\frac{1}{8}$. This means the bonds are selling for $\frac{1}{8}$ of 1 percent lower or in this case $1.25 less for a $1,000 bond than at the close of the previous market day.

BUYING AND SELLING BONDS

Bonds may be purchased through almost any broker listed in the yellow pages of the local telephone directory under "Stock and Bond Brokers." A typical commission is $10 for the purchase or sale of a bond with a face value of $1,000. While these commissions are comparatively small, they should be taken into account when calculating yield or value of a bond.

Investment Decisions

Since the market value of a bond fluctuates with the market until maturity, care must be taken not to confuse the value or yield of a bond at the present time with the value or yield at the time of purchase, or any other time. For example, should a $1,000 bond with a 7.9 percent coupon be sold in order to invest in some alternative of equal quality but yielding a 10 percent return? The answer depends not on the fact that the bond originally cost $1,000 but on the current market price and resulting yield to maturity. Whatever has been spent previously is gone, and is classified as "sunk cost." Sunk cost has little or no bearing on current yield or yield to maturity (an exception may occur where credit for tax losses is available). The following example illustrates the principal of sunk costs when comparing a bond currently owned with an alternative investment.

Example 17.4

Assume the current date is February 15, 1981. A friend of yours had $1,000 to invest 5 years ago this month, so she purchased a Consolidated Edison Company (Con Ed) bond for $1,000 ($990+$10 commission). The bond paid 7.9 percent annually by means of the usual coupons redeemable for $79 per year every February 14. The maturity date on the bond is February 14, 2002. The friend now has an opportunity to invest any amount in a mutual fund that is currently paying 10 percent on whatever is invested. The friend asks your advice on whether to sell the bond and invest the proceeds in the mutual fund at 10 percent. You look in the paper and find the following listing for the current selling price of the bond. An additional expense of $10 commission is involved if the bond is sold. Assume the mutual fund is the same quality (risk) investment as the Con Ed bond.

Bond Listing: Con Ed 7.90 02 12.5 25 $63\frac{3}{8}$

a. What is the current yield for this bond?
b. What is the yield to maturity for an investor buying this bond at today's price as listed above plus $10 commission?
c. Would you advise the friend that the yield on the bond is about 7.9 percent and the bond should be sold and the proceeds reinvested in the mutual fund at 10 percent?
d. Calculate the yield to maturity of the bond from the friend's point of view, as a current *owner* of the bond and advise on the basis of comparative yields whether or not to keep the bond or reinvest in the mutual fund.

SOLUTION

a. The current yield may be read directly from the market quote as 12.5 percent or may be calculated as $7.9/63.375 = 12.47$ percent.

b. The yield to maturity for a new investor is calculated by trial and error as in rate of return problems in Chapter 10.

	$i=12\%$	$i=13\%$
cost of purchase $P_1 = (-\$63\frac{3}{8} \times 10) - \$10 =$	$-\$643.75$	$-\$643.75$
annual coupons $P_2 = +79\,(P/A, i, 21) =$	$+\ 597.40$	$+\ 561.02$
maturity value $P_3 = +1,000\,(P/F, i, 21) +$	92.56	$+\ \ 76.80$
	$+\$\ 46.21$	$-\$\ \ 5.93$

Interpolation yields

$$\frac{46.21}{46.21+5.93} \times 1\% + 12\% = 12.9\% \text{ yield to maturity}$$

c. No. The yield on the bond at the time of purchase *was* about 7.9 percent but it no longer is. It currently is higher than the alternative 10 percent available, so the bond should be retained. The difference in the $1,000 price paid originally for the bond and the currently available price of $623.75 (net after payment of commission) is lost. If interest rates go down part of the loss could be recovered. But if the decision is made on the basis of the current market price, then the bond should be retained for the best yield.

d. The friend's point of view should see a bond with a present value of $623.75 (net after paying $10 sales commission), yielding $79 annually for another 21 years, and $1,000 lump sum at the end of 21 years. All other income and expense have occurred in the past and should be disregarded (except where tax losses or gains may be involved). The rate of return for the friend may be calculated similarly to that in part (b) above as follows.

		$i = 12\%$	$i = 13\%$
cash value (net) $P_1 = (-\$63\frac{3}{8} \times 10) + \$10$	$=$	$-\$623.75$	$-\$623.75$
annual coupons $P_2 = +79(P/A, i, 21)$	$=$	$+ 597.40$	$+ 561.02$
maturity value $P_3 = +1,000(P/F, i, 21)$	$=$	$+ 92.56$	$+ 76.80$
		$+\$ 66.21$	$+\$ 14.07$

Extrapolation yields

$$\frac{14.07}{(66.21 - 14.07)} \times 1\% + 13\% = 13.3\% \text{ yield to maturity}$$

In other words, the friend, if she sells, will lose $1,000 - 623.75 = \$376.25$ and will be able to invest only $623.75 in the mutual fund earning 10 percent. But, her $623.75 *value* in the bond will earn 13.3 percent at maturity, so she is better off by keeping the bond.

BEFORE- AND AFTER-TAX COMPARISONS

When comparing investment opportunities, investors are most concerned about the after-tax income or rate of return rather than income before taxes. Tax payments are really just another normal expense involved in gaining income and the gross income is not nearly as important as how much is left over after all expenses (including taxes) are paid. When considering investment opportunities in bonds, yields of tax exempt municipals routinely compete with taxable corporates on the basis of yield after taxes. The following example illustrates such a comparison after the cost of tax payments are taken into consideration.

Example 17.5 (*After-tax rate of return*)

Assume the year is 1981 and your firm has some idle cash that should be kept fairly liquid but at an acceptable after-tax interest rate. Your boss is particularly interested in two alternative bonds and asks you to calculate the after-tax yield to maturity on each of them.

The first is an American Airlines $10\frac{7}{8}$, 92 bond rated BBB, currently selling at $95\frac{1}{4}$. The second is a tax exempt Florida Turnpike bond 7.10, 2005 bond rated BBB and selling for 97.

Your firm is in the 38 percent tax bracket (capital gains are taxed at $0.4 \times 38\% = 15.2\%$) and expects to pay a $10 commission on each bond purchased. The bonds are assumed to have a $1,000 redemption value with no commission paid upon redemption.

SOLUTION

The after-tax yield is found by the customary trial, error, and interpolation method. Both bonds have coupons that come due semiannually, but a reasonably close result may be obtained by treating the income more simply on an annual basis. For a rough estimate helpful in approximating the first trial, i, find the current yield and adjust it for appreciation or depreciation upon redemption. In this case the current market price is only slightly under the redemption value, so the yield to maturity should be only slightly under the current yield. For convenience the problem may be worked in terms of the percentage values given in the market quotes, rather than in terms of actual cost and income (which are simply the percentages multiplied by 10).

Florida Turnpike Bond. Current yield $7.1/(97+1)=7.2$ percent. Since the yield to maturity is probably less than 8 percent but greater than 7 percent, use 8 percent for a first trial i.

$$\text{cost} = P_1 = \qquad\qquad\qquad\qquad -98.00$$

$$\text{interest income} = P_2 = 7.1(P/A, 8\%, 24) = \qquad +74.75$$

$$\text{redemption income} = P_3 = (100 - 0.152 \times 2.00)(P/F, 8\%, 24) = \underline{+15.72}$$

$$-7.53$$

Since the negative sign on the result indicates that the actual return is less than 8 percent, the next trial is made at 7 percent.

$$P_1 = \qquad\qquad\qquad\qquad -98.00$$

$$P_2 = 7.1(P/A, 7\%, 24) = \qquad\qquad +81.43$$

$$P_3 = (100 - 0.152 \times 2.00)(P/F, 7\%, 24) = \underline{+19.65}$$

$$+3.08$$

Interpolation then yields:

$$\frac{3.08}{3.08 \mid 7.53} \times 1\% + 7\% = \underline{\underline{7.3\%}} \text{ yield to maturity}$$

Since this is a tax free bond (the interest payments are not taxable), the before-tax yield to maturity is almost the same as the after-tax yield. Only the 2-point capital gain is taxed, and that tax is payable in the year of redemption of the bond.

American Airlines Bond. Current yield $10.875/(95\frac{1}{4}+1)=11.30$ percent. This is a considerably greater before-tax current yield than the Turnpike bond. The first trial i is estimated by finding the current yield after taxes and adjusting as necessary. The tax rate is 38 percent so the after-tax income is 62 percent of the coupon amount. Thus, current after-tax yield $0.62 \times 10.875/96.25 = 7.01$ percent. The yield to maturity will be slightly more, since the redemption value less capital gains tax is still slightly more than the purchase price plus commission. Calculating the yield using a first trial i of 7 percent leads to the following.

$$\text{cost} = P_1 = \hspace{5cm} -96.25$$

$$\text{interest income} = P_2 = 0.62 \times 10.875(P/A, 7\%, 11) = \hspace{1cm} +50.56$$

$$\text{redemption income} = P_3 = [100 - 0.152(3.75)](P/F, 7\%, 11) = \underline{+47.24}$$

$$+ 1.55$$

A second trial at 8 percent produces the following results.

$$P_1 = \hspace{4cm} -96.25$$

$$P_2 = 0.62 \times 10.875(P/A, 8\%, 11) = \hspace{1cm} +48.14$$

$$P_3 = [100 - 0.152(3.75)](P/F, 8\%, 11) = \underline{+42.64}$$

$$- 5.47$$

Interpolating yields

$$\frac{1.48}{1.55 + 5.47} + 7\% = \underline{\underline{7.2\%}} \text{ after-tax yield to maturity}$$

Discussion: The comparative after-tax yields of the two bonds are:

Florida Turnpike $= 7.3\%$

American Airlines $= 7.2\%$

The difference is not large, but the tax exempt Florida Turnpike bond does yield a better after-tax return. The ratings on the two bonds are the same, so their relative risks are approximately the same. Therefore, choose Florida Turnpike.

SUNK COSTS, BUY-SELL DECISIONS

Investors in bonds are dealing with a constantly fluctuating market although the changes in the bond market are not usually as rapid and large as changes in stock or commodity markets (Figure 17.3). Sometimes the interest rates decline, with consequent rises in the average price of bonds, and other times the reverse happens. Astute investors frequently try to upgrade their portfolios by selling current holdings and buying more attractive issues. The decision to sell or buy is usually based on a number of factors, such as the quality of the bond (fiscal soundness, marketability of product, amount and type of bond collateral pledged, etc.) as well as the income and yield. When comparing the yields on current versus prospective holdings, a question sometimes arises concerning the true value of the current holding. If the investor paid 96 for a corporate bond five years ago (95 market + 1 commission), and the current market value of the bond is 84, which figure should be used for current value? Actually the current value is really the amount of cash that can be realized in exchange for the bond (see Example 17.4d). In most cases this is the current market value, less commission, plus or minus any tax considerations. Where a capital gain is involved income taxes must be paid (or credited). In the example at hand, the loss in value normally will result in a tax credit benefit. Thus, assuming a taxable $1,000 corporate bond with a market value of 84, the investor would receive $840 less a $10 commission. Since the bond is taxable, a capital loss credit may be available. The original purchase price of the bond does not enter into the calculation except in considering the tax loss or gain. The loss in value is considered as a "sunk cost" and should not be confused with the value of the bond to the investor. When comparing rates of return between this bond and a prospective replacement, the current cash value calculated above should always be employed.

SUMMARY

This chapter presents the cost analysis of bonds, both municipal (tax exempt) and corporate. A bond is simply a long-term public debt of an issuer that has been marketed in a convenient denomination, usually through a broker. The bond has a face-redemption value, together with a series of interest payments usually designated by a nominal interest rate of the face value, and a maturity date upon which the bond may be redeemed for its face value.

There is an active bond market in which bonds are bought and sold each business day. Depending upon a number of variables including the demand, money supply, bond rating and yield on competing investments, the bonds are traded for more or less than their face value. Thus, the actual interest earned on an investment in bonds may be less, or greater, than the interest paid on the face amount.

Bonds are commonly accepted as a means of acquiring needed funds for capital improvements for both government and private industry.

A.3) $-5,500+350(\%,7,20)$
$+5000(\%F,i2,20)$

$try\ i=6\% \Rightarrow NPW= 73$

$i=6.13\%$

$i=7\% \Rightarrow NPW=-500$

PROBLEMS FOR CHAPTER 17, BOND FINANCING

Problem Group A

Given values for four of the five variables P, A, F, i, n, find the fifth.

A.1 A $5,000 bond with 8 percent annual coupons for $400 will mature in 20 years. It can be purchased for $4,300.
  a. Find the current yield.
  b. Find the yield to maturity.

A.2 Same as Problem A.1 except the 8 percent coupons are paid semiannually ($200 every six months).

A.3 A coupon bond is purchased for $5,500. Coupons are clipped from the bond and cashed in at the bank once a year for $350. (The bond has a face value of $5,000 with 7% coupons.) At the end of 20 years the last coupon is turned in at the bank for $350 and the bond is cashed in for $5,000. What is the before-tax rate of return?

A.4 An investor desires a municipal bond yielding 6 percent to maturity. A friend has one with a $2,000 face value, maturing in ten years and with 20 more semiannual coupons for $50 each coupon. What price should he offer to obtain a 6 percent yield? (Use $n=20$.)

A.5 A bond is in default and is not paying interest. It has a face amount of $10,000 and 20 annual coupons, each for $575 still attached ($5\frac{3}{4}$% on the $10,000 bond).
  a. The bond is supposed to mature in 20 years. It is rumored that the interest coupons will not be honored until the end of the 20-year period. At that time the project will be refinanced and all previous interest payments will be paid off, with interest on the interest payments compounded from the dates due, at the $5\frac{3}{4}$ percent rate. The principal will also be paid in full at that time. If an investor is willing to accept the rumored payment schedule as true, how much should he offer to pay for one $10,000 bond if he desires 10 percent compounded return on his investment over the 20-year period?
  b. If the $10,000 bonds are offered for sale now for $4,000 each and the above sequence of events occurs leading to a complete payoff, what yield to maturity would the investor realize?
  c. Assume the $10,000 bonds described in (a) continue in default until ten years *after* the maturity date and then pay off in full with compounded interest at $5\frac{3}{4}$ percent on all amounts from the dates due. If the selling price now is $4,000, what is the yield (rate of return) until pay-off date?

A.6 A port authority toll bridge is currently yielding net revenues of $100,000 per year, increasing by $10,000 per year. The bridge is old and needs replacement, and will be out of service for two years during reconstruction. Revenue bonds will be sold to finance the replacement. If the bonds draw an interest rate $i=7$ percent, what is the maximum amount of bonds that can be sold now and paid off within 25 years from now assuming the full amount of net revenues is used to pay off the bonds? Assume the existing bridge is taken out of service now, and the first income for repayment of bonds occurs at the end of year 3 in the amount of $130,000, increasing by $10,000 per year thereafter.

A.7 Same as Problem A.6 except that while the bonds show an interest rate of 7 percent, investors want to receive 9 percent on their investment.

A.8 A transportation authority wants to finance the construction of a bridge by borrowing money on bonds. How much can they borrow if the estimated toll receipts occur as listed below? The bonds are expected to have a 7 percent interest rate, and be paid off by the end of the twenty-fifth year. O & M costs will begin at EOY 3 at $40,000 per year and increase by $2,000 per year each year through EOY 25.

EOY	TOLL RECEIPTS
1	none (bridge under construction)
2	none
3	$ 15,000
4	$ 35,000
5–7	(increase $20,000/yr)
8	$115,000
9	$120,000
10–24	(increase $5,000/yr)
25	$200,000

A.9 A $10,000 bond with 7 percent coupons payable annually was purchased four years ago for $9,800 and sold today for $9,500.
   a. What before-tax rate of return was realized on this investment by the original purchaser?
   b. If the new purchaser holds the bond to maturity (eight more years) what before-tax rate of return will be realized?

Problem Group B

Compare alternatives. Sunk costs.

B.1 Ten years ago a client of yours paid $1,000 for a 5 percent bond (annual coupons of $50 each) that will mature 20 years from the date of purchase. She now would like to reinvest (sell that bond at market value) in another bond that sells for $920, has coupons of $95 per year, and will mature 15 years from now. The bond that she presently owns has a market value of $620. In addition, the brokerage fee on each sale or purchase is $10 per bond lump sum regardless of the amount. Advise your client on the most profitable alternative before any tax considerations are taken into account.

B.2 Your firm intends to invest some of its funds into bonds and is particularly interested in two bonds both issued by the same organization. The after-tax yield to maturity on both is 7.96 percent. The only difference is bond A has a maturity date 3 years from now and Bond B matures in 25 years.
   a. What advantage(s) does bond A have over B?
   b. What advantage(s) does B have over A?

B.3 Four years ago you purchased a computer science (Comp Sci) bond for $987.00 including commission. The bond has coupons redeemable for $60 per year and the bond is expected to be worth $1,000 upon maturity in 1994. Assume that the anniversary date on the latest coupon just happened to coincide with yesterday's date, so it will be a full year before the next coupon comes due. Assume that the maturity date on the bond will also coincide with yesterday's date but in the year of maturity. You now have an opportunity to invest any amount in a mutual fund that is currently paying 10 percent on whatever is invested and must decide whether or not to sell the bond and invest the proceeds in the mutual fund at 10 percent. You look in the paper and find the following listing for the current selling price of the bond. An additional expense of $10 commission is involved if the bond is sold. Assume the mutual fund is the same quality (risk) investment as the bond.

> Bond listing:
> Comp Si   6s 94 10.2 59...

B.4 An investor has a choice of investing $9,500 in an eight-year certificate of deposit yielding 9 percent compounded monthly (0.75% per month), or investing in an unrated $10,000 bond yielding 8 percent payable semiannually ($400 each six months) and maturing in eight more years. He can purchase the bond for $7,800 today.
   a. Compare the rates of return for the two investments.
   b. Discuss the relative merits of the two investments, including (1) the fact that the certificate of deposit cannot be cashed in early without paying a substantial penalty, (2) if the bond is purchased, what happens to the difference in money between the two investments ($1,700 in this case), and (3) the unknown risk associated with purchasing an unrated bond.

Problem Group C

Tax considerations.

C.1 An investor is seeking a corporate bond with at least a 7 percent yield to maturity after taxes. She is offered a corporate bond with a face value of $5,000 which she may buy at 80 (for $4,000). The bond will mature in ten years and has 8 percent coupons ($400/yr). The investor expects to pay an annual income tax of 30 percent on the interest income and an annual intangible tax of 0.1 percent on the face value of the bond. In addition, the difference between the $4,000 purchase price of the bond and the $5,000 redemption amount is taxable at the time of redemption at capital gains tax rate of 40 percent of the gain taxed at the ordinary rate. Does the after-tax yield exceed 7 percent? Show proof.

C.2 In an effort to put idle cash to work, your client purchased ten Gen Inst bonds four years ago today, at 89+1 for commission. The bonds each have a face value of $1,000 and the coupons pay interest of 5 percent once a year on an anniversary date coinciding with yesterday's date. The bonds will be redeemed on that same date 20 years from now. The bonds are currently selling at the following market price. If the bonds are sold, an additional cost of $10 per bond will be incurred. Gen Inst 5s xx 6.9 $72\frac{3}{4}$.

Your client now has an opportunity to invest in tax-free municipal bonds that have a yield to maturity of 6 percent.

Your client pays an income tax of 40 percent on current income including bond interest payments, plus a capital gains tax on any net gain between purchase price and selling cost. If he sells for less than his purchase price, he is entitled to a tax credit against other income at the same capital gains (now a capital loss) rate.

a. What is the current yield for this bond before taxes?
b. Find whether or not your client should sell his current holdings (ten bonds) and reinvest the funds in the tax-free municipals. Do this by determining whether a higher rate of return is obtained by leaving the cash invested in the bonds or selling the bonds and investing in municipals at 6 percent (calculate net return after paying commission costs). Assume either bond would be held until maturity.

C.3 Calculate the after-tax rate of return for the original investor of Problem A.9 if he has an effective tax rate of 27.2 percent.

A.9|

$P = 15\,000\,(P/A, 7\%, 23)\,(P/F, 7\%, 2) + 20\,(P/G, 7\%, 6)\,(P/F, 7\%, 2)$

$+ 100\,(P/A, 7, 18)\,(P/F, 7, 7) + 5\,(P/G, 7, 18)(P/F, 7, 7)$

$- 40\,(P/A, 7, 23)\,(P/F, 7, 2) + 2\,(P/G, 7, 23)\,(P/F, 7, 2)$

$P = 687,656$

Chapter 18
Home Ownership and Mortgage Financing, Owning Versus Renting

MORTGAGE FINANCING FOR HOME OWNERSHIP

Currently about 2 million housing units are constructed annually at a cost of almost $100 billion (comprising over one-third of the total construction market and about 3% of the nation's entire gross national product).

In the early days of our country's history, when the business of mortgage lending was in its horse-and-buggy stages, mortgages were difficult to obtain and prospective home buyers were expected to make down payments of about 50 percent or more of the purchase price. In addition the life of the mortgage was typically shorter, say five years, and the amount borrowed was repayable on the due date in one lump sum. Consequently, as few as one-third of American homes were owner-occupied (compared to about two-thirds now) and mortgage foreclosures were not uncommon.

Federal Housing Administration (FHA) and Veterans Administration (VA)

The mortgage lending industry has progressed and enlarged considerably since those early days. A major turning point occurred in 1934 when the Federal Housing Administration (FHA) was conceived by Congress to insure mortgages against default. Since 1934 the FHA has written mortgage loan insurance aggregating over $200 billion, covering over 10 million homes, more than a million living units in multifamily projects, and over 35 million property improvement loans. The FHA is one of the very few government agencies that is entirely self-supporting. It derives its income from fees, insurance premiums, and investments, and its insurance reserves are well over a billion dollars. The FHA does not build houses or lend money. It acts only as an *insurer* of privately made loans from approved lenders. Any FHA-approved mortgage lender, such as a bank, savings and loan association, or other financial institution can make insured mortgage or home improvement and rehabilitation loans.

A minimum down payment is required by FHA in order to qualify for an insured mortgage loan. The amount of down payment is revised and updated from time to time. The current (became effective 1980) schedule of down payments required by FHA, based on the purchase price of the house, is

AMOUNT OF LOAN	DOWN PAYMENT
Less than $25,000	3%
Between $25,000 and $67,500	$750 + 5% of amount over $25,000
Greater than $67,500	$2,875 + 100% of amount over $67,500

For example, the down payment required on a $20,000 house is just 3 percent of $20,000 or $600 down payment. The down payment on a $30,000 house is

3% of first $25,000	=	$ 750
5% of next 5,000	=	250
total down payment required =		$1,000

On a $70,000 house the down payment is calculated as

3% of first $25,000	=	$ 750
5% of next 42,500	=	2,125
all of remaining 2,500	=	2,500
purchase price = $70,000		$5,375 = down payment required

Several years after the establishment of FHA, the Veterans Administration (VA) was given a similar charge to guarantee loans for the nation's 30,000,000 veterans who served in the Armed Forces. The VA does not require any down payment on home purchases by qualified veterans.

At present (1980) the VA is authorized to compensate a mortgage (lender) for 60 percent of losses of up to $17,500 per mortgage. For example, assume a veteran purchases a $40,000 house with no down payment and assumes a $40,000 VA mortgage. Then for some reason the veteran defaults and the mortgagee (lender) forecloses. If the foreclosure sale brings in $35,000 for the mortgagee, the VA will pay $5,000 to the mortgagee to make up the full $40,000 owed on the mortgage. If the house brought only $20,000 at the sale, the VA would pay a maximum of only $17,500 and the mortgagee would lose the remaining $2,500 if he could not recover from the mortgagor-veteran. The creation of the FHA and VA had a very stabilizing influence on the industry and gave lenders more confidence in mortgages as an investment. Over the years this brought a greatly increased inflow of mortgage loan funds into this financial market area since the FHA insured against loss through foreclosure. (An extra $\frac{1}{2}$% of the unpaid balance is charged the borrowers to finance the insurance fund. If a mortgage is seriously in default, the mortgage may be purchased from the mortgage holder with money from this fund. The FHA then forecloses if necessary and resells the property.) One effect of FHA and VA on the housing market has been to make housing available to many who could previously not afford the high down payments. The average down payment in the latter 1960's was about 14 percent, while more recently it has dropped to about 7 percent. An increasing number of homes are sold with no down payment at all. In 1963 10 percent of all new homes sold with mortgages were sold with no down payment, whereas more current data show a rise of up to 17 percent.

By far the largest number of home mortgage loans are termed conventional loans in which the borrower usually pays 20 to 25 percent of the purchase price as a down payment and borrows the remainder from major lending institutions such as savings and loan associations. Since the mortgage on the property is the only protection the lender has, borrowers must demonstrate some ability to repay the loan.

The type of financing for new homes throughout the United States is given in Table 18.1. The FHA and VA insure mortgage loans obtained by home owners and developers through conventional lending channels. In addition the Government National Mortgage Association (GNMA) purchases mortgages insured by FHA and VA so that mortgage lenders may have additional cash to loan on more mortgages. The Federal National Mortgage Association (FNMA), a private-public corporation, also buys mortgages at auction from private mortgage lending institutions in periods

Table 18.1. TYPE OF FINANCING, ALL NEW HOUSES, 1971 THROUGH 1977

	PERCENT DISTRIBUTION						
	1971	1972	1973	1974	1975	1976	1977
FHA-Insured	25	17	8	7	9	6	7
VA-Guaranteed	9	10	8	8	8	8	8
Conventional	49	59	69	69	58	67	68
Farmers Home Adm.	(*)	(*)	(*)	(*)	7	5	4
Cash	17	14	15	17	18	15	14

SOURCE: Bureau of the Census, U.S. Department of Commerce.
*Withheld because estimate did not meet publication standards on basis of sample style.

of short money supply. These organizations were created by Congress to insure a healthy and efficient mortgage market since so much of our nation's new housing and commercial building is dependent upon mortgage financing.

Sources of Funds for Mortgage Loans

The primary sources of mortgage funds loaned in a typical year is shown below. A total of $46.3 billion was loaned on mortgages that year with the percent loaned by each group as shown:

savings and loan	52%
commercial banks	21%
federal sponsorship	14%
mutual savings associations*	8%
life insurance companies	3%
finance companies	2%
others	<1%

MORTGAGES IN THE INVESTMENT MARKET

There is an active market for buying and selling of mortgages. The professionals who do this as a business usually call themselves mortgage bankers or mortgage brokers. Mortgage brokers simply act as middlemen between borrower and lender. Mortgage bankers on the other hand offer a complete line of service to the investor. They will find and screen the property owner or potential buyer in need of a mortgage, lend the money, sell the mortgage to the investor and service the mortgage for him or her. Servicing includes collection, bookkeeping, following up on delinquent accounts, carrying escrow accounts for insurance and taxes, and periodically checking on the condition of the property. Typically about one-quarter of the mortgage banker's gross income comes from origination fees (front end charges or discount points paid by the borrower), and one-half comes from service fees to the investor. All of these services greatly assist in bringing money into the mortgage money market. The greater the supply of money available, the lower the interest rate and the easier it is for the prospective home buyer to obtain a mortgage loan.

WHAT IS A MORTGAGE?

A real estate mortgage and note is simply a promise to repay borrowed funds using real estate as collateral. They are normally long-term notes (20

---

*Mutual savings associations are very similar in nature to savings and loan institutions. They are confined for the most part to the northeastern states. They are state chartered and typically hold about 70 percent of their assets in mortgages.

to 35 years are common). In event of default the mortgage lender has a claim against the property and typically has recourse to court action requiring sale of the property at public auction to satisfy that claim.

A mortgage involves two basic items, (1) the note, and (2) the mortgage itself.

1. The note provides evidence of the debt. It describes the dollar amount, the interest rate, and terms and dates of repayment.
2. The mortgage gives the lender an interest in the property until the note is paid off. In 30 states (including Florida), the mortgage constitutes a lien on the property and goes with the property until paid off regardless of transfers of ownership. In the other 20 states the mortgage transfers title to the lender until the note is paid off.

Both the note and the mortgage must be properly executed in order to constitute a complete legal document.

Mortgages are usually freely transferable by either party. Mortgages are *not* transferable only if so stipulated in the agreement between the lender and borrower. In the absence of such a stipulation the mortgage lender is free to sell the mortgage, and the borrower is free to sell the property and let the new owner take over the mortgage with the property without permission of the other party. However, in case of default, each person who has previously assumed the mortgage is liable for any loss on the mortgage. The holder of the mortgage may look to each person who has sold the property subject to the mortgage as an individual guarantor of the mortgage amount.

SECOND AND THIRD MORTGAGES

A property owner may owe to more than one lender simultaneously, using the same property as collateral for all the loans. A mortgage securing a second loan is a second mortgage. Third and more mortgages can be taken out; the number is only limited by each subsequent lender's faith in the ability of the borrower to repay. Each successive loan is subsidiary to the one before it. In case of default the last lender may offer to take over all obligations and payments to the prior lenders in exchange for a quit-claim deed from the owner. For instance the third mortgage holder can offer to take over payments to mortgage holders numbers 1 and 2, in hopes of salvaging something from his interest. The owner clears up delinquent debts on the property plus salvaging a sagging credit rating. In case of foreclosure the first priority usually goes to payment of taxes (delinquent and current) and any other government claims. Then (provided there are no prior liens) the first mortgage holder gets all balance due (together with compounded interest on delinquent payments) before the second mortgage holder gets anything. The third is in the same position with respect to second. Consequently, at a sheriff's foreclosure auction sale the first mortgage holder is usually interested in bidding the price up high enough to cover the amount due him. The second mortgage holder may take over the bidding at this point if he feels the value of the property exceeds the amount of the first mortgage and he is able to raise the cash to pay off the first mortgage balance (often by refinancing the property).

Assume, for example, the first mortgage balance is $10,000 and the second mortgage is $5,000. The second mortgage holder feels the property is worth just $14,000 due to some recent deterioration. The highest bid during the course of the auction is $12,000. The second mortgage holder has the option of letting it go to this bidder, who would pay the auctioneer $12,000. Out of this the first mortgage holder gets $10,000 (assuming that taxes and other priority liens have been paid) and the second mortgage holder would net $2,000 after the first mortgage was paid. The second mortgage holder would logically bid up to $14,000 in expectation of either getting the $4,000 or more in cash if someone else is the successful bidder, or at least getting the property worth $14,000 in exchange for $10,000 to pay off the first mortgage in case he is the highest bidder. If no one else bids higher, his only obligation is to pay off (or refinance) the first mortgage $10,000 balance. Any auction sales price over the $15,000, of course, results in the overage going to the owner, providing there are no other liens on the property.

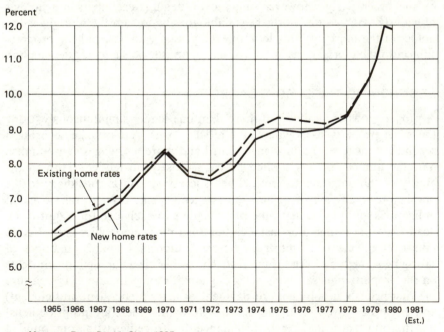

Mortgage Rates Double Since 1965
Chicago–Year-end graph from National Association of Realtors traces average interest rates from 6 to 12 percent. During last three years, average rate for new and existing home mortgages have been the same.

Figure 18.1   Average effective interest rates for new and existing home mortgages. (SOURCE: National Association of Realtors, December, 1979)

Foreclosure does not free the borrower of any unpaid obligation. If the sale brings less than the mortgage owed, the borrower is still personally liable for the difference.

MORTGAGE TERMS, *i* AND *n*

Residential first mortgages typically have a life of from 20 to 35 years with the majority being 25 to 30 years. Second mortgages typically run 10 years although any mutually agreeable life is possible. The interest rate depends upon supply and demand of mortgage money and the risk. For many years first mortgage interest rates hovered typically between 5 percent and 6 percent, with periodic dips or peaks within 1 percent on either side. Then in the mid-1960's the interest rate began a gradual rise. In recent years mortgage interest rates peaked at about 16 percent and began a gradual decline, modified by cyclical ups and downs (Figure 18.1).

INCREASING THE YIELD

Sometimes lending institutions participate in various government-sponsored lending programs that involve ceilings on interest rates for mortgages. If the current mortgage market interest rate is higher than the government ceiling, the lending institution may either stop lending under that program or utilize some other method for reaching the market rate. Methods currently in use to increase the yield above the ceiling include (1) mortgage discounts and other "front end" charges, and (2) nonassignment provisions.

1. *Mortgage discounts* are called "points" if they are expressed in terms of a percentage of the amount of the loan. For instance, if a lender requires "3 points" as a condition of making a $30,000 mortgage loan, he wants 3 percent or $0.03 \times \$30,000 = \$900$, in *addition* to the down payment. Actually he wants the borrower to agree to repay $30,000 plus interest in exchange for a loan of $29,100, (or repay $30,900 for a loan amount of $30,000). The points may be added to the balance due or deducted from the amount loaned depending on the policy adopted by the lender.

Under FHA regulations, the *buyer* is prohibited from paying more than the current interest rate set by FHA. This prohibition applies to all extra charges including points. The FHA however does not prohibit the *seller* from paying points. Therefore the seller customarily pays the points on sales involving FHA mortgages in markets where the prevailing interest rate is higher than FHA. The rule-of-thumb equivalency between points and interest rates is: 8 points equals approximately 1 percent interest. (This is fairly accurate for mortgages of 25 to 30 years but loses accuracy rapidly with shorter time periods.) For instance, assume a $30,000 mortgage is needed. The specified FHA interest rate is $8\frac{1}{2}$ percent and the mortgage market rate is $9\frac{1}{4}$ interest. The seller of the property is required to pay the bank about 8 points per each 1 percent, or in this case, 6 points for the $\frac{3}{4}$ percent difference between FHA and market rates. Six points amounts to $0.06\times\$30,000=\$1,800$ that the seller pays in this case. Some sellers, foreseeing this cost, raise the selling price of the property to compensate before putting it on the market. In either case the true market value is determined when the seller and the buyer, who each have a reasonably full knowledge of other offerings and who are competing with others in the marketplace, finally reach a mutually agreeable bargain. An example of a real estate transaction involving a discounted mortgage follows.

Example 18.1

A buyer desires to purchase a $33,000 house with $3,000 down. The mortgage lender will give an FHA $7\frac{1}{2}$ percent mortgage for 30 years but since the market is $7\frac{7}{8}$ percent this requires 3 points from the seller as a condition of granting the loan.

    a. How much is the monthly payment? (Add $\frac{1}{2}\%$ for FHA insurance, so $i_n=8\%$.)

    b. What is the actual rate of interest that the lender receives?

SOLUTION

The mortgage amount to be repaid is $\$33,000-\$3,000=\$30,000$. The amount loaned is $0.97\times\$30,000=\$29,100$. The payments and receipts of each of the three parties is noted as follows.

    Buyer (mortgagor-borrower)

pays	$ 3,000	cash down payment (d.p.)
assumes	30,000	mortgage note payable to lender (mortgagee)
	$33,000	total cost of d.p. plus mortgage to buyer

    Seller

receives	$ 3,000	cash down payment from buyer
receives	29,100	cash from lender (mortgagee)
	$32,100	total cash received from buyer and lender

    Lender (mortgagee)

gets	$30,000	mortgage note payable in monthly installments by buyer (mortgagor)
gives	29,100	cash to seller
nets	$ 900	3 points discount

a. The monthly payments are

$$n = 30 \text{ yr} \times 12 \text{ months/yr} = 360 \text{ months}$$

$$i = 8/12 - 0.667\% - 0.00667/\text{month}$$

$$A = P\left(\frac{i(1+i)^n}{(1+i)^n - 1} \right)$$

$$= 30,000 \left(\frac{0.00667(1.00667)^{360}}{(1.00667)^{360} - 1} \right) = \underline{\underline{\$220.21/\text{month}}}$$

Note: A method of solving $A = P\{i(1+i)^n/[(1+i)^n - 1]\}$ for high values of n and low values of i is given in Chapter 4.

b. The cash flow of the lender is shown in Figure 18.2.

$$220.21 = 29,100(A/P, i, 360) = 29,100\left(\frac{i(1+i)^{360}}{(1+i)^{360} - 1} \right)$$

The actual interest rate the lender receives is, by interpolation

try $i = 0.007$, $A = +220.21 - 29,100(0.007618) = -1.48$

try $i = 0.0068$, $A = +220.21 - 29,100(0.007449) = +3.43$

By interpolation

$$i = 0.0068 + \frac{3.43}{4.91} \times 0.0002 = 0.0069 = 0.69\%/\text{month}$$

$$i_e = (1.0069)^{12} - 1 = 0.08653 = \underline{\underline{8.7\%}}$$

Approximate calculation: The mortgage payment may be conveniently approximated by using $\frac{1}{12}$ of the annual payments and the tables, as follows.

$$A = \$30,000 \underset{0.088828}{(A/P, 8\%, 30)} /12 = \$2,665/12 = \underline{\underline{\$222/\text{month}}}$$

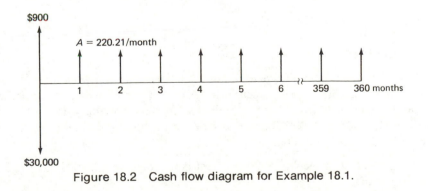

Figure 18.2 Cash flow diagram for Example 18.1.

Notice that the loss of accuracy in this case is less than 1 percent. While perhaps not acceptable for accounting purposes, this accuracy is usually more than required for problems in cost analysis. Likewise the actual effective interest rate the lender receives may be approximated conveniently using 12 times the monthly payments and the tables as follows.

$$\frac{\text{annual payments}}{\text{amount loaned}} = \frac{\$ \, 2{,}665}{\$29{,}100} = 0.091575$$

Interpolation yields

$(A/P, i, 30)$	i
0.088828	8%
0.091575	
0.097337	9%

$$i = 8\% + \frac{0.091575 - 0.088828}{0.097337 - 0.088828} \times 1\% = 8.3\% \text{ effective interest rate}$$

Notice the loss of accuracy in this case is

$$\frac{8.7 - 8.3}{8.7} = 0.046 \quad \text{or} \quad 4.6\%$$

2. *Nonassignment of Mortgage*: *Effect on yield*: Under this provision the mortgage is assigned to the original borrower only. If he sells the property, the mortgage must be paid off in full. This has two advantages for the lender. First, if the old mortgage interest rate is lower than the current rates, the lender gets to terminate a low interest loan and lend the money out on a new mortgage at higher current interest rates. Second, if the old mortgage was discounted when the loan was made, then the effective interest rate received by the lender is substantially increased. The following example illustrates this situation.

Example 18.2

Two years ago a home buyer took out a $20,000 mortgage providing for 30 annual repayments with 9 percent interest on the unpaid balance. Terms for the mortgage included 4 discount points so the buyer actually only received $0.96 \times \$20{,}000 = \$19{,}200$. She now decides to sell the house and pay off the balance due on the mortgage.

a. Find the effective rate of interest if the loan were kept the entire 30 years.

b. Find the interest rate if the loan were paid back in two years.

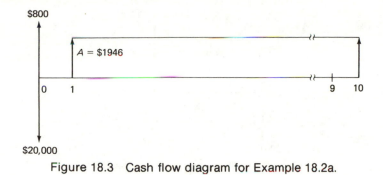

Figure 18.3   Cash flow diagram for Example 18.2a.

SOLUTION

a. The annual payments are

$$A = 20,000 \, (A/P, 9\%, 30) = \$1,946$$
$$0.0973$$

The lender's cash flow diagram is shown in Figure 18.3. Given A, P, and n, then solve for i to find the 30-year discounted mortgage loan rate.

$$A/P = (A/P, i, 30) = 1,946/19,200 = 0.10135$$

By interpolation,

i	$(A/P, i, 30)$
10%	0.1061
	0.10135
9%	0.0973

$$i = 9\% + \frac{0.10135 - 0.0973}{0.1061 - 0.0973} \times 1\% = \underline{\underline{9.5\%}} \text{ 30-yr discounted rate}$$

b. To determine the interest rate if the loan is paid off after two years, first calculate the balance due after two years as the present worth of $30 - 2 = 28$ payments of $1,946. Thus,

$$P = 1,946 \, (P/A, 9\%, 28) = \$19,686$$
$$10.1161$$

If the mortgage loan is paid off in two years, the lender's cash flow diagram appears in Figure 18.4.

$$P = -19,200 + 1,946(P/A, i, 2) + 19,686(P/F, i, 2) = 0$$

Solving for i by interpolation yields

	$i = 11\%$	$i = 12\%$
$P_1 = -19,200$	$-19,200$	$-19,200$
$P_2 = +1,946(P/A, i, 2)$	$+ 3,336$	$+ 3,289$
$P_3 = +19,686(P/F, i, 2)$	$+15,977$	$+15,694$
	$+ \quad 113$	$- \quad 217$

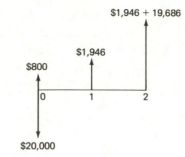

Figure 18.4   Cash flow diagram for Example 18.2b.

By interpolation,

$$i = 11\% + \frac{113}{113 + 217} \times 1\% = \underline{\underline{11.3\%}}$$

Comment. By paying the note off early (after two years in this case) the lender receives a substantially higher interest rate on her investment (raises from about 9.5% to 11.3%) due to the discounting of the note (4 points in this example).

INCOME TAX BENEFITS FROM HOME OWNERSHIP

In order to encourage home ownership the U.S. Congress has permitted certain expenses to be deducted on income tax returns. The most important of these to the average home owner are the deductions for interest payments and real estate taxes. Since the greater part of the payments on a new mortgage consist of interest payments (note Example 18.2b), and the average family moves into a home with a new mortgage approximately once every five years, this income tax deduction is significant to the average home owner. In addition, should the home owner be fortunate enough to sell the home for more than was paid for it (a common situation in inflationary times), additional tax concessions are available. Instead of paying income tax at regular rates, the profit is taxed as a capital gain, providing it has been held for more than one year and qualifies as capital gains. In addition there is a provision in the tax law which permits the full amount of the sale price to be reinvested in another home thereby deferring any tax on the gain for as long as the taxpayer owns a qualified home.

AD VALOREM TAX

In addition, some states, including Florida, permit a homestead exemption to real estate taxes to owner-occupied homes. In Florida the exemption is scheduled to go to $25,000 of assessed valuation, whereas in Texas the exemption is $3,000. This means that a home in Florida valued at $65,000 would be taxed at only $40,000 if it qualified for homestead exemption. If the tax millage rate is at 30 mills, the real estate (ad valorem) tax with homestead exemption is $0.03 \times \$40,000 = \$1,200$ instead of the $0.03 \times \$65,000 = \$1,950$ tax on a home not occupied by the owner, for a saving of $750 per year. In order to qualify for homestead exemption in Florida or Texas an owner must live in the home as a permanent residence on January 1 and notify the tax assessor by March 1 of each year.

Ad valorem taxes may be levied by any authorized level of government including cities, counties, school boards, and special tax authorities. In Florida there is a state-administered ceiling of 10 mills on the county *and* city, and 8 mills on the school board. The voters acting in referenda can vote themselves additional ad valorem taxes for special projects (bond issues) if they so desire. Thus for the city dweller in Florida the maximum tax without referenda is 28 mills plus special tax authorities. As a practical matter the average level is frequently just above 30 mills since worthy bond projects frequently meet with voter approval despite the increased tax required. A mill is $\frac{1}{10}$ of a cent, and refers to the tax on a dollar's worth of assessed value. Thus, if the millage rate is 30, a property assessed at $100,000 pays $0.03 \times \$100,000 = \$3,000$ taxes.

COMPARING MORTGAGES AT DIFFERENT INTEREST RATES

Home buyers sometimes encounter situations where either the existing first and second mortgages may be assumed or the property may be refinanced. A comparison of the two alternatives may be made by comparing the combined interest rate of the two mortgages to the interest rate of the proposed replacement. For example, suppose a house has a present first mortgage at 6 percent plus a second mortgage at 10 percent. The mortgages could be assumed by the buyer or he could refinance at 8 percent. Which is the most economical alternative? The following example illustrates a solution to this type of problem.

Example 18.3

The data on the two mortgages (mtg.) is provided in the table below.

MTG.	BALANCE REMAINING	INTEREST RATE	MONTHLY PAYMENTS	NUMBER OF PAYMENTS REMAINING
First	$10,860	6%	$120.57	120
Second	25,472	10%	244.86	243

Should the new owner refinance at 8 percent?

SOLUTION

The equivalent interest rate may be determined by finding the common i at which the present worth of the monthly payments equals the sum of the balance remaining on the two mortgages.

The sum of the balances is

first mtg.	$10,860
second mtg.	25,472
sum of first and second mtgs.	$36,332

Since the 10 percent mortgage is the larger of the two the resulting interest rate will be greater than midway between 6 percent and 10 percent, or greater than 8 percent. Therefore, try 9 percent for a first trial.

The present worth of the first and second mortgages at 9 percent is the PW of the remaining payments, or

first mtg. $P_1 = \$120.57(P/A, 9/12\%, 120)$ $= \$\ 9,518.00$

second mtg. $P_2 = \$244.86\ (P/A, 9/12\%, 243)$ $=\ 27,335.38$
$$11.63679$$

PW of first and second mtgs. at common i of 9% $= \$36,853.38$

Since the PW of the income to the lender is higher than the $36,332 PW of the loan, the actual return is higher than 9 percent. Therefore, make another trial at 10 percent and interpolate.

first mtg. $P_1 = \$120.57(P/A, 10/12\%, 120)$ $= \$\ 9,124$

second mtg. $P_2 = 244.86(P/A, 10/12\%, 243)$ $=\ 25,472$

PW of first and second mtgs. at common i of 10% $= \$34,596$

Interpolation yields

9%	36,853	36,853
desired combined i	36,332	
10%		34,596
	521	2,257

$$\frac{521}{2,257} \times 1\% + 9\% = 9.23\% = \underline{9.2\%}$$

The effective combined interest rate of the two mortgages is 9.2%. Since this is greater than the refinance interest rate of 8 percent the new owner should *refinance*. Note that discount on refinance has not been considered.

Reviewing the preceding steps in the example, the combined interest rate may be found by

Step 1. Find the monthly payments, A_1 and A_2.

Step 2. Find the sum P_1 and P_2 of the unpaid balance remaining on the mortgages.

Step 3. Select a logical trial i and find the present worth P_3 and P_4 of each of the series A_1 and A_2 using the same trial i. Find $P_3 = A_1(P/A, i_{trial}, n_1)$ and $P_4 = A_2(P/A, i_{trial}, n_2)$.

Step 4. By trial and error and interpolation find the i at which

$$P_3 + P_4 = P_1 + P_2.$$

This is the combined effective interest rate of the two mortgages.

EFFECT OF DIFFERENT MORTGAGE LIVES

When the life of the two mortgages are different $(n_1 \neq n_2)$ the combined rate changes with time. It gradually approaches the dominant i as the shorter mortgage is paid off. In the example case, after 120 months the first mortgage is paid off and the effective i obviously becomes 10 percent, the i of the second mortgage. If the two mortgages have the same life $(n_1 = n_2)$ then the effective rate remains the same throughout the life of both mortgages.

AMORTIZATION SCHEDULES

An amortization schedule is simply a tabulation of the periodic repayments A, required to pay off an amount P, over n periods with interest at i percent. The amortization schedule usually includes a listing of the due dates, amount of each payment due, amount of interest paid, amount of principal paid, and the total amount of principal remaining after each payment is made. For instance, the amortization schedule for a $30,000 mortgage payable monthly over 30 years with i at 8 percent per year appears as

(1)	(2)	(3)	(4)	(5)
				UNPAID PRINCIPAL
		INTEREST	PAYMENT ON	REMAINING
DATE DUE	PAYMENT DUE	PAYMENT	PRINCIPAL	$30,000.00
July 1, 19___	$220.21	$200.00	$20.21	29,979.79
Aug. 1	220.21	199.87	20.34	29,959.45
Sept. 1	220.21	199.73	20.48	29,938.97
Oct. 1	220.21	199.59	20.62	29,918.35

The significance of the figures in each column is discussed in turn.

• *Column (1) Date Due.* This typically begins about 30 days after the loan is made or "closed" but is usually arranged for mutual convenience. Interest is paid in arrears at the end of each time period rather than in advance at the beginning of each period. Thus the interest paid on July 1 is owed for use of the money during the month of June. If the closing were, say May 20, then the interest for the last 10 days of May would probably be prepaid at the time of closing. Then the interest paid on July 1 would be calculated the same as for every other month as $(8\%/12)\times$(unpaid principal). For July that amounts to $(0.08/12)\times\$30,000=\$200$.

• *Column (2) Payment Due.* This is calculated as $A=P(A/P,i,n)$ or in this case

$$A=30,000(A/P,8/12\%,360)=\$220.21$$

• *Column (3) Interest Payment.* This is calculated as noted above in column (1).

• *Column (4) Payment on Principal.* This is simply the portion of the monthly payment left after the interest payment is deducted, or $\$220.21-\$200.00=\$20.21$ for July 1.

• *Column (5) The Unpaid Principal Remaining.* This begins as the full amount of the mortgage and is reduced each month by the payments of principal from column (4). It eventually reduces to zero at the end of 30 years.

For commercial users, amortization schedules are typically run by computer at a nominal cost. They are useful when comparing alternatives, calculating taxes, and buying or selling properties under mortgage.

OUTLINE OF COMPARISONS: RENTING AN APARTMENT TO OWNING A HOME

Throughout much of the remainder of our lives each of us must periodically choose between renting and buying. In addition, advice is often requested from friends or clients regarding the relative merits. The following may be used as a checklist when calculating numerically the present worth or annual cost comparisons.

1. Renting—advantages.
   a. No down payment required (but first and last months rent plus damage deposit may be required).
   b. Maintenance usually taken care of by landlord. Often there is no yard to mow. Major facilities such as pool, sauna, tennis courts, game room-laundry facilities are all maintained by the management.

   c. No direct real estate taxes and insurance charges (although the landlord naturally must charge enough rent to cover these items).

   d. Housing cost is a known amount predetermined for the life of the lease (1 to 3 years). It is *x* dollars per month with no worries about unexpected bills for leaky roofs or faucets.

   e. In place of an equity in a house, a renter may wisely invest the amount of the down payment saved and have a large sum at the end of 30 years. For instance, $3,000 invested at 8 percent for 25 years = $20,538. In addition, if there is say a $25 per month saving in the total rent costs over ownership costs, and if this could be invested at 8 percent, it would result in a balance of about $22,000 after 25 years. However, if inflation continues at its present pace, the probabilities are remote of renters having lower monthly costs than owners over more than the first few years.

   f. Economies of scale realized by landlord in building and maintaining a large number of units are hopefully passed on to the tenant.

   g. Mobility is high. If the need or desire to move occurs there are usually only relatively small penalties involved in breaking a lease (typically one or two months rent).

2. Renting–disadvantages.

   a. No build up of equity.

   b. No right to alter the premises without landlord approval.

   c. Inflation brings increases in the rent charged (although it also should bring increases in salary sufficient to pay it; also the down payment saved and invested in some inflation hedge should also increase in value).

3. Owning–advantages.

   a. Inflation tends to raise the resale value of the property.

   b. Mortgage payments remain unchanged (while rents go up), unless escalation charges are permitted.

   c. Ownership equity increases to 100 percent over the life of the mortgage. (However a 100% ownership of a depreciated property may lack significant value.)

   d. Income tax deductions are allowed for all expenditures for interest and taxes. During the first few years, most of the mortgage payment is applied to interest payment, so a large proportion of the housing cost is tax deductible.

   e. Has no landlord and thus pays nothing extra for landlord's profit (although the landlord's profit could well consist of just the difference between retail prices paid by the home owner and the commercial discount prices of building and maintenance paid by the landlord).

4. Owning–disadvantages.

   a. Requires a down payment (none required for renter, but he or she

            may have to deposit first and last month's rent plus damage
deposit). Down payment involves loss of interest that could result
from investment of this money.

    b. Monthly housing costs are not definitely known since all mainte-
nance and repair (including unexpected repairs) is the responsibility of the home owner.

    c. If deflation should occur, the home owner could be burdened
with high mortgage payments.

SUMMARY

Mortgage financing is an event familiar to millions of Americans who own
their own home. Through such federal agencies as the FHA and VA, loans
are guaranteed and down payments are reduced to levels that make home
ownership an attainable goal for almost everyone.

    A mortgage has two parts: (a) a note that describes the terms of the
debt, and (b) the mortgage itself which gives the lender an interest in the
property until the note is paid off.

    Mortgages vary in length of time for payout, with most first mortgages
running from 20 to 35 years. Some mortgages carry fixed interest rates but
varying rates predominate. With the rising cost of money, lending institutions have resorted to "front end" charges to increase their yield. One such
charge is a *mortgage discount*, termed "points," expressed in terms of a
percentage of the amount of the loan. For example, if a lender requires 3
points on a $50,000 loan, he or she is requiring $1,500 in addition to the
down payment. The points may be added to the balance due or deducted
from the amount loaned. Some mortgage terms may restrict the home owner
from selling the mortgage to another party (through an assumption agreement). If the property changes hands, refinancing may be required if
stipulated in the agreement (usually called a "due on sale" clause).

    One of the main economic advantages to home ownership is the
reduction in income taxes through deductions for taxes and interest on the
mortgage. In deciding whether to sell or rent a home, this advantage should
be compared to other advantages and disadvantages.

PROBLEMS FOR CHAPTER 18, HOME OWNERSHIP
AND MORTGAGE FINANCING

Problem Group A

Find the monthly payments, the balance due on mortgages, effective i, combined i.

A.1 A developer advertises a house for sale at $25,000 with 5 percent down. A
    25-year conventional mortgage is available at 8 percent providing the buyer is

willing to pay 4 points. (Add 4% to mortgage balance.)
a. What are the monthly payments? (Use the equation, with $n = 300$.)
b. What is the actual interest rate? (Approximate, using end of year payments.)
c. What is the actual interest rate? (Use equation, with $n = 300$.)

A.2 Assume you purchased a home five years ago at a sales price of $15,000. You paid 5 percent down and obtained a 30-year 7 percent mortgage. You paid no points. You now have decided to sell it and find the market value is $20,000. You offer to sell to a prospective buyer for 5 percent mortgage balance and the selling price less the down payment. The second mortgage has an $n = 10$ yr, and $i = 10$ percent. (To simplify the calculations, assume end of year payments on both mortgages.)
a. What are the payments on the first mortgage?
b. What is the balance due on the first mortgage?
c. What is the amount of the second mortgage?
d. What are the payments on the second mortgage?

A.3 Assume that five years ago you purchased your present residence under the following terms:

$$\text{price} = \$20,000$$

$$\text{down payment} = \$2,000$$

$$\text{your share of closing costs} = \$500$$

Mortgage terms are 8 percent, 30 years. Your $2,000 down payment, plus the mortgage cover all the costs, including a 5 point discount on the mortgage note. Therefore the mortgage is for

price	$20,000
closing	500
total cost	$20,500
less down payment	− 2,000
amount loaned by mtg.	$18,500
5 point discount	× 1.05
amount of mtg. principal owed	$19,425

Now (five years after purchase) you are ready to move. You estimate the house is now worth $28,000. You put it up for sale for $28,000, asking $2,000 down plus closing costs, and a second mortgage ten years at 10 percent. Find:
a. The amount of annual payments on the first mortgage.
b. The balance due on the first mortgage, 5 years after your purchase of the house.
c. The total amount required to be financed by the second mortgage.
d. The annual payments on the second mortgage.
e. If you would like to sell the second mortgage for cash, and the expected rate of return is 12 percent for investors in this type of paper at present, how much cash can you get for it?

A.4 A house is for sale for $40,000. No down payment is required as a VA mortgage is available at $8\frac{1}{2}$ percent, $n = 30$ years with 6 points discount. The seller is required to pay the discount on the mortgage. He is perfectly willing to pay the discount, but says he will sell you the same house for $37,600 if you buy the house for cash. Find the actual interest rate, i, assuming the $37,600 is the actual value of the house.

A.5 A house is for sale with the following mortgages.

	PERCENT	UNPAID BALANCE	TIME REMAINING
First mortgage	4	12,540	12 yr
Second mortgage	8	8,650	20 yr

a. What is the actual combined interest rate?
b. If the two mortgages can be consolidated into one mortgage for $21,190 at 6 percent for 20 years: (i) Will the annual payments rise or fall? (ii) Will the interest rate rise or fall from the level found in (a)?

Note: Use end of year annual payment for a close approximation.

A.6 Six years ago your client purchased a house for $30,000. She paid 10 percent of the purchase price as down payment in cash, and financed the remainder with a 25-year mortgage at 7 percent. (As a simplifying approximation, assume end of year payments throughout this problem.)

a. What are the annual payments on the mortgage?
b. What is the balance left to pay on the principal after six annual payments have been made?

Now after six years, your client wants to sell the house. She thinks it will sell for $50,000 if she can obtain FHA financing on it.

c. What down payment is required of the new buyer? (See text for FHA down payment requirements.)
d. Subtract the down payment from the purchase price and find the amount of mortgage. This is the amount of mortgage that will be insured by FHA, providing that the house and the buyer qualify.
e. Current FHA regulations stipulate that the buyer cannot pay more than $8\frac{1}{2}$ percent interest on an FHA mortgage, but the bank needs to get 9 percent or they won't make the loan. Approximately how many points is the bank asking for?
f. Assuming that the mortgage will be calculated on the basis of the amount found in (d) above, and that the payments will be calculated by multiplying that amount times the $(A/P, 8\frac{1}{2}, 30)$, who pays the points, and how much do they pay?
g. The buyer is having trouble coming up with the full down payment, so your client offers to take back a second mortgage for half the down payment. She offers terms of 8 percent, and ten years. What are the amounts of the EOY payments on the second mortgage?
h. Your client needs cash now, so she intends to sell the second mortgage to an investor who agrees to buy it if it is discounted so as to yield 15 percent rate of return. How much cash will the investor offer your client for the mortgage?

Problem Group B

Find the equivalent annual or monthly costs. Compare alternatives.

B.1 You need to buy a house and have narrowed the available selection to two which are both equally desirable. Which is the more economical alternative given the following characteristics?

a. Costs $40,000 with a $10,000 down payment. Balance of $30,000 payable over 20 years at 9 percent.

b. Costs $45,000 with $10,000 down and mortgage at 8 percent, 20 years.

B.2 A young married couple asks your advice on selecting a home. They have narrowed the choices down to two, a mobile home, or a custom built home. They already own a lot zoned for either. (For estimating purposes use end of year annual payments instead of monthly.) ($i = 6\%$.)

a. Buy a mobile home, 1,400 ft² at $7.50/ft². The down payment is $500 and the mortgage is ten years at 10 percent. The resale value in ten years is expected to be about $4,000. Annual taxes, $30 for a license plate.

b. Buy a custom built home, 1,400 ft² at $22.00/ft². The down payment is $2,000 and the mortgage is 30 years at 8 percent. The resale value of the house over the near term is estimated as the difference between the replacement value (which is expected to increase about 5%/yr), and the depreciation (which for a 50-year life, straight line, is about 2% decrease per year). The net result is a 3 percent compounded increase per year in resale value. Annual taxes 30 mills on a $20,000 assessed valuation.

1. Over the ten-year period, which has the lowest equivalent annual cost?

2. Which has the lower equivalent annual cost if the custom built house appreciates at a net rate of 6 percent compounded?

B.3 You are being assigned to a distant city for a five-year period. You find you can either rent a suitable house to live in on a five-year lease for $350 per month or you can buy the same house for $35,000. The average ownership costs over the five-year period are estimated as

Real estate taxes are 3 percent of purchase price per year.
Maintenance is $50 per month.
Insurance is $100 per year.
Rebate on income tax is $400 per year.

a. If you were able to cash in some 9 percent bonds and pay cash for the house ($i = 9\%$), what would the minimum sales price have to be in five years to reach the break-even point with renting?

b. If you were able to buy the house with no down payment and get an 8 percent, 30-year mortgage loan, what would your break-even selling price be in five years? ($i = 8\%$ for the mortgage, 9% for everything else.)

Note: Assume all end of year annual payments and no capital gains or other taxes except as listed in the problem.

B.4 A solar hot water heating system costs about $400 new with about $20 per year maintenance costs. An electric hot water heater of the same capacity costs about $80 new with about $12 worth of electricity per month (use $144/yr). A competitive gas hot water heater costs $120 new with about $8 worth of gas per month (use $96/yr). Any of the three can be selected for a new residence now being designed. The mortgage will be at 8 percent. The anticipated life of each system is 15 years. Which is the most economical (a) for the home owner; (b) for the developer?

B.5 A prospective house buyer has a choice between gas and electricity for heat, A/C, hot water, and kitchen. He gets estimates that the gas installation will cost about $3,000 new plus $0.003/1,000 BTU while electric costs about $2,000 plus $0.011/1,000 BTU. Estimated life of each is 15 years with mortgage at 7 percent. How many BTU's per year must be used at the break-even point between gas and electric?

Problem Group C

Arithmetic and geometric gradients.

C.1 You are called as a consultant to determine the lower cost alternative between two types of housing over a representative five-year period. Assume end of year payments as a convenient approximation in all cases. Assume further that your clients habitually invest and borrow at 10 percent.

Alternate A: Purchase a 1,200 ft² home.

purchase price	@ $20/ft² plus $6,000 for the lot
closing costs, excluding "points"	$600
mortgage	$28,000 @ 8%, 30 yr
points required to obtain mortgage	6 (6% of $28,000 added to the other closing costs)

Expenses associated with the home are

Ad valorem taxes are 30 mills on $30,000 assessed valuation.
Taxes remain constant over the five-year period.
Maintenance is $600 per year, increasing $50 every year.
Insurance is $100 per year, increasing $5 every year.

Resale value of the house increases with inflation at an estimated 6 percent per year compounded. The net resale proceeds to the owner is reduced by a 6 percent sales commission to the real estate broker less an additional $200 for the seller's share of closing costs. The owner is assumed to be in the 25 percent income tax bracket. The tax deductible items are (a) ad valorem taxes, and (b) interest payments on the mortgage. In addition any capital gain (difference between selling price and buying price minus any selling and buying costs) is taxable at 10 percent upon sale of the house.

Alternate B: Rent at $300 per month, increasing $20 per month every year. For simplicity, assume all payments are made at EOY.

Chapter 19
Investment Property

LAND INVENTORY

For purposes of discussion, real estate investment property may be subdivided into (1) undeveloped and agricultural land, and (2) developed property. The United States contains about 2.26 billion acres of land area. As yet, less than 3 percent of this area has been developed for urban and transportation uses. About 80 percent of the total land area is used for crop land, pasture, range, and forest land. Another 5 percent is used for recreational, wildlife, and other special uses. About 13 percent is essentially unused land, including desert, tundra, and swamps. In 1981 three-fifths of the land area of the United States was in private ownership, and two-fifths was owned by federal, state, and local governments. Much of the privately held undeveloped and agricultural land will eventually be developed in response to the increasing public demand for more residences, shopping centers, increased industrial production, and so forth. Currently about 1.2 million additional acres of land each year are undergoing development. Meanwhile many of the owners of the remaining undeveloped land hope the value of the land is increasing rapidly enough to warrant the cost of ownership.

FEATURES OF INVESTMENT REAL ESTATE

Undeveloped Land as an Investment

Investors in undeveloped land typically incur continuing costs while waiting for a profitable opportunity to sell or develop the property. These costs include

1. Loss of use of the capital equivalent of the market value. This can be quantified as the present worth of the market value, less sales commissions, closing costs, and capital gains taxes, if any.
2. Annual real estate taxes.
3. Annual maintenance costs, if any. The property may require construction and maintenance of access trails, fences, culverts, and so forth.

The income in most cases consists of the anticipated profit at the time of resale, but may also include small amounts for timber, hunting rights, crop rental, and so forth. A typical example of a cost analysis for an investment in undeveloped land follows.

Example 19.1

An investor group holds 3,000 acres of undeveloped land which they purchased at $1,000 per acre five years ago. The land is assessed at $500 per acre and the tax rate is 20 mills so they have paid $10 per acre per year in taxes. At the end of the fourth year they put in some access trails at a cost of $50,000. What must they sell it for at the end of the fifth year to realize a before-tax rate of return of 15 percent on their investment?

SOLUTION
Use future values, referred to the time of resale of the property.

Future value of cost

purchase price, $F_1 = 3,000$ acres
$\times \$1,000/\text{acre} \ (F/P, 15\%, 5)$ $= -\$ 6,034,100$

taxes, $F_2 = 3,000 \text{ acres} \times \$10/\text{acre}/\text{yr} \ (F/P, 15\%, 5) = -\$ \ 202,300$

access road, $F_3 = \$50,000(F/A, 15\%, 1)$ $= -\$ \ \ 57,500$

selling price required after 5 yr for 15% return $= \ \$ 6,293,900$

If this same group had been able to acquire the property with a small down payment their selling price could be much lower (or their rate of return much higher), as illustrated by the next example.

Example 19.2

Assume for instance, the group of investors were able to acquire the property in Example 19.1 for 20 percent down, paying interest on the balance at 10 percent, with no payments on the principal due until the end of five years.

$$\text{down payment} = 0.20 \times \$1{,}000/\text{acre} \times 3{,}000 \text{ acres} = \$600{,}000$$
$$\text{balance borrowed} = 0.80 \times \$1{,}000/\text{acre} \times 3{,}000 \text{ acres} = \$2{,}400{,}000$$
$$\text{annual interest payments} = 0.10 \times \$2{,}400{,}000 = \$240{,}000$$
$$\text{annual taxes} = 3{,}000 \text{ acres} \times \$10/\text{acre} = \$30{,}000$$

Find the selling price to realize 15 percent compounded on the $600,000 investment.

SOLUTION

down payment, $F_1 = \$600{,}000(F/P, 15\%, 5)$ $= -\$1{,}206{,}800$

taxes and interest, $F_2 = \$270{,}000/\text{yr}\ (F/A, 15\%, 5) = -\$1{,}820{,}400$

access road, $F_3 = 50{,}000(F/P, 15\%, 1)$ $= -\$\ \ \ 57{,}500$

future value of investment *excluding* note $= \$3{,}084{,}700$

  +balance owed on note (interest has been paid) $=$ $\$2{,}400{,}000$

selling price required after 5 yr for 15% return $=$ $\$5{,}484{,}700$

Note this selling price is $809,200 *less* than is required without the financing.

Developed Property

Developed investment property can be further subdivided into

1. Residential
2. Commercial
3. Industrial
4. Governmental-institutional-miscellaneous

The largest dollar volume is in residential investment property, mostly apartment houses.

Residential rental property (apartments and single family homes) provides housing for about 70,000,000 Americans as well as investment income for many millions of owners.

As an investment, rental real estate has several features that set it apart from other types of investment.

CHANGES IN LAND VALUE

Appreciation

Real estate typically (but certainly not always) increases in value with time. One news report recently told of homes in South Florida selling for 40 percent more than they did two years previously. The average price of homes in the area reportedly increased over the previous year by 15 to 20 percent. The national average for increased home prices recently was reported as 9.5 percent per year.

Depreciation

All structures have finite lives. Some day practically every structure will be demolished. At that time its value will be reduced to the value of the salvageable materials if any. Recognizing this the Internal Revenue Service (IRS) permits a depreciation allowance for all improvements that have a determinable economic life. For buildings an allowance of 15 years is typical. Depreciation is a recognized cost of doing business, so an annual depreciation allowance is permitted by IRS for the theoretical cost of depreciation (with little regard to actual market depreciation or appreciation). When the property is sold, however, taxes are levied on any difference (capital gain) between depreciated book value and actual sale price. The depreciation allowance then amounts only to a temporary tax shelter rather than an avoidance of taxes. See Chapters 15 and 16 for treatment of taxes and depreciation.

Combined Appreciation and Depreciation

The combined effect of appreciation and depreciation on the market price of a property may be approximated by noting the effect of each component individually and then combining the two together.

The appreciation effect is caused by rising replacement costs (primarily from inflation), or more realistically by the increasing costs of constructing similar new structures. Buyers seeking a given type of structure may either build one at a cost of construction which typically increases with each passing year, or buy an existing one at a competitive price which tracks the rising cost of construction. Indeed, if the existing structure could stay perpetually new (and marketable, that is with no obsolescence problems), the amount of annual appreciation would equal the amount of annual increase in the cost of new construction. This annual increase in construction costs can be reasonably approximated by some of the popular construction cost indexes, for instance the Building Cost Index published weekly by McGraw-Hill, Inc., in the *Engineering News-Record*. As an example, if this index were expected to rise by an average of 7.2 percent compounded annually over the next five years, the replacement value of a property could be expected to increase by a factor of $(1+0.072)^5$ according to the compound amount equation of Chapter 2. Putting this equation into more general terms applicable to appreciation we have

$$R = P(1+i_i)^m \tag{19.1}$$

where

$R =$ the appreciated replacement value of the structure when it is m years old

$P =$ present value when $m =$ zero years of age

$i_i =$ appreciation-inflation rate in construction costs expressed as an annual compound amount in decimal form

$m =$ age of the structure

Unfortunately, however, the value of the structure does not keep pace with the value of new construction since the structure is not perpetually new. It gradually deteriorates until at the end of the economic life the structure has essentially zero value regardless of how much the replacement value has increased.

To predict a property value that decreases according to the straight line depreciation method, the following equations may be used (see Chapter 16).

$$V_m = P\left(1 - \frac{m}{N}\right) \tag{19.2}$$

where

$V_m =$ the depreciated book value at the end of m years

$N =$ the entire allocated economic life of the structure in years

$P =$ purchase price of the property

When the structure is new, the depreciated value equals the present value $(V_0 = P)$ at the beginning of the economic life. At the end of the economic life N, the depreciated value equals zero.

The combined effect of appreciation in replacement value and depreciation of original value can be approximated by combining the two factors affecting the purchase price (appreciation and depreciation) to obtain:

$$RV = P(1+i_i)^m\left(1 - \frac{m}{N}\right) \tag{19.3}$$

where $RV =$ the future market value at the end of m years. In reality the actual market value will be determined by willing buyers striking bargains with willing sellers, both sides having a reasonable knowledge of alternative opportunities to buy and sell. Since the feelings of these two unknown groups are difficult to predict for any given future date, this equation may provide a simplified but useful method of approximating future market values which are affected by appreciation and depreciation. The following example illustrates the use of this method.

Example 19.3

A home was purchased 5 years ago for \$42,000, which consisted of a house valued at \$36,000 on a lot valued at \$6,000. If the average rate of appreciation for property in this area were 12 percent, and the house could be expected to last 40 years with a zero salvage value at EOY 40, what would the market value of the house and lot be now?

SOLUTION

Remember that land does not depreciate, therefore the land and the house should be treated separately and summed.

$$\text{land, } R_{c1} = \$6,000(1+0.12)^5 = \$10,600$$

$$\text{house, } R_{c2} = \$36,000(1+0.12)^5 \left(\frac{40-5}{40} \right) = \$55,500$$

$$\text{total} = R_{c1} + R_{c2} = \$66,100$$

TAXES ON REAL ESTATE INVESTMENTS

The various levels of government tax real estate investments in several ways:

1. Real estate taxes are assessed annually by local government (county, school board, city) on the actual value of property. Tax assessors determine the value of the property as fairly as possible. They frequently use a formula including square footage, number of bathrooms, general condition, and so forth. In Florida the tax assessors are required to assess at full value. This typically averages out to about 80 percent to 85 percent of the actual sales price. The tax assessors are allowed to adjust the sales price downward to reach a lower assessed value to compensate for sales commission, financing costs, and so forth. From a practical viewpoint, assessors have found that lower assessments bring fewer complaints than higher ones.

2. Income taxes are paid on the taxable annual proceeds from real estate investments, just as on any other investment. Depreciation allowances frequently shelter a significant share of this income until time of sale of the property. Trading the property instead of selling it can frequently extend the life of the tax shelter.

3. Only 40 percent of capital gains are taxed at the taxpayer's ordinary tax level (see Chapter 15). There are three basic requirements for income to qualify as capital gains.

   a. The income must occur as a profit resulting from the sale of an asset, or from a written instrument representing an asset (such as stocks or bonds).

b. The asset must have been held for more than one year after the purchase date.

c. Regardless of the time held, the sale of the asset or instrument must not be part of the regular business of the taxpayer. Taxpayers lose their claims to capital gains if they are dealers and have held the assets primarily for sale to customers. In determining whether real property is held primarily for sale to customers the IRS provides a prescribed list of tests involving continuity of sales and sales-related activities over a period of time, and frequently of sales as opposed to isolated transactions, and so on.

Generally, developing and selling land to builders and other individuals is considered as engaging in the business, and profits are taxed as ordinary income. On the other hand, if property is purchased, developed, held for several years, and then sold in one piece, the profit will probably be treated as a capital gain.

Another exception to capital gain occurs in the case of a depreciated asset. Land of course cannot be depreciated, but the improvements on the land (buildings, pavement, etc.) may be depreciated. These depreciation allowances are income tax deductible as a cost of doing business. However, if the property is sold at a profit, then part of the allowance for depreciation may be taxable. More specifically, if the taxpayer has claimed *accelerated* depreciation (more than allowed by straight line), then the excess of accelerated depreciation over straight line depreciation recaptured in the sale will be treated as ordinary income. The gain attributable to *other than* the recapture of excess depreciation is treated as capital gain, provided the real property has been held for more than one year.

OWNER'S VIEWPOINT, INVESTMENT RETURN

The three variables that most influence a rental investment are (1) leverage, (2) mortgage interest rate and pay out period, and (3) depreciation schedule. Each is discussed in turn.

Leverage

The amount of borrowed money available to help finance the investment is referred to as the amount of leverage, and is common to many other types of investments as well. (A familiar example is margin trading on the stock or commodity exchange.) Rental properties can typically be mortgaged to 75 percent or 80 percent of their purchase price. Thus, it is not unusual for an investor to purchase a $1,000,000 apartment complex by paying $200,000 cash and arranging for an $800,000 mortgage to finance the remainder. Most investors prefer high leverage. The next example illustrates the difference in rate of return which can occur with high leverage.

Example 19.4

An apartment house can be constructed which is estimated to earn a profit of $20,000 per year before income taxes. The total project cost is estimated as $200,000 which may be paid either (a) all in cash, or (b) cash investment of $40,000 and a mortgage for the $160,000 balance at 8 percent, annual payments of interest only, with principal due at some future date. Which is the more profitable in terms of before-tax rate of return? What tax advantages are there to mortgage financing?

SOLUTION

a. The rate of return on the cash transaction is

$\dfrac{\$\,20,000}{\$200,000} = 10\%$, assuming the $200,000 value of the property neither increases or decreases

b. With an 8 percent mortgage, the interest payment is $0.08 \times \$160,000 = \$12,800$ per year. The net income before-tax then is $\$20,000 - \$12,800 = \$7,200$.

The before-tax rate of return is $\$7,200/\$40,000 = 18$ percent return on $40,000 investment.

In this simplified example (with no depreciation, capital gains, recapture, or investment tax credit), the after-tax rate of return is found by simply multiplying the before-tax rates by $(1-t)$. Thus, if the owner is in the 30 percent tax bracket, the after-tax rates of return are:

$$\text{cash purchase } (1-0.3) \times 10\% = \underline{\underline{7\%}}$$

$$20\% \text{ down payment } (1-0.3) \times 18\% = \underline{\underline{12.6\%}}$$

Thus, the additional leverage results in a significant increase in the rate of return. Of course, the risk is greater with high leverage, since the income may decline during economic recessions, while the mortgage payments fall due regularly.

Mortgage Interest Rate and Payout Period

Small changes in interest rates can have a large influence on profitability. If the interest rate in Example 19.4 were raised to 9 percent the effect would be

mortgage	=	$-\$160,000$
net income before interest	=	$+\quad 20,000$
interest payment @ 9%	=	$-\quad 14,400$
net income after interest	=	$+\quad 5,600$

$$\text{percent return on } \$40,000 = \frac{5,600}{40,000} = \quad 14\%, \text{ instead of the previous } 18\%$$

Depreciation Schedule

Depreciation is a tax deductible cost of doing business. The IRS generally permits a 15-year recovery period on depreciable real estate. The accelerated recovery schedule is based on the 175% declining balance method with an automatic change to the straight line method to maximize the deduction. The depreciation cost may be deducted from income during years of ownership but upon sale of the property, any differences between depreciated book value and sale price is taxed as either regular income or capital gains. At worst, the depreciation deduction is equivalent to an interest-free loan from the government until the property is sold.

REQUIRED RENTAL RATE DERIVED

At times, estimated data on costs and desired rate of return are given for a proposed rental project. Then the feasibility of the proposal is checked by deriving the rental rate required and comparing to actual rental rates currently charged in the local rental market. Rental rates for the proposed project, of course, should be competitive with those charged by current local units of the same type and location. The next example illustrates a simplified before-tax version of such a feasibility study.

Example 19.5

A ten-unit apartment house can be constructed for $200,000. The mortgage company will furnish an 80 percent mortgage at 8 percent for 25 years. Operating costs are estimated at $3,000 per year. If 90 percent occupancy is expected, what annual rent is required to return 15 percent on the investment the first year before any income tax considerations and with zero depreciation?

SOLUTION

$$\text{amount invested} = 0.20 \times \$200,000 \qquad\qquad = \$40,000$$

$$15\% \text{ return required} = 0.15 \times 40,000 \qquad\quad = \quad 6,000$$

$$\text{interest payments first year} = 0.08 \times \$160,000 = \quad 12,800$$

$$\text{operating costs} \qquad\qquad\qquad\qquad\qquad = \quad \underline{3,000}$$

$$\text{income required} \qquad\qquad\qquad\qquad\qquad = \$21,800$$

$$\text{annual rent per unit} = \$21,800/(10 \times 0.90) \quad = \$2,422/\text{yr}$$

AFTER-TAX DETERMINATION OF PROJECT FEASIBILITY

Taxes are a normal cost of doing business for most investors, and must be taken into account when calculating net rates of return. The next example illustrates a normal study of a proposed rental project over a two-year period, and seeks to determine a profitable selling price at the end of the two-year period. However, the same techniques may be applied to solve for most other unknown variables related to the problem. Projects may be examined over longer periods of time by repetitions of the same procedures for as many years as desired. These calculations are readily adaptable to a computer program and if many projects are to be examined, this procedure is recommended. The computer can range all of the relevant variables and put out a large amount of significant data concerning the feasibility of each project in a relatively short time.

Example 19.6

A client requests you to examine the feasibility of a proposed ten-unit office rental project with the following estimates of costs and income.

total purchase price
(incl. $20,000 land, $180,000 building) = $200,000

mortgage   = $180,000
   $[A = 180,000(A/P, 10\%, 30 \text{ yr})$ = \$ 19,094/\text{yr}]$

initial cash investment required
   including closing costs   = $ 20,000

straight line depreciation schedule
   $180,000/15 yr   = $ 12,000/yr

rental income, 10 units × $225/month
   × 12 months × 90% occ.   = $ 24,300/yr

operating expense   = $ 10,000/yr

Your client is in the 35 percent tax bracket, and any losses incurred in the real estate rental investment may be deducted from other income for tax purposes.

    Your client expects to sell at the end of the second year of ownership (after the second mortgage payment has been made). What is the minimum selling price if her goal is 20 percent profit on the investment compounded annually, including capital gains or recapture tax?

SOLUTION

The solution is tabulated in Table 19.1, resulting in an *NFW* of $27,324. The sale price of the property after allowance for the payment of capital gains tax (*CGT*) is calculated as follows:

FW at EOY 2 of all cash flows, gain or loss = $ 27,324

Balance due on mortgage at EOY 2   = <u> 177,703</u>

Selling price (net before considering *CGT*) =   205,027

Depreciated book value of property   = − 176,000

Capital gain included in selling price   = $ 29,027 + *CGT*

Table 19.1 SOLUTION FOR EXAMPLE 19.6

THE AMORTIZATION SCHEDULE FOR THE FIRST TWO YEARS

END OF YEAR	END OF YEAR PAYMENT	INTEREST	PAYMENT ON PRINCIPAL	MORTGAGE BALANCE
0				$180,000
1	$19,094	$18,000	$1,094	178,906
2	19,094	17,891	1,203	177,703

$$CGT = \$29,027 \times \{[1/(1-0.4 \times 0.35)] - 1\} \quad = \$ \quad 4,725$$

Selling price (net before CGT) $\qquad = \quad 205,027$

Selling price (including CGT) to obtain
  a ROR of 20% on investment $\qquad = \$209,752$

A table such as that on page 416 may be constructed and the net cash gain or loss after taxes for each year determined in the right-hand column.

SUMMARY

Real estate investments can become complex, and this chapter addresses just a representative group of the many factors influencing such investments. These factors include credit terms for mortgage financing, depreciation, capital gains, recapture, and after-tax rate of return. If better information is lacking, the effect of appreciation and depreciation on real property can be approximated by

$$RV_m = P(1+i_i)^m \left(1 - \frac{m}{N}\right)$$

PROBLEMS FOR CHAPTER 19, INVESTMENT PROPERTY

Problem Group A

Find the required rental income to fulfill stated objectives. No income taxes.

A.1 Your client is interested in purchasing a small ten-unit office complex. The purchase price is $500,000 with 10 percent down payment required. In addition to the down payment, closing costs and front end charges require an additional cash outlay of $2,000. The mortgage is for 30 years at 10 percent. The operating costs including property taxes are estimated at 40 percent of the gross rental income. The resale value in 30 years is estimated at $700,000. Approximate the following answers by using end of year annual payments and income. Your institutional client pays property taxes but no income taxes.
  a. How much monthly rent will he need to charge each of the ten tenants to develop a 15 percent rate of return before taxes over the 30-year life of the building if he stays 100 percent rented?

CASH FLOW OUT	TAX DEDUCTIBLE COSTS	TAXABLE INCOME (OR LOSS)	INCOME TAX PAID (OR SAVED)	CASH FLOW INCOME	AFTER TAX NET CASH FLOW THIS YEAR
FIRST YEAR					
Operating expense $10,000	Operating expense $10,000	Rent $24,300	Taxable	Rent $15,700	$29,795 Cash in
Mtg. interest 18,000	Mtg. interest 18,000	Costs (40,000)	Tax rate	×0.35 Income tax (5,495)	29,094 Cash out
Principal 1,094	Depreciation 12,000				
Cash out $29,094	Costs $40,000	Taxable income ($15,700)	Tax ($5,495)*	$29,795	($ 701) First year
SECOND YEAR					
Operating expense $10,000	Operating expense $10,000	Rent $24,300	Taxable	Rent $15,591	$29,457 Cash in
Mtg. interest 17,891	Mtg. interest 17,891	Costs (39,891)	Tax rate	×0.35 Income tax (5,457)*	29,094 Cash out
Principal 1,203	Depreciation 12,000				
$29,094	$39,091	(($15,591)	($5,457)*	$29,457	($ 363) Second year

*The tax shown in parenthesis indicates a rebate of the amount in ( ) due to deductible expenses in excess of taxable income. Then the future worth at EOY 2 of each year's net profit or loss is determined and summed as follows:

F_1 = future worth of initial cash investment at EOY 2

−$20,000 $(F/P, 20\%, 2)$ = −$28,800

F_2 = future worth of net cash gain or loss for EOY 2,

$701 $(F/P, 20\%, 1)$ = + 841

F_3 = future worth of net gain or loss at EOY 2 = + 363

total FW at EOY 2 of all cash invested = −$27,595

b. More realistically, he expects to average 85 percent occupancy and to be able to raise his rents by 5 percent per year. What should his beginning rent be?

c. What is the first year cash flow under the conditions of (b)?

A.2 You have just completed a preliminary estimate for a proposed warehouse. The total estimated first cost for the project is $1,000,000 with a resale value after 30 years of $200,000. An investor is interested in having you design and build it for her if she can get a 10 percent preincome tax return on her investment. How much annual rent must she charge (in excess of property taxes and operating expenses) in order to obtain a 10 percent return on her investment?

Problem Group B

Find the rental rate for a given after-tax rate of return.

B.1 Your client is interested in building and operating a small shopping center of 100,000 ft^2 and asks you to calculate the rental rate needed per square foot in order to make a 15 percent rate of return after taxes on all funds invested. Your client is in the 45 percent tax bracket for ordinary income. The future costs and income for the project are estimated as follows.

EOY	
0	Purchase land for $100,000.
1	Finish construction of $2,000,000 worth of buildings and other improvements depreciable over a 15-yr life by straight line depreciation method, with zero salvage value at the end of 15 yr. Begin rental and operation of the center, but rental income and operating costs are not credited or debited until EOY 2.
2	First year's operating and maintenance costs charged at EOY 2=100,000/yr. First year's income deposited after income taxes are paid at a tax rate of 45% net income (where net income=gross income−O&M−depreciation).
3–6	Same as EOY 2. (O&M costs continue at $100,000/yr.)
6	Sell the shopping center for original cost plus appreciation estimated at 4%/yr compounded. The difference between the depreciated book value and the sales price is treated as a capital gains (40% of the gain taxed at ordinary rate).

a. Draw a cash flow diagram indicating the time and amount of each cash flow occurrence.

b. Find the rental rate in terms of dollars per square foot each year to obtain 15 percent return after taxes on all funds invested.

Problem Group C

Find the present worth or future value (find the selling price or net price).

C.1 A land subdivider advertises 5-acre parcels at $2,000 per acre (total= $10,000/parcel) on some rural acreage he is subdividing. He advertises 6 percent ten-year mortgages with no down payment. In order to raise cash, he is having to resell these mortgages to investors, and due to the risk involved, the investors are requiring a 20 percent interest return.

a. How much is the subdivider actually getting for the land in dollars per parcel after discounting the mortgages (before taxes)?

    b. Assume the subdivider bought the land for $1,000 per acre and paid $100 per acre in legal and surveying fees. He is in the 45 percent tax bracket and pays ordinary income tax on all net profit. How much is the subdivider netting per parcel *after* taxes?

C.2 A client purchased some undeveloped land ten years ago for $10,000. She has paid an average of $300 per year taxes and other expenses on the property. She now has a customer interested in purchasing the land.

    a. She asks you what the minimum selling price should be in order to make a 20 percent profit *before* income taxes on all funds invested in the land.

    b. She is in the 30 percent tax bracket (and has been for the last ten years) and pays ordinary rate taxes on 40 percent of the capital gain. What is the selling price in order to make 20 percent profit *after* income taxes?

C.3 A land owner wants to sell 100 acres of land. He has two offers, and asks your advice on which is the better offer. He paid $100 per acre for the land 15 years ago and has been paying property taxes of $300 per year ever since. He has been in the 30 percent tax bracket the entire time. The terms of each offer to purchase are

Offer A. $1,000 per acre cash now. The gain on the sale is eligible for capital gains treatment.

Offer B. A real estate salesperson offers to sell the land for him in 5-acre parcels for $2,000 per acre ($10,000/parcel) for a 10 percent commission "off the top" ($200/acre commission taken out of the down payment). The salesperson suggests terms of $1,000 down and a mortgage for nine additional end of year annual payments of $1,000, no interest charged. The land owner finds he could sell these mortgages for cash providing he will discount them to yield a 15 percent rate of return. Assume that with these terms the land would sell quickly and he could get his cash reasonably soon. The gain on the sale is taxed as ordinary income.

C.4 The developer of a 360-unit seafront condominium asks you, her consultant, to calculate how much she should sell her condominium units for. You work up the following time and cost schedule. All costs and income are credited at the end of the month in which incurred.

TIME	
0	Land purchase $1,000,000
Months 1 through 6	Plan and design the project. Designer is paid $600,000 in equal monthly installments of $100,000 at the end of each month.
Months 7 through 18	Construct the project. Contractor is paid $500,000 at the end of each month for 12 months, plus an additional $1,000,000 retainage paid to the contractor at the end of the twelfth month of construction (total of $7,000,000).

TIME		
Months 19 through 36		Sales of the 360 units are expected to progress at the rate of 20 units per month. To aid sales promotion, the developer wants to start prices low and raise them 1% per month over the 18-month sales period. Sales promotions and commission=10% of sales price.
EOM		
12	$ 50,000 for	Ad valorum taxes, insurance, and other income tax deductible items.
24	$200,000 for	Same
36	$200,000 for	Same

Main office overhead chargeable to this project=$10,000 per month for 36 months.

The developer pays 1.5 percent per month interest on all funds used to finance the project. She desires to make $3,000 per unit profit after taxes on each unit sold. She is in the 42 percent tax bracket. *No* capital gains are involved. Assume taxes are paid in monthly installments when due.

a. What must the selling price be for the first 20 condominiums sold in the nineteenth month of the project life (credited at the end of the nineteenth month)?

b. What must the selling price be for the last 20 condominiums in the thirty-sixth month?

C.5 Your are interested in investing in a warehouse that is currently for sale. The following data are available. (Assume EOY payments):

1. Current rental income is $100,000 per year with an estimated increase in rental income of 5 percent per year.
2. Estimated sales price for warehouse in 6 years is $1,400,000.
3. Estimated property tax, maintenance, and insurance is $40,000 per year with an estimated increase of 5 percent per year.
4. A 15-year mortgage for $750,000 is available at 10 percent with equal annual payments.

Find the maximum price you can pay to net a 20 percent rate of return before income taxes on your actual cash investment (down payment).

C.6 Same as Problem C.5 except add the following assumptions.

5. Assume the following tax deductions are available as a part of the investment and that you are in the 42 percent tax bracket. (a) $50,000 per year for depreciation of warehouse; (b) $40,000 property tax, maintenance, insurance and interest on mortgage, with an estimated increase of 5 percent per year.
6. Assume also you will pay 42 percent tax on 40 percent of the capital gains of $500,000 upon selling the property in 6 years. Find the maximum price you can pay now to net an after-tax rate of return of 15%.

Problem Group D

Find the rate of return, i, *before* taxes.

D.1 An 80 acre tract of land is available for $100,000 cash. Upon inquiry the owner says she will accept as an alternative a down payment of $20,000 and six additional end of year annual payments of $20,000. To what interest rate would the alternative payment plan be equivalent (before taxes)?

D.2 A client asks your opinion on a proposal to construct a warehouse rental property in a local industrial park. The estimated terms are shown below. Your client specifically wants to know the anticipated rate of return before taxes if the property is sold after ten years.

total first cost of project	$2,000,000
first year gross income	240,000/yr
annual increase in income each year	12,000/yr/yr
first year O&M costs	100,000/yr
annual increase in costs each year	5,000/yr/yr

estimated change in resale value is

$$R_c = P(1+i_i)^m \left(\frac{N-m}{N} \right)$$

where

$P =$ the cost new

$N =$ the anticipated life of the structure (40 yr in this case)

$i_i =$ annual % increase in replacement cost expressed as a decimal (0.07 for this case)

$m = 10$ yr

D.3 A friend bought a 40-acre farm eight years ago and wants to sell the remainder of it at the present time. He gives the figures shown below and asks what before tax rate of return he will realize on the entire transaction.

EOY			
0	Purchase price	−	$65,000
1−5	Annual taxes and O&M	−	8,000/yr
1−5	Annual income	+	12,000/yr
5	Sell 25 acres at $3,000/acre	+	75,000
6−8	Annual taxes and O&M	−	6,000/yr
6−8	Annual income	+	8,200/yr
8	Sell remaining 15 acres at $4,000/acre	+	60,000

D.4 A client needs a preliminary estimate of before-tax rate of return on a proposed high rise apartment named Plush Manor (balconies overlooking the sea for every apartment) with 100 apartment units costing $30,000 per unit. Rental rate per unit is set at $600 per month with 90 percent occupancy expected. Taxes, maintenance, and other operating costs (excluding depreciation) are estimated at $2,100 per year for each unit regardless of whether occupied or not. Life expectancy is 20 years with 50 percent salvage value at the end of that time. What is the rate of return before income taxes?

Problem Group E

Gradient, sunk cost.

E.1 Your client has located a 2,000 acre tract of land suitable for development which is available for $1,000 per acre. She has visions of developing a country club type residential community around an 18-hole championship golf course. The tract would contain 2,000 lots averaging $\frac{1}{2}$ acre each, with the remaining land used for golf course, lakes, open space, and right-of-way. She feels she can sell 200 lots the first year and continue at that same rate of 200 lots per year for ten years. The sales strategy is to start the prices as low as possible the first year and increase them 20 percent per year. (For example, if the first year's price is 1.0, then the price the second year is 1.20, the third year is 1.44, and the fourth year is 1.728, etc.) The following development costs are anticipated as of the *end* of year listed.

YEAR	COST	ITEM
0	$2,000,000	Purchase of land (begin land sales)
1	2,000,000	Site grading, roads, utilities, taxes, etc.
2	2,000,000	Construct golf course, tennis, etc., maintenance
3	200,000	Maintenance and taxes, additional development
4	220,000	Same
		Increase 20,000/yr/yr
10	340,000	Same

Your client wants to make 18 percent before income taxes on her invested capital, and so will charge the project at 18 percent interest on all of her funds invested therein. How much should each residential lot sell for (a) the first year; (b) the tenth year? Assume that all income received from sale of lots is credited to the project as of the end of the year in which the lot is sold.

E.2 A proposed shopping center is under consideration with the following costs and income estimates.

Original cost of land and buildings complete is $4,000,000
    (this will be financed by $1,000,000 owner's capital and $3,000,000 from a mortgage at 8% and 30 years with P&I payments made annually).
Annual O&M costs are $200,000.
Extra repair costs of $100,000 per year are expected in years 6 through 12.
At the end of 12 years the owner expects to pay off the balance of the mortgage in one lump sum and sell the center for $5,000,000.

Rental income the first year should net $500,000 and should raise a certain amount per year after that. What arithmetic gradient increase in rents is necessary for the owner to make 15 percent compounded return on his cash investment, not considering taxes?

E.3 Thinking at the time that it was a good investment, a friend of yours purchased a 10 percent, six-year mortgage on a lot in a new development three years ago. The borrower made prompt and regular monthly payments up to six months ago. The last payment received reduced the balance owed on the mortgage principal to $3,875. Now after checking into the matter, apparently there is no hope of the payments being resumed, but the lot upon which the mortgage was issued can be foreclosed. Because of some unusual legal problems involved with this particular development, the cost of foreclosure is estimated at $1,500. Due to poor market conditions, the lots in this subdivision have a current cash market value of $990.

a. How much did the friend invest in the mortgage if he paid the base amount when new?

b. Do you advise your friend to foreclose? Why?

E.4 Your client has an opportunity to purchase 670 acres along the Suwannee River for $1,200 per acre. She asks your advice about marketing the land in 1-acre vacation-home sites. You investigate the site and estimate there are a total of 500 good home sites, with the remainder used for roads, common recreation areas, and green space. After consultation with realtors in the area you estimate lot sales at the rate of 100 lots per year for five years. As is customary you intend to start the lots at a relatively low price and raise the price by 20 percent every year. (For example, if the first year's price is 1.0, then price during the second year is 1.2, the third year the price is 1.44, etc.) The following development costs are anticipated as of the *end* of the year listed.

YEAR	COST	ITEM
0	$804,000	Purchase the land (begin land sales).
1	932,000	Install roads, water, and sewer. *Begin sale of lots*, but income is credited at EOY 2.
2	120,000	Maintenance, taxes, additional facilities (boat ramp, tennis, etc.)
3	140,000	Same Increase by $20,000/yr/yr
6	200,000	Same. Sell last lots. Turn common areas over to Homeowners Association or the county.

Your client wants to make 15 percent rate of return on all funds invested in the project.

a. How much should the first 100 lots sell for in terms of dollars per lot?

b. How much should the last 100 lots sell for in terms of dollars per lot?

Assume that all income from the lots is credited to the project at the end of the year in which the lots are sold.

E.5 Your client has just located some land for sale which might make a good location for a new shopping center. He asks you to calculate the rate of return using the following estimates of cost and income. (*Hint*: Find *i* between 10% and 20%.)

EOY		COST	INCOME
		(ALL FIGURES ARE $\times$ 1,000)	
0	Purchase the land	$- 100	0
1	Complete the design	- 80	0
2	Complete construction	- 1,000	0
3	Operating expense and income	- 200	+ 100
4	Operating expense and income	- 210	+ 150
5	Operating expense and income	- 220	+ 200
6	Operating expense and income	- 230	+ 250
7	Operating expense and income	- 240	+ 300
8	Operating expense and income	- 250	+ 350
9	Operating expense and income	- 260	+ 400
10	Operating expense and income	- 270	+ 450
10	Resell shopping center		+ 3,600

Chapter 20
Utility Rate Studies

RATE STUDIES

A utility rate study is simply an analysis of the utility's finances in which the cost of service to each class of customer is determined and compared to the rates paid by each class. Utilities usually try to set rates so that each class of customer pays for its own fair share of costs. Most of these rate studies are carried out by engineers who are familiar with both the utility system and with the techniques of cost analysis.

Rate studies will vary in level of sophistication and detail depending upon the complexity of the system, the requirements of the system's policy board, and the performance and capability of the consultant. A method that is adequate for a small utility having only one class of customer, and therefore a very simple rate schedule, will serve as the first example. Later, a second example will deal with problems of more complex utility systems.

Costs fall basically into two categories.

1. Ownership costs, also called fixed costs, capital costs, plant costs, installation costs, and so forth.
2. Operating costs, including maintenance, repairs, customer accounts, and so forth, also called variable costs.

These costs may be further subdivided for more complex studies as desired. The allocation of these costs into rates is typically done by charging a flat base rate per customer to cover the cost of constructing and owning the utility and connecting it to the customer's property. This is sometimes called a "ready to serve" charge. In the simplest case this base rate amounts to the total monthly ownership costs divided by the total number of customers. If a water system has ownership costs equivalent to $128,000 per month with 32,000 customers, then the flat rate per customer is $128,000/32,000=$4.00 per customer.

The operating costs are covered by charging an additional unit cost per unit of product used. If the operating costs are $80,000 per month to operate a water system selling 320,000 kgal (1 kgal=1,000 gallons*) of water per month, then the variable rate is $80,000/320,000=$0.25/kgal.

Therefore, the rate for this utility system having only one class of customer is

1. $4.00 per month base charge to each customer regardless of consumption to cover ownership costs plus
2. $0.25/kgal for each 1,000 gallons consumed by the customer per month. Thus, a customer using 6 kgal in one month receives a bill for $5.50, and a customer using 10 kgal is billed for $6.50.

MORE THAN ONE CLASS OF CUSTOMER

As utility systems increase in size they frequently are required to service more than one type of customer. The question then arises, should the large customer be charged the same rate as the small customer, or should customers be divided into classifications with different rates? Following the original premise that each customer should pay its own fair share, an analysis can be made to determine the actual costs attributable to each customer class. The system's costs are subdivided into appropriate categories, such as ownership costs and operating costs, and each class of customer pays a fair share of each.

ELECTRIC UTILITY SYSTEMS RATE STUDIES

An electric utility is similar to other rate making situations in that there are ownership costs and operating costs, but they are called by different names, demand costs and energy costs.

Demand Costs

The cost of ownership usually is called "demand" costs, since it relates to the amount of generating capacity in kilowatts (kW) which may be required

*The symbol k is the first letter of the word kilo, meaning 1,000. The current usage of M (the Roman letter for 1,000) is not used here because of the confusion that arises when M is incorrectly assumed to be one million.

(demanded) by the system or a class of customer. The electric utility cannot store its product and therefore installs capacity needs to exceed system demand by a safe margin. The margin should be sufficient to account for generator outages during peak load, and for growth until a new generating unit is ready to come on line. In effect, each customer requires a certain number of kilowatts reserved from the system's total kilowatts of generating capacity to be ready to serve the customer on demand, at the flick of a switch. These kilowatts of capacity can be shared, of course, with off-peak customers. But for all practical purposes, a certain number of kilowatts of generating capacity are standing by, reserved and ready on demand for the personal use of each customer at the time of peak load. Part of the reserved kilowatts of demand capacity is used daily, while another significant part is used only a few hours each year, either at the summer peak (late afternoon of a hot summer day) or the winter peak (usually a cold winter morning). The remaining part (about 10–15%) of the reserved kilowatts is needed for emergency reserve to be used only during breakdown or maintenance downtime for the rest of the system. Thus, reasonably, each customer should expect to pay a fair share of the demand costs, which include all of the capital costs for purchase and installation of all plant equipment, in addition to the costs of maintaining the plant and equipment in ready-to-serve standby condition. Demand costs usually are calculated in terms of

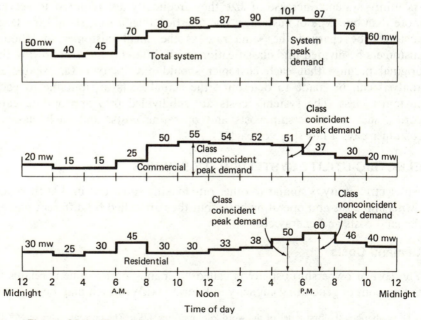

Figure 20.1 Example of power demand for two customer classes.

New York University

Interdepartmental Communication

Journal of
Engineering Valuation & Cost Analysis

dollars per kilowatt. The nomenclature used to describe system and customer-class demand follows.

1. System peak demand is the maximum demand (kW) imposed on the system, sometimes called system coincident peak demand, or simply coincident peak.
2. Class noncoincident peak demand is the maximum demand of a customer class (in kW) which does not occur simultaneously with the system peak.
3. Class coincident peak demand refers to the maximum coincident demand (kW) of any one class of customers.

Examples of the variation and types of demand are given in Figure 20.1.

Energy Costs

The cost of operating the system and producing kilowatt-hours (kWh) of electrical energy are called energy costs. These are the costs for turning the generators including labor, repairs, supplies, maintenance, and a base charge for fuel. Since fuel costs have risen so rapidly, a separate fuel escalation charge is added later to compensate for the incremental fuel costs above the base fuel cost. Energy production costs are normally treated as a constant unit cost (in terms of $/kWh), although any given generator can usually produce energy for slightly less cost at optimum load than at other loads. Energy production costs usually are assumed to be the same in dollars per kilowatt-hour for each class of customer.

Customer Costs

In addition to the familiar owning (demand) and operating (energy) costs, electric utility costs usually involve one or more additional cost categories not directly related to the kilowatts of demand nor the kilowatt-hours of energy consumed. For instance, "customer" costs include billing costs and meter reading costs and any other costs related directly to servicing the customer. Since there is negligible difference in the expense of billing a large customer compared to a small customer, each customer can equitably share a prorated portion of the cost of customer service. Customer costs usually are calculated first in terms of dollars per customer per year, but then may be billed to the customer either as a flat monthly charge per customer, or prorated over the first few kilowatt-hours each month and charged in terms of dollars per kilowatt-hour.

Revenue Related Costs

Another cost category sometimes encountered is costs related to revenue, such as the gross receipts tax passed on to state government. Usually this charge first is calculated as a percent increase, and then applied evenly to all charges, including both flat monthly charges as well as unit dollar per kilowatt-hour charges.

Basic Approach to Electric Rate Making

An electric rate schedule should pass two important tests.

1. It must return adequate revenue to cover all costs of the system.
2. It should treat each class of customer on the same equitable basis, fairly apportioning the cost of serving to the customer for whose benefit the cost is incurred.

To accomplish this, all the costs of the electric utility system are first allocated to an appropriate cost category, such as demand costs, energy costs, customer costs, and revenue related costs. Then each cost category is totaled and apportioned out appropriately to each class of customer, as illustrated in the following example.

Example 20.1 (*Electric utility problem*)

A municipally owned (no income taxes) utility system has three classes of customers (residential, commercial, and bulk power). The bulk power customer purchases directly from the transmission line, and does not use the distribution system. The following data on the system are provided:

1. Customers. There are 20,000 residential customers, 2,000 commercial customers, and 1 bulk power customer.
2. Expenditures are $14,000,000 per year consisting of the cost items listed in Table 20.1.
3. Class peak demand. The residential and commercial customers peak simultaneously with the system peak, while the bulk power customer is at 80 percent of its customer class peak when the system peaks. The customer class peaks and the system coincident peaks are listed below.

	CLASS NONCOINCIDENT PEAK DEMAND	CLASS COINCIDENT PEAK DEMAND
Residential	84,000 kW	84,000 kW
Commercial	40,500 kW	40,500 kW
Bulk power	50,500 kW×0.80=	40,400 kW
System coincident peak demand		164,900 kW

4. Energy sales.

	ANNUAL TOTALS	AVERAGE MONTHLY/CUSTOMER
20,000 residential customers	354.08×10^6 kWH	1,475 kWH
2,000 commercial customers	206.02×10^6 kWH	8,584 kWH
1 bulk power customer	159.00×10^6 kWH	13,250 kWH
Total sales	719.10×10^6 kWH	

5. Book value (BV) of the system.

		PERCENT OF VALUE
Generating and transmission	$45,750,000	79.8
Distribution system	11,580,000	20.2
Total BV of electric utility	$57,330,000	100.0

6. Total expenditures for the system are $14,084,000 per year, summed from the individual cost items listed in Table 20.1.

Table 20.1. SUMMARY STATEMENT OF EXPENDITURES FOR FISCAL YEAR FOR EXAMPLE 20.1.

POWER PRODUCTION EXPENSE ELECTRIC GENERATION		COST CATEGORY
a. Operations	$ 3,198,000	Energy
b. Maintenance	351,000	Demand
c. Transmission expense	160,000	Demand
c. Distribution expense	511,000	Demand
c. Engineering expense	148,000	Demand
d. Customer accounts	281,000	Customer
e. Sales promotion expense	56,000	Demand
f. System security	21,000	Demand
g. Administrative and general		
Employee pensions and benefits	508,000	50% demand 50% energy
State utility tax	224,000	Revenue
Remaining A & G	940,000	Demand and revenue
h. Transfer to bond, PIF, general government	6,187,000	Demand
Subtotal	$12,585,000	
Fuel escalation charges collected separately on a flat cost-of-energy per kWH basis ($/kWH)	1,499,000	
Total expenditure for the fiscal year	$14,084,000	

Determine equitable rates to charge each customer for electric service.

SOLUTION

The problem is to determine what fair share of the systems cost should be borne by each of the three customer classes. A rate schedule can then be constructed which will adequately apportion the costs to the appropriate customer classes.

The first step is to examine each of the cost items in Table 20.1 and allocate each item to one or more of the four cost categories: demand, energy, customer, and revenue. Systematic examination of each of these items in the summary statement of expenditures leads to the allocations shown in Table 20.2 and discussed below.

a. The entire expenditure entitled "operations" under "power production expense electric generation," is related to operating rather than owning the system and so is allocated to the *energy* category. All costs incurred in actually turning the generators are considered energy related costs.

b. The expenditure entitled "maintenance" under "power production expense electric generation" is necessary to keep the plant in standby ready-to-serve condition, and so is a cost allocated to ownership or *demand*.

c. Transmission expense, distribution expense and engineering expense are all incurred due to the installation rather than operation of the utility and so are logically considered ownership costs and charged as *demand* related costs.

d. Customer accounts is charged to the third category of costs, the *customer related costs*.

e. The sales promotion expense of $56,000 is usually justified by its proponents as necessary to increase the load factor for the whole system which reduces the ownership cost per customer. Therefore, it is attributed to ownership and charged to the *demand* category.

f. System security is required whether the generator turns or not, so is attributable to *demand*.

g. Administrative and general. The largest single item here is $508,000 for employee pensions and benefits. This is logically divided between *demand* and *energy* on the basis of the approximate proportion of personnel servicing each category, assumed here as 50-50. Another large item, $224,000 for state utility tax is charged as a percent of the customer's bill and passed on to the state, and so is *revenue* related.

h. The annual equivalent costs for depreciation, financing and other capital investment costs are reasonably approximated by the current annual transfer item in the summary. The transfer includes the annual payment for principal and interest on electric revenue bonds, plant improvement fund, disaster fund, and general government (in lieu of taxes). Some of these transfers are required under the bond covenants. The transfer to general

Table 20.2. SUMMARY OF COST ALLOCATIONS FOR EXAMPLE 20.1.

	DEMAND RELATED COSTS BASED ON kW OF CP DEMAND	ENERGY RELATED COSTS BASED ON kWh OF ENERGY CONSUMED	CUSTOMER RELATED COSTS BASED ON THE NO. OF CUSTOMERS	REVENUE RELATED COSTS BASED ON $ REVENUE RECEIVED
Power production expense, electric generation, operations, maintenance	$ 351,000	$3,198,000*		
Transmission expense	160,000			
Distribution expense	511,000			
Engineering expense	148,000			
Customer accounts			$281,000	
Sales promotion expense	56,000			
System security	21,000			
Administrative and general Employee pensions and benefits	254,000	254,000		
State utility tax				$224,000
Remaining A&G	554,000			386,000
Transfer to bond, PIF, general government	6,187,000			
	$8,242,000	$3,452,000	$281,000	$610,000

*Does not include $1,499,000 in fuel escalation charges which are billed and collected separately.

government in lieu of taxes compensates city government for property that would generate tax revenue if privately held since it doubtless generates as much general government costs (police and fire protection, road and traffic) as equivalent privately held property. These costs constitute justifiable charges against all electric customers on an equitable basis and properly are ownership costs allocated to *demand*.

Summary of cost allocations: Based on the foregoing discussions a summary of cost allocations may be derived from the summary statement of expenditures (Table 20.2).

This table plus the fuel escalation charges add up to the actual $14,084,000 spent on the system as shown below.

demand (fixed plant costs)	=	$8,242,000
energy (operating costs)	=	3,452,000
customer accounts	=	281,000
revenue related	=	610,000
fuel escalation (billed and collected separately)	=	1,499,000
	=	$14,084,000

Assumptions for allocation of costs to customer's class rates: After allocating the actual system costs to the four cost categories, the costs must still be allocated from the four cost categories to the customer's rates by classes.

In arriving at an allocation of costs to rates, the following are usually assumed unless the utility's policy-making board has adopted policies to the contrary.

1. Each new class of customer is entitled to be charged the pooled costs of the system rather than the incremental costs of serving the new customer class. Under this system the new capacity and any new fuel source cost (for instance, the incremental costs of high priced oil versus limited supply of lower priced gas) of serving the new customer class raise the average cost of all old customers. (While the pooled-cost concept is widely accepted, the contrary concept of charging impact fees to large new customers who cause large incremental cost increases to the system certainly is worth consideration and discussion.)

2. There is a reasonably uniform load factor among customers within a given class and therefore all customers within a class equally share the demand costs of the class.

3. Fixed costs for the systems are allocated on the basis of demand, which in turn is derived by any one of several methods in current use.

Three common demand allocation methods: Demand related costs are usually allocated to classes of service in an average cost of service study using one of the three methods described below.

1. *Peak responsibility method.* Demand costs are allocated to classes of service based on the proportion of class contribution to peak at the time of the system coincident peak (*CP*). This method charges demand costs to customer classes that create peak demand and requires peak demand-meeting facilities to be installed.

2. *Noncoincident demand method.* Demand costs are allocated to classes of service based on the proportion of the summation of the class noncoincident peak (*NCP*) demand. This method charges demand

costs to customer classes based on facility requirements necessary to serve the class maximum demand and gives credit for diversity benefits.

3. *Average and excess demand method.* Demand costs are allocated to classes of service on the basis of the extent of facilities as well as contribution to system peak demands. The maximum demand portion of demand related costs is allocated to customer classes on the basis of peak responsibility or noncoincidental group demands. The balance of costs is allocated to customer classes on the basis of peak responsibility or noncoincidental group demands. The balance of costs is allocated on the basis of average demands utilizing system load factors.

Only the first method, called the peak responsibility method, will be illustrated here. The load curves for the system are shown in Figure 20.2, subdivided into the three customer classes described. Under this method the three customer classes will share the costs of peak demand in proportion to the ratio of class coincident peak demand to the system coincident peak demand.

$$\text{residential} = \frac{84,000 \text{ kW}}{164,900 \text{ kW}} = 0.509$$

$$\text{commercial} = \frac{40,500 \text{ kW}}{164,900 \text{ kW}} = 0.246$$

$$\text{bulk power} = \frac{40,400 \text{ kW}}{164,900 \text{ kW}} = \underline{0.245}$$

$$\text{total} = \hspace{3cm} 1.000$$

The annual cost for demand totals $8,242,000 for the generation, transmission, and distribution parts of the system. However, the bulk power customer uses only the generation and transmission parts of the system and should not be charged for distribution costs. Therefore, the calculations are separated.

Generation and transmission account for 79.8 percent of the book value of the system (distribution accounts for the remaining 20.2%), and thus all customers using generation and transmission should pay for 79.8 percent of the demand cost, or $0.798 \times \$8,242,000 = \$6,577,116$. Each customer class is charged according to its proportion of peak load.

residential $= \$6,577,116 \times 0.509 = \$3,347,752$

commercial $= \$6,577,116 \times 0.246 = \ \ 1,617,971$

bulk power $= \$6,577,116 \times 0.245 = \ \ \underline{1,611,393}$

$$\$6,577,116$$

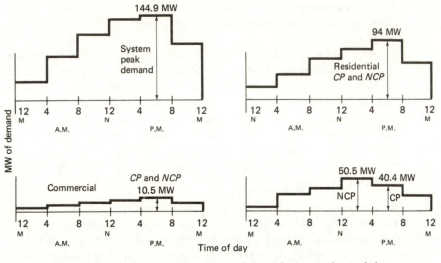

Figure 20.2 Graph of customer demand versus time of day.

The demand costs of the distribution system are not used by the bulk power customer and so should be charged only to residential and commercial customers. Their cost allocations in proportion to the remaining peak demand (after deducting bulk power peak) is

demand cost attributable to the distribution system
$$= 0.202 \times \$8,242,000 = \$1,664,884$$

$$\text{residential} = \$1,644,884 \times \frac{84,000 \text{ kW}}{84,000 + 40,500} = 1,123,295$$

$$\text{commercial} = \$1,664,884 \times \frac{40,500 \text{ kW}}{84,000 + 40,500} = \underline{541,589}$$
$$\$1,664,884$$

There are two basic alternative methods of billing these demand charges to the customers, plus any number of combinations of the two methods.

Alternative method 1, lump sum monthly demand charge billing: Bill each customer a fixed lump sum demand charge monthly. Each class of customer has a different fixed sum, but all customers within a class are treated as if they had the same demand requirements on the system. Using this system the monthly bills would be calculated as follows.

	DEMAND CHARGE
Residential	
Generation and transmission	$3,347,752
Distribution	1,123,295
Total demand charge to residential ÷	$4,471,147 ÷ (20,000 customers × 12 months)
(Number of customers	= $18.63/customer/month
× months/yr)	
Commercial	
Generation and transmission	$1,617,971
Distribution	541,589
Total demand charge to commercial ÷	$2,159,560 ÷ (2,000 customers × 12 months)
(Number of customers	
× months/yr)	= $89.98/customer/month
Bulk power	
Generation and transmission	$1,611,393/12 months = $134,284/month

With this method the large all-electric residence pays the same demand charges as the small cottage using gas for cooking, heat, and hot water. Obviously this system favors the higher-than-average-demand customers within each customer class.

Alternative method 2, demand-proportional-to-energy charge allocated to kilowatt-hours: The second method assumes that demand is proportional to energy consumption. Advocates of this method reason that the residence consuming twice the average energy in kilowatt-hours must have twice the peak load demand. This of course assumes that the load curves all have the same profile shape and vary only in magnitude. The annual demand charge is obtained by dividing demand cost by the energy consumption kilowatt-hours. The resulting charge in terms of dollar per kilowatt-hour is billed to the customer on the basis of kilowatt-hours consumed. Electric rates calculated by this method would be

residential (demand charges for generation, transmission and distribution)/(energy consumed in kWh/yr)
$4,471,047/354.08 × 10^6 kWh = $0.01263/kWh

commercial (demand charges for generation, transmission and distribution)/(energy consumed in kWh/yr)
$2,159,560/206.02 × 10^6 kWh = $0.010482/kWh

bulk power (demand charges for generation and transmission)/(energy consumed in kWh/yr)
$1,611,393/159.00 × 10^6 kWh = $0.010135/kWh

This method favors those customers who have higher than normal peak demand compared to their energy consumption. This type customer is described as having a low "load factor."

The load factor for each customer is simply defined as the "amount of power purchased by each customer relative to their peak load requirement."

Thus, for example, a 100 percent load factor indicates a customer is using peak load 100 percent of the time. A 40 percent load factor may indicate the full kilowatt-hour peak load is used an average of 40 percent of the time, or 40 percent of the kilowatt-hour peak load is used all the time (except for a short peak), or any other combination.

The load factor for the whole class of residential customers in this example problem is

$$\frac{354.08 \times 10^6 \text{ kWh}}{8{,}760 \text{ hr/yr} \times 84{,}000 \text{ kW}} = 0.481 = 48\%$$

Customers with a lower than average load factor would be favored by alternative method 2, while those with higher load factors would pay a little more than their fair share.

Energy cost allocation to customers' rates: After allocating the demand costs to customers' monthly billing rates, the allocation of energy costs is relatively simple. The total cost of producing energy ($3,452,000) is divided by the total energy sold (719.10×10^6 kWh) and a cost in dollars per kilowatt-hour is determined. For the example problem this yields

$$\$3{,}452{,}000/719.10 \times 10^6 \text{ kWh} = \$0.0048/\text{kWh}$$

This charge is the same for all customers in all classes since it costs the same in dollars per kilowatt-hour to turn the generator for the small customer as for the large. There are energy losses in the system, so the denominator is the net energy sold at the meters rather than the gross energy generated.

Customer costs allocated to customer rates: The customer costs are assumed to be about the same for one customer as for another. So each customer regardless of customer class is billed the same monthly fixed charge to cover customer costs. The amount of that charge per customer per month is calculated as the (total customer costs)/(total number of customers$\times$12 months/yr), or

$$\frac{\$281{,}000/\text{yr}}{(20{,}000 \text{ residential} + 2{,}000 \text{ commercial} + 1 \text{ bulk customer}) \times 12 \text{ months/yr}}$$
$$= \$1.07/\text{month/customer}$$

Revenue costs allocated to customer rates: The revenue costs are based on the amount of gross revenue (less fuel adjustment charges) collected by the system and are simply passed through to the state tax collector. The amount is calculated as a percentage and is the same for all customer classes. The percentage is calculated as

$$\frac{\text{amount of tax } (\$610{,}000)}{\text{gross revenue less fuel adjustment } (\$14{,}084{,}000 - \$1{,}499{,}000)}$$
$$= 0.0485, \quad \text{or} \quad 4.85\%$$

Rate schedule: All of these charges (demand, energy, customer, revenue) are added together to make up a rate schedule, shown as follows.

Using demand method 1:

Residential
demand $18.63/month
energy $ 0.048/kWh
customer $ 1.07/month
revenue   multiply × 1.0485

To obtain a usable billing rate for residential customers, the two monthly rates are combined ($18.63/month + $1.07/month = $19.70/month) and multiplied by the revenue multiplier. Also the energy charge is multiplied by the revenue multiplier to yield the following billing schedule of rates.

monthly base total = $19.70/month × 1.0485 = $20.66/month

energy total = 0.0048 × 1.0485 = $0.00503/kWh

A typical 1,000 kWh monthly bill totals to

$20.66/month + $0.00503/kWh × 1,000 kWh = $25.69/month

Commercial
demand $89.98/month
energy $ 0.0048/kWh
customer $ 1.07/month
revenue   multiply × 1.0485

Using the same process as for residential rates, the commercial rates are established as follows.

monthly base total = $91.05/month × 1.0485 = $95.47/month

energy total = $0.00503/kWh

A typical 10,000 kWh bill totals to

$95.47/month + 10,000 kWh × $0.00503/kWh = $145.77/month

Bulk Power
demand $134,284/month
energy $ 0.0048/kWh
customer $ 1.07/month
revenue   multiply × 1.0485

monthly base total = $134,284/month × 1.0485 = $140,797

energy total = $0.00503/kWh

A typical 13.25×10^6 kWh bill totals

$140,797 + 0.00503/kWh × 13.25×10^6 = $207,444/month

If the alternate method 2 for demand allocation is used, the resulting monthly rates are as follows.

Residential
demand $0.01263/kWh
energy $0.0048/kWh
customer $1.07/month
revenue   multiply×1.0485

monthly base=$1.07/month×1.0485=$1.12/month
  energy total=($0.01263+$0.0048)×1.0485=$0.01828/kWh

A typical 1,000 kWh monthly bill totals to

$1.12/month+0.01828×1,000=$19.40/month

The 1,000 kWh monthly bill is lower than the average 1,475 kWh, so the resulting monthly bill is lower using demand method 2.

Commercial
demand $0.010482/kWh
energy $0.0048/kWh
customer $1.07/month
revenue   multiply×1.0485

monthly base=$1.07/month×1.0485=$1.12/month
  energy total=($0.010482+$0.0048)×1.0485=$0.016023/kWh

A typical 10,000 kWh monthly bill totals to

$1.12/month+$0.016023×10,000=$161.35/month

The 10,000 kWh monthly bill is higher than the 8,584 kWh average, so the monthly charge is higher using demand method 2.

Bulk Power
demand $0.010135/kWh
energy $0.0048 kWh
customer $1.07/month
revenue   multiply×1.0485

monthly base=$1.07/month×1.0485=$1.12/month
  energy total=($0.010135+$0.0048)×1.0485=$0.015659/kWh

A typical $13.25×10^6$ kWh monthly bill totals to

$1.07/month+$0.015659×13.50×10^6$ kWh=$207,483/month

The $13.25×10^6$ kWh bill is the average for this bulk power customer. Using method 2, the monthly bill is higher in the fifth significant figure because of round-off in the rates.

Additional Fuel Adjustment Charges

Before the late 1960s when fuel prices began their rapid escalation in cost, fuel oil cost about $2 per barrel with little fluctuation over long periods of time. With productivity increases and greater efficiencies in electric generating equipment, periodic rate reductions were common. Whenever a rate revision was in order, public hearings were held and in due course the rates were appropriately changed. After fuel prices began their rapid rise it became impractical to have a rate hearing every time the price of fuel went up. In addition the utilities received unpredictable allocations of cheaper price-regulated gas interspersed with higher foreign oil, so the fuel cost varied from month to month. As a result, an automatic fuel adjustment pass-through charge is allowed, so that all fuel costs over the equivalent of about $2 per barrel for oil are billed to the customer separately. Thus, if the excess cost of fuel (over the $2 per barrel base cost) amounts to $875,000 for a given month and the amount of energy sold that month to all customers amounts to 60,000,000 kWh, then the fuel adjustment charge to every customer is

$$\frac{\$870,000/\text{month}}{60,000,000 \text{ kWh}/\text{month}} = \$0.0145/\text{kWh}$$

There is no revenue percentage added to this fuel adjustment charge.

SUMMARY

Utility rate studies are often carried out by engineers who are familiar with both the utility system and with techniques of cost analysis. Rates are generally calculated in two parts—a base rate which is usually a fixed cost per month for the service, regardless of the amount of product used and a cost per unit of product consumed. Added to these costs are costs for fuel adjustment.

In formulating electric rates there are several critical points requiring sound engineering judgment and experience.

1. First, sound judgment and familiarity with the system are required to properly allocate the electric utility system costs to the four cost categories of demand, energy, customer, and revenue. High allocations to demand favors the off peak customer if the peak responsibility method is used, and favors low-peak-compared-to-load-factor customers when most other methods are used. A high allocation to energy raises rates to the bulk power customer and lowers them to residential and commercial. A change of $1,000,000 from demand to energy (8% change) raises bulk rates about 4 percent and lowers residential about 1 percent. A change in coincident peak of 17 percent from 100 percent of noncoincident peak to 83 percent of noncoincident peak lowers the bulk power rate 11 percent.

2. Second, good local data should be used when determining the load curves for each customer class. These load curves are needed for determining class demand peaks and the resulting allocations of system demand costs. Local data can be obtained by means of special recording meters located on representative sample customer locations, or by installing recording meters on designated substations that serve representative customer classes. If no other means is available, data from neighboring utilities systems may be used, but obviously constant monitoring of local data is preferred.

PROBLEMS FOR CHAPTER 20, UTILITY RATE STUDIES

Problem Group A

Determine the rate.

A.1 A small city requests that you make up a rate schedule for them adequate to just cover their costs for a water utility system now being constructed. The system will have 8,000 customers. The whole system is financed with 20-year special assessment bonds with $i = 8$ percent. Assume the life of the system is 20 years with $500,000 resale value at the end of that time. They estimate costs and consumption as follows.

Capital costs	$3,000,000
Operating costs	
Maintenance	$120,000/yr escalating $5,000/yr
Operations	40,000/yr escalating $4,000/yr
Consumption	1,000,000 gal/day

a. Find a rate that will pay for the system *without* increases. The rate should include: (i) a flat monthly charge, plus (ii) a unit charge in dollars per 1,000 gal consumed.
b. Find a rate that will pay for the system *with* increases. Begin with (i) a flat rate of $2.00 per month, plus (ii) a unit charge of $0.25 per 1,000 gal.

A.2 Same as Problem A.1 except the system will have, initially, 6,000 customers each consuming 125 gal per day and the number of customers will increase by 500 customers per year.

A.3 A new sewage treatment plant has an anticipated life of 20 years after which it should be replaced. The local government prudently decided to charge enough extra in their rate structure to finance the new plant 20 years from now. Taking into account inflation and other factors the best guess is the new plant will cost $20,000,000 at a time 20 years from now. The current plant will process 150,000,000 gal per month at the start. This will increase by 1,000,000 gal per month for the next 20 years.

Assume that the sewage billing will be on the basis of water used since the sewage is not metered but the water is. It is estimated that for every gallon of water metered, 0.7 gal goes into the sewer. How much should be charged per 1,000 gal of water billed in order to pay for the new sewage treatment plant 20 years from now? Nominal annual $i = 7$ percent, $n = 240$.

A.4 The county commission requests your help in setting fares for a proposed bus transportation system currently under consideration. The following data are provided.

cost per bus, new	$60,000/each
expected life	20 yr
interest	7%
salvage value at the end of 20 yr	$10,000
O & M costs (including overhead)	$1.00/mi the first year, increasing by $0.10/mi each year thereafter

The buses will run 12 hours per day, 6 days per week, 52 weeks per year at an average speed including stops of 10 mph. Each bus is expected to carry an average of 80,000 passenger trips per year, and the annual revenue is calculated as the fare per passenger times the 80,000 passenger trips per year.

As a simplifying assumption, all O & M costs and fare receipts are accounted for at the *end* of the year in which they occur.

a. If the fare ($/passenger) is constant over the 20-year period, what fare should be charged each passenger in order for the bus system to break even at the end of 20 years?

b. If the average passenger rides 2 mi per trip, what is the equivalent cost per passenger mile? (*Hint:* Divide $/passenger by 2 mi/passenger.)

c. If travel by private auto costs $0.15/mi now and is expected to increase by $0.015/mi every year for the next 20 years, compare the costs per passenger-mile to determine if it would be less expensive to move people by car or by bus for this particular bus system. (Compare the annual equivalent $/passenger-mile for each alternative.)

d. For the bus system, if fares are $0.35 the first year, what increase in fares per year (arithmetic gradient) is required to let the system break even at the end of 20 years? (Find answers in $/each fare/yr.)

e. If fares are $0.35 each year and increases are limited to $0.05 per fare per year, find the present worth of the subsidy required per bus.

f. Assume the federal government offers to pay for the cost of the new buses, so that the county receives them free of charge, if the county will pay for O & M costs. How much subsidy is the federal government providing in terms of equivalent dollars per fare for each passenger? (Assume the fare and equivalent subsidy is constant over the 20-yr period.) In other words, how much does the government subsidy amount to in terms of dollars per passenger assuming that the subsidy per passenger does not change over the 20-year life of the system?

Problem Group B

Is the rate adequate?

B.1 As consultant to a small municipal utility (income tax exempt) you are requested to report on the feasibility of a proposed water main extension. The following data have been supplied.

$i = 7\%$

Expected life $= 25$ yr

Extension length is 10,000 ft of 8 in. i.d. water line

Cost of materials and installation is $8.10/ft

Current monthly billing rate for general residential customers is $0.75/1,000 gal

Average current consumption is 10,000 gals/month/customer

Current cost to utility are $0.45/1,000 gal for O&M and $0.30/1,000 gal for P&I (amortize capital costs)

The proposed water main will serve a subdivision where lots average 100 ft frontage on both sides of the main. Deducting 25 percent for open space and street intersections yields an average density of one metered customer for each 62.5 ft of water main on an average. Full development of the subdivision is expected to take five years and to occur at the rate of about 20 percent per year. Thus, the first year only 20 percent of the ultimate number of customers will need water, the second year there will be 40 percent, and so on. (For simplification, use end of the year convention, with 20 percent of full receipts occurring at the end of the first year, 40 percent at the end of the second, etc. Assume the utility installs the full 10,000 ft of line at the beginning of the first year.)

a. Will the current schedule of amortization charges be adequate to pay for capital costs of this installation?

b. If not, how much is the shortage or surplus?

c. If the income from the amortization charge is not enough to make the P&I payments on this installation, and the difference is to be made up as a one time hook-on capital facilities charge, how much should this charge be per meter if the meters were installed at the *beginning* of the first year? (*Hint*: Find the *PW* of the shortage and divide by the total number of customers.)

d. How much should be the capital facilities charge for meters installed at the *end* of the first year? [*Hint*: Each new hook-on should pay accrued interest from the EOY 0, or $F = P(F/P, i, n)$.]

B.2 A small municipality owns its own electric utility and sells electricity to customers within the urban area. Since the generator is subject to downtime (both scheduled and unscheduled) the city commission has decided to intertie with another larger neighboring system. They ask you to determine whether or not the intertie will result in a net profit or loss to their system. They provide you with the following data. Assume $i = 10$ percent.

Existing System

generator size	20 MW rated capacity
load factor	50% (actually generates only 50% of the product of the rated MW × total h/yr)
average sales price to customer	$0.05/kWh
average fuel, operations & maintenance cost	$0.03/kWh
overhead and administration cost	$500,000/yr
annual payments on $5,000,000 first cost of 20-MW plant $(A/P, 10\%, 40)$	$511,300/yr
average downtime	3%

Proposed Intertie
>   original cost $300,000
>   estimated life of intertie   40 yr (no salvage value)
>   maintenance costs of intertie $15,000/yr
>   cost of electricity purchased
>   through the intertie $0.04/kWh

Assume the intertie plus the existing generator will provide electricity 100 percent of the time.

a. Find the equivalent annual cost of the intertie (including the additional costs and income) and tell whether it is a net income or a net cost to the system.

b. Same as (a) except assume the following costs and incomes escalate at the rate of 5 percent per year compounded (geometric gradient) over the 40-year life of the intertie: (i) average sales price to customer; (ii) average fuel, operations, and maintenance cost; (iii) overhead and administrative costs; (iv) maintenance costs of intertie; (v) cost of electricity purchased through the intertie.

Note: More information is provided in the problem than is needed for the solution.

Problem Group C

Find the rate of return.

C.1 A private utility company is considering the feasibility of servicing a large new subdivision. You are provided with the following income and cost projections and are requested to find the before-tax rate of return, *i*. (*Hint*: Try 10%.)

EOY	EOY CAPITAL COST	EOY NET OPERATING INCOME
0 Begin design	− $ 50,000.00	
1 Begin construction	− 300,000.00	
2 Complete construction		
commence operations	− 500,000.00	
3 Operations underway,		
subdivision in		
growth stages		−$100,000.00
4		− 60,000.00
5		− 20,000.00
6		+ 20,000.00
7		+ 60,000.00
8		+ 100,000.00
9 Subdivision fully		
developed (change		
of gradient)		+ 140,000.00
10		+ 145,000.00
11		+ 150,000.00
(Increase $5,000.00/yr		
through thirty-fifth year)		
35		+ 270,000.00
35 Salvage value		+ 500,000.00

Problem Group D

Find the errors in a proposal for a rate increase.

D.1 A utility company bases proposed rates on the following statement of costs. Point out the errors in the statement and correct them.

capital recovery of investment of $500,000 @ 6%, $n=40=$	\$ 33,230
operating costs for maintenance supplies, labor =	75,000
bond interest @ 6% =	30,000
overhead costs =	10,000
depreciation =	12,500
major overhaul costs every 10 yr=$50,000/10 =	5,000
amortization of bond issue over 20 yr=$500,000/20 =	25,000
federal income taxes, state income taxes, local	
property taxes =	50,000
	\$245,730

Problem Group E

Electric utility rate making.

E.1 Use the data below to develop an electric utility rate schedule that will balance all the estimated costs with revenue. Assume the large power customers do not use the distribution system. Omit the revenue cost category.

NUMBER OF CUSTOMERS	CUSTOMER CLASSES	ANNUAL CONSUMPTION (kWh)	NONCOINCIDENT CLASS (kWp)	COINCIDENT PEAK ÷NONCOINCIDENT PEAK
32,900	Residential	340,354,000	83,950	1.00
3,500	Commercial	262,508,000	72,000	0.90
8	Large power	56,148,000	17,325	0.80
	Total	659,010,000		

System peak demand 162,610 kWp

ALLOCATED ANNUAL SYSTEM COSTS

Demand	$21,297,000/yr
Energy	3,294,000
Customer	891,000
Total	$25,482,000/yr

BOOK VALUE OF THE SYSTEM

Generating	$53,022,000
Transmission	12,835,000
Distribution	23,171,000
Total value	$89,028,000

Find the schedule of rates for each class of customer. Use method 2 for allocating demand costs to the customer's electric rates.

E.2 Same as Problem E.1 except use demand method 1.

Chapter 21
Replacement Analysis

ECONOMIC LIFE

Most economic investments, with the notable exception of land, have three distinct "lives": (1) the actual useful life, (2) the depreciation life, and (3) the economic life. Life here is defined as the period of time the investment is used by one owner.

1. *Actual useful life*, as the name implies, is the period over which the investment is actually used. This may or may not coincide with the depreciation life or the economic life.
2. *Depreciation life* is the anticipated period of time the investment is expected to be used and the investment depreciates from new cost to salvage or resale value over this time period. For example, a power company may plan on switching from natural gas to coal in their boilers, necessitating a change in burners. If this change is to take place in six years, then the depreciation life of the existing gas burners is six years. Or a contractor may depreciate a bulldozer over a five-year period expecting to sell it at that time. However, if the equipment is found to still be quite serviceable after five years, it may not be replaced until a later time. Thus the actual useful life

may exceed the depreciation life and is dependent upon the specific conditions surrounding the use of an investment.

3. *Economic life*, on the other hand, is a period of time defined as *the life (period) for which the net annual worth (cost) is a minimum.* A simple example illustrates this principle.

Example 21.1

An automobile is purchased for $8,000. Anticipated annual maintenance costs and salvage values are shown below.

YEAR	ANNUAL MAINTENANCE COSTS ($)	END OF YEAR SALVAGE VALUE ($)
1	0	6,000
2	1,000	5,000
3	2,000	4,000
4	3,000	3,000
5	4,000	2,000

For an $i = 15$ percent, what is the economic life?

SOLUTION

YEAR (n)	ANNUAL WORTH OF PURCHASE $8,000(A/P,15\%,n)$	ANNUAL MAINTENANCE $1,000(A/G,15\%,n)$	ANNUAL WORTH OF SALVAGE VALUE $F(A/F,15\%,n)$	NET ANNUAL WORTH
1	− $9,200	− $ 0	+ $6,000	− $3,200
2	− 4,920	− 464	+ 2,324	− 3,060 ←
3	− 3,504	− 908	+ 1,152	− 3,260
4	− 2,800	− 1,326	+ 600	− 3,526
5	− 2,384	− 1,728	+ 296	− 3,816

A plot of the net annual worth (costs) is shown in Figure 21.1. The economic life is two years, which may be different from either the actual useful life or the depreciation life.

In Example 21.1 the annual cost of owning the investment (annual cost of purchase less the annual worth of salvage, or resale) decreases as the life increases. However the annual worth (cost) of maintenance increases each year. Summing these three factors leads to an economic life in which the annual cost is a minimum (or conversely the annual worth is a maximum if income is involved).

If at the end of every two years another auto could be purchased for the same price ($8,000) and *if* all other costs were the same, then the lowest

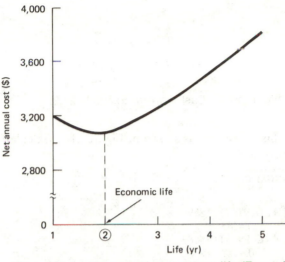

Figure 21.1   Net annual cost versus life (Example 21.1).

cost alternative is to replace with a new auto every two years. Since costs of new replacement autos usually rise over time, then the new auto should not be purchased until the projected annual cost of the existing auto (defender) exceeds those of the new replacement (challenger). As the defender ages, a prestige or status cost may be added if relevant to the actual problem situation.

With the passing of time, current costs and technologies often change. As a result the economic life of an investment may undergo a periodic reevaluation. One recurring question that arises upon reevaluation is what value to place on an investment that is several years old. *The only relevant value is the current market value.* The original cost, present book value obtained from some arbitrary depreciation method, or present replacement cost from a comparable piece of equipment has *no* bearing on the present value to be used in an analysis of economic life (except for calculating taxes). The following example illustrates this point.

Example 21.2

An economic analysis of the costs associated with the use of a blueprint machine reveal it was purchased ten years ago for $7,200 and lately has been requiring a great deal of maintenance to keep it running. Because of improved technology the machine has a zero market value today or any time in the future. Anticipated annual maintenance costs are estimated to be $500 next year and increase by $200 per year for each year thereafter. What is the economic life of the printing press, for $i = 9$ percent?

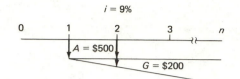

Figure 21.2 Cash flow for Example 21.2.

SOLUTION

The cash flow of this investment is shown in Figure 21.2. Calculation of net annual worths (costs) are:

n	ANNUAL COSTS
1	$500
2	$500 + 200\,(A/G, 9\%, 2) = \$596$
	0.4785
3	$500 + 200\,(A/G, 9\%, 3) = \$689$
	0.9427

Note the original purchase price has nothing to do with the problem and clearly the net annual costs of this machine are increasing each year. Therefore the economic life is *one year*. This does *not* mean that a new machine should be purchased at the end of one year. It simply indicates that *if* another used machine is available with lower equivalent annual costs, then it is cheaper to replace with the second used machine than to continue to operate the first machine with increasing annual costs.

From the preceding problem it can be stated that *whenever the annualized costs are increasing from time zero the theoretical economic life is always one year.* Also note that the original purchase price is irrelevant, as it is a sunk cost.

MAINTENANCE LEVEL

In a similar manner to economic life, there often is a maintenance level that either maximizes profit or minimizes cost. Obviously wear and tear accumulation occurs when maintenance programs are at a level insufficient to prevent this accumulation. Some maintenance programs are the result of considerable thought, calculation, and planning. Other maintenance programs just happen on the spur of the moment whenever someone becomes concerned enough to do the maintenance. Considering maintenance as the independent variable and profit (or rate of return or similar measure) as the dependent variable, a characteristic graph of the relationship between the two is shown in Figure 21.3.

Ordinarily the curve is nearly flat on top with decline in net profit with either too high or too low a level of maintenance. To determine the optimum level of maintenance for a particular capital investment, estimates of income and costs using different types of maintenance programs should be made and a measure of profitability derived. Gross profit and intangibles should also be evaluated. A simplified example follows.

Figure 21.3   Net profit versus maintenance level.

Example 21.3

A recreation facility is being designed for use by the public. At this point two maintenance programs are being considered with costs and income as shown.

	MAINTENANCE LEVEL A	MAINTENANCE LEVEL B
First cost	$100,000	$100,000
O & M costs	50,000/yr	13,000/yr
	(High level of maintenance)	(Low level of maintenance)
Expected life	20 yr	5 yr
	(Long life)	(Life shortened by poor level of maintenance)
Income	60,000/yr	60,000/yr decreasing 10,000/yr

Which level of maintenance should be selected if $i = 8$ percent?

SOLUTION
 Plan A

 annual first cost, $A_1 = -100,000 \, (A/P, 8\%, 20) \quad = -\$10,190$
            0.1019

 annual income, O & M, $A_3 - A_2 = 60,000 - 50,000 = \underline{+\;10,000}$

                 $-\quad\;\;190$

 Plan B

 annual first cost, $A_1 = -100,000 \, (A/P, 8\%, 5) \quad = -\$25,050$
            0.2505

 annual income $- A_2 = 60,000 - 10,000 \, (A/G, 8\%, 5) = +\;41,535$
               1.8465

 annual O & M costs $= A_3 = -13,000 \qquad\qquad = \underline{-\;\;13,000}$

                $+\;\$3,485$

In this case the lower maintenance level and consequent shorter life and declining income actually yields a greater profit on an annualized basis.

However, in evaluating the gross profit and intangibles, note that the life of the investment is only five years, and *no* consideration has been given to the possible changes in first costs as replacements are purchased.

Also, in considering intangibles, the lack of maintenance, while more profitable over the short term could lead to long-term erosion of public confidence and good will, and decrease in value of related investments (surrounding property values) and other costs to owner, investor, and community.

Maintenance costs may also depend upon the quality of product. An engineer planning structures or roads is frequently confronted with a choice between the alternatives of (A) high first cost with resulting low maintenance requirements or (B) low first cost and high maintenance. The variables to be evaluated include:

1. Estimate of maintenance costs for each level of first cost quality. Will wage rates and prices of maintenance supplies rise in the future? (Yes favors A, with low maintenance requirements.) If yes, how rapidly? (Anticipated near-term rises favor alternative A, with low maintenance requirements.)
2. Estimate of expected life span for each level of quality.
   a. Is this life adequate to accomplish the objective?
   b. If future replacements are necessary will the replacement cost rise? If so, how much?
3. Source of funds.
   a. Capital funds to purchase new equipment may be difficult to acquire, while funds for maintenance may be approved routinely.
   b. If privately owned, tax considerations may favor maintenance (favors alternative B, causing less drain on limited capital, higher tax deductible maintenance costs).
4. Tax considerations. Engineering investments involving a tradeoff capital costs versus O&M costs are frequently influenced by tax considerations. The basic reason is that for tax purposes maintenance costs are usually deductible in the year in which they occur, while capital costs usually must be deducted as they are amortized over the entire life of the investment. Thus when faced with a choice between alternates of the same equivalent before-tax cost, tax considerations usually weigh in favor of the maintenance alternative.

An example of capital costs versus maintenance analysis follows.

Example 21.4

A water service line is required for a new subdivision. Estimates on the costs of the alternative installations follow. Funding may be obtained through utility revenue bonds currently selling at 6 percent interest.

ALTERNATIVE	LIFE	CAPITAL COST	ANNUAL MAINTENANCE
Proposal A	10 yr	$100,000 Replacement increases $50,000 each 10 yr	$10,000 increasing $5,000/yr for 10 yr until replacement
Proposal B	40 yr	$300,000	$10,000 increasing $1,000/yr

No salvage value is anticipated in either case. Which of the two proposals is more economical if owned by a tax exempt municipality?

SOLUTION

Find the present worth of each proposal and compare. (Equivalent annual cost comparisons could be made equally well.)

Proposal A

$$\text{first cost } P_1 \qquad\qquad\qquad\qquad\qquad = \$100,000$$

$$\text{replace in 10 yr } P_2 = F(P/F, 6\%, 10) = \$150,000 \times 0.5584 = \quad 83,760$$

$$\text{replace in 20 yr } P_3 = F(P/F, 6\%, 20) = 200,000 \times 0.3118 = \quad 62,360$$

$$\text{replace in 30 yr } P_4 = F(P/F, 6\%, 30) = 250,000 \times 0.1741 = \quad 43,525$$

$$\text{maintenance/yr } P_5 = A(P/A, 6\%, 40) = 10,000 \times 15.046 = \quad 150,460$$

$$\text{maintenance gradient } P_6 = G(A/G, 6\%, 10)(P/A, 6\%, 40)$$
$$= 5,000 \times 4.024 \times 15.046 \qquad\qquad = \underline{\quad 302,583}$$

$$\text{total present worth of 40 yr of service} \qquad = \$742,688$$

Proposal B

$$\text{first cost } P_1 = \text{first capital cost} \qquad\qquad = \$300,000$$

$$\text{maintenance } P_2 = A(P/A, 6\%, 40) = \$10,000 \times 15.046 = \quad 150,460$$

$$\text{maintenance gradient } P_3 = G(P/G, 6\%, 40) = \$1,000 \times 185.957 = \underline{\quad 185,957}$$

$$\text{total present worth of 40 yr of service} \qquad\qquad \$636,417$$

The totals indicate proposal B is the preferred alternative under these circumstances. By way of analysis, inspection of the calculations indicates the areas of high cost for each proposal. For proposal A, the increasing maintenance cost is a large factor. If some means were found to reduce this rapidly rising cost to the level of maintenance for proposal B, proposal A could become the least costly. The sum of present worths of capital investments for four successive installations of proposal A totals $289,645, appreciably below the $300,000 for proposal B. Even though the cash flow sum of capital expenditures for A total $289,645, compared to $300,000 for B, the timing of the expenditures brings the present worth of A to less than

B. Other comparisons of capital investment cost versus maintenance costs can be found in installations such as

CAPITAL INVESTMENT	MAINTENANCE OR OPERATING EXPENSE
Insulation	Annual heating and cooling costs
Larger duct work	Janitorial costs
Built-in vacuum cleaner system	Reduced maintenance and longer life
More durable hardware and fixtures of all types	
More durable exposed surfaces, floor, ceiling, walls, outside and in	

FUNCTIONAL OBSOLESCENCE

In these times of rapidly changing technology examples of functional obsolescence are common. Piston engines have made horses and mules functionally obsolete for many applications, while jet engines and turbines have in some cases replaced piston engines. Several companies have stopped making slide rules since electronic calculators became available. In many industries machines and methods that were the best available five to ten years ago are no longer even competitive. With the rapid rate of technological change in mind some engineers laconically circulate the observation that "if it works, its obsolete." In a surprisingly large number of industries this is quite close to the truth. Whatever has been done can be improved upon, and given time and resources, will be.

In addition to functional obsolescence caused by technological advancement, there is a second type brought about by changing tastes or styling.

Vance Pachard, author of "The Wastemakers" expressed the concern of a lot of thinking people when he wrote "I have no quarrel with legitimate technological advancement, but rather seriously question the morality and economic validity of a system that in order to exist, depends upon artificially shortening the useful life of products." Change of auto body styles, clothing styles, and even building architecture styles without some functional improvement is subject to legitimate criticism and should be discouraged as a waste of money, time, and other scarce resources.

REPLACEMENT ANALYSIS

When considering whether or not an existing machine should be replaced, one of the alternatives is usually to continue using the existing machine. To do this, the economic life of the existing machine should be determined. As discussed earlier, the economic life is that period in which the annual costs are minimized (or annual profits are maximized). This annualized worth should be compared to the annual worth of the replacement. In reality the question is "*Should the existing machine (defender) be replaced (by a challenger) now, or continued in service for at least one more year?*" This question *only* considers the situation for the *next year*. The reasons are obvious; costs vary, technology varies, needs change. Therefore a longer analysis period is usually a waste of effort. If the decision is to keep a defender for another year, then a new replacement analysis can be made after one more year. If a challenger replaces a defender then the challenger's economic life is used in the analysis and it can become a defender at any time in the future. The following example illustrates the method.

Example 21.5

An existing automobile can be sold now for $4,000, or if kept, will decrease in value by $1,000 each year. Maintenance costs of the automobile are anticipated to be $1,200 next year, increasing by $1,400 per year thereafter. The company can replace this automobile with the new one, costing $8,000, and described in Example 21.1. Should the existing automobile be kept for one more year ($i = 15\%$)?

SOLUTION
The first step is to calculate the economic life of the defender, based on *annual costs*. Thus,

YEAR	ANNUAL COST OF INVESTMENT $4,000 $(A/P, 15\%, n)$	ANNUAL COST OF MAINTENANCE $1,200+1,400$ $(A/G, 15\%, n)$	ANNUAL WORTH OF SALVAGE $F(A/F, 15\%, n)$	NET ANNUAL COSTS
1	−$4,600	−$1,200	$3,000	−$2,800
2	−$2,828	−$1,852	930	−$3,750
3	−$2,014	−$2,070	288	−$4,196
4	−$1,610	−$3,056		−$4,666

As the annual costs are increasing each year the economic life of the defender is one year and its annual costs are $2,800 *the next year*. Comparing the annual costs to the challenger (Example 21.1) reveals the challenger has an annual cost of $3,060 which is $260 more than the defender.

The answer is to keep the defender one more year and then reevaluate.

In Example 21.5 there is a tendency to select the challenger, since the defender's annual costs would be $3,750 for the next two years as compared to the challenger's annual costs of $3,060 over the same period. This conclusion is *not* correct because

1. By keeping the defender for one more year, $260 less costs are incurred.
2. If the situation does not change (changes do occur), then after one year the challenger could be selected, resulting in lower annual costs over the next three years.

There is also a tendency to conclude that the challenger should be purchased one year from now. That also is not valid logic since circumstances may change during the next year requiring a new evaluation at that time. A mental bias regarding the outcome of future decisions should not be made before a current study is completed. Therefore all that can be stated with confidence concerning Example 21.5 is that the defender will be kept for *one more year* and then reevaluated.

Example 21.6

An existing airport handles an average of 300 airplanes per day at an annual cost for operating and maintenance of $200,000. The land and existing facilities could be sold on the open market for approximately $3,000,000. A new airport has been proposed using a more efficient layout and equipment that could handle the same traffic for O&M costs of $100,000 per year. The new airport, estimated to cost $4,000,000 including land facilities, would be financed with municipal bonds at 6 percent. Both airports have an anticipated life of 40 years after which the resale value of the existing airport is estimated at about $15,000,000 and of the proposed airport about $10,000,000.

SOLUTION
Since the annual O&M costs for each facility are constant (no gradient) the minimum annual cost for either initial investment would be for them to last as long as possible, that is, the economic life equals the useful life. In both cases this life is 40 years. Since the lives are the same, comparison betwe n challenger and defender can be made by any of the standard methods. The present worth method is used in this example.

Present worth of defender

annual O&M costs $P_1 = \$200,000\,(P/A,6\%,40) = -\$3,009,000$
$$15.046$$

market value of defender $P_2 = \$3,000,000 \qquad = -\$3,000,000$

resale value $P_3 = \$15,000,000\,(P/F,6\%,40) \qquad = +\$1,458,000$
$$0.0972$$

total present worth of defender $\qquad = -\$4,551,000$

Present worth of challenger

annual O&M costs $P_4 = \$100,000\,(P/A,6\%,40) = -\$1,505,000$
$$15.046$$

capital cost of challenger $P_5 = \$4,000,000 \qquad = -\$4,000,000$

resale value $P_6 = \$10,000,000\,(P/F,6\%,40) \qquad = +\$\ 972,000$
$$0.0972$$

total present worth of challenger $\qquad = -\$4,533,000$

The conclusion on the basis of this analysis is that there is no significant difference between the alternatives. These data contain no justification for a move. Therefore, unless there are additional arguments to be considered, the present location (defender) should be retained.

Most professional problems are more complex and involved than Example 21.6. However, the same basic approach of life-cycle costing is used regardless of the complexities.

Example 21.7

Example 21.6 is further complicated by an annual 10 percent increase in air traffic expected to level off when the volume reaches 1,000 planes per day. The existing airport can handle up to 440 planes per day with no increase in cost. With an additional runway costing $1,000,000 and another $50,000 per year in operating costs, it can handle a total of up to 1,000 planes per day.

The proposed new airport could be built to handle 1,000 planes per day for $5,000,000. In addition, the larger commercial jets cannot use the existing airport but could use the new facility, which would bring better schedules and an estimated $100,000 per year worth of new business into the community. This new business figure should increase at a rate of about $10,000 per year. However, the new airport would be located further out and the increased distance and time is expected to cost the 100,000 annual users (the public), increasing about 10,000 users per year, an additional $0.50 in time and operation costs for each user.

With these additional considerations in mind, make a new analysis of the challenger-defender contest.

SOLUTION
The same basic approach is followed and the present worth of each alternative determined (since the two alternatives have the same lives).

Defender: It is necessary first to find how long the existing airport can handle the traffic without a new runway.

$$300 = 440(P/F, 10\%, n)$$

$$n = 4 \text{ yr until a new runway is needed}$$

The present worth of this cost today is

$$P_7 = \$1,000,000 \, (P/F, 6\%, 4) = \$792,100$$
$$0.7921$$

The present worth now of an additional 36 years of O&M costs, which start four years from now is

$$P_8 = 50,000(P/A, 6\%, 36)(P/F, 6\%, 4)$$
$$= (50,000)(14.621)(0.7921) = \$579,100$$

Defender summary of present worth of existing airport and improvements is as follows.

annual O & M costs, existing runway	$P_1 = -\$3,009,000$
market value of defender	$P_2 = -\$3,000,000$
resale value in 40 yr	$P_3 = +\$1,458,000$
new runway 4 yr from now	$P_7 = -\$\ 792,000$
O & M costs of new runway	$P_8 = -\$\ 579,000$
	$-\$5,922,000$

The present worth of the *challenger* may be listed as follows.

PW of new business brought to community
$$= P_{10} = \$100,000\ (P/A, 6\%, 40) = \$1,505,000$$
$$15.046$$

PW of gradient increase in new business
$$= P_{11} = 10,000(P/G, 6\%, 40) = 10,000(A/G, 6\%, 40) = \$1,860,000$$

PW of increased travel costs
$$= P_{12} = 100,000 \times \$0.60\ (P/A, 6\%, 40) = \$903,000$$
$$15.046$$

PW of gradient increase in travel costs and people
$$= P_{13} = 10,000 \times \$0.60(P/G, 6\%, 40) = \$1,116,000$$

PW of new airport is simply the new increased cost
$$= P_{14} = \$5,000,000$$

Challenger summary of *PW* of proposed new airport (P_{14} replaces P_5).

capital cost of new airport	$P_{14} = -\$5,000,000$
resale value in 40 yr	$P_6 = +\ \ \ 972,000$
O & M costs	$P_4 = -\ \ 1,505,000$
new business to community	$P_{10} = +\ \ 1,505,000$
gradient increase in new business	$P_{11} = +\ \ 1,860,000$
increased travel costs	$P_{12} = -\ \ \ \ 903,000$
gradient increase in travel costs	$P_{13} = -\ \ 1,116,000$
	$-\$4,187,000$

Conclusion. Under the conditions designated, the challenger new airport is the least cost alternative. The existing airport should be sold and the new one constructed.

SENSITIVITY ANALYSIS

Estimators and planners are constantly being called upon to predict future costs. Obviously this is a hazardous occupation. Many variables must be considered, each of which can influence the results. For instance, future operating costs will be strongly influenced by both fuel costs and labor costs. These are difficult enough to predict for the short term, but many estimates require long-term predictions. To increase the accuracy of the results, some estimators perform a sensitivity analysis. This consists of

1. Listing the variables most likely to affect the future cost figures being estimated.
2. Determining a probable range over which these variables may traverse.
3. Determining the effect on the future cost figures being estimated of the variables ranging over their probable range.

If a future cost figure is significantly affected by the ranging variable then the cost estimate is said to be very sensitive to that variable. Obviously the important variables merit special attention and analysis, whereas the variables to which the cost is relatively insensitive will merit a somewhat lower expenditure of time and effort. Sensitivity analyses may be further refined by estimating a probability basis for ranging the variables (for example, using a 90% probability). To illustrate, if the future unit costs of a certain earthmoving fleet are being estimated, labor costs are certainly one important variable. After careful research and consideration, an estimator could perhaps estimate that labor costs will rise a mean of 8 percent per year over the service life of the fleet with a 90 percent probability that the labor costs will rise not less than 6 nor more than 10 percent. The effect of the range of possible cost increases on the unit cost of earth moved could be calculated, combined with the effect of other variables, and compared with the cost of an alternative method of earthmoving. At least an informal sensitivity analysis should be performed on the life cycle costs of every major investment. Formal studies are time consuming and typically involve computer programming to perform the many repetitious calculations. However, these sensitivity analyses frequently yield critically important results not otherwise obtainable.

A few of the variables involved in determining the most economical method of generating electricity are discussed in the following example.

Example 21.8

Assume that you are planning to build an electric generating plant and are choosing between a nuclear energy plant or a fossil fuel plant.

ITEM	NUCLEAR	FOSSIL
Capital	$120,000,000	$50,000,000
Fuel	1,000,000/yr	5,000,000/yr
Operating expense	10,000,000/yr	10,000,000/yr

Assume further that money could be borrowed through bonds at 6 percent interest rate and the life of both plants would be about 40 years. Finding the present worth (cost in this case) of both alternatives, the P/A factor is 15.046, therefore, the nuclear plant has a total present cost of $285,092,000, while the fossil fuel plant has a present cost of $275,276,000.

The variables involved in each plant in this simplified version are: (1) the capital investment, (2) the fuel, (3) the operating expenses, and (4) the life. Analyze each in turn.

SOLUTION

1. *Capital investments.* First, how good are the estimates? Fossil fuel plants have been built for many, many years. Innovations occur from time to time but the basic design is familiar to a large number of professional electric power plant designers. Nuclear plants on the other hand are relatively new and there is not a great deal of experience in the area. Consequently, the reliability of estimates of nuclear plant costs is much less than that of fossil fueled plant costs. Therefore, the possibility of an overrun on capital costs for nuclear plants is greater than that for an overrun on fossil fuel plants.

2. *Fuel.* Nuclear plants require only a small amount of fuel. This fuel is at present available only from the government, therefore, the price is fairly predictable and the market fairly well established. Contrasting this, fossil fuel is available on a rapidly moving market. Prices rise and fall as a result of supply and demand. Recent estimates of fossil fuel reserves have been alarmingly low. As a result, prices of fossil fuel have begun to rise. In addition, at present it is difficult to predict the requirements for pollution control on fossil fuels. The price of fossil fuels, therefore, is quite variable particularly when considering a 40-year economic life span.

3. *Operating expenses.* We may assume that these consist of at least 50 percent labor. The pay demands of labor vary from year to year. The ability of labor to enforce these demands likewise vary and investigation of the relative labor requirements of each type of plant certainly would be in order in a sensitivity analysis of these two plants.

4. *Life.* While the useful life of complex plants such as these is difficult to accurately estimate at best, the rapidly changing energy and environmental influences make useful life very difficult to predict. New pollution control measures, mandated changes in operating procedures or fuels, and changes in the demand for power may force a large change in useful life.

SUMMARY

Replacement analysis involves two separate, yet interrelated, economic analyses. First, the economic life of the existing unit (termed defender) must be determined, then the following question answered. Should the existing unit (defender) be replaced (by a challenger) *now*, or continued in service for at least *one more year*? The economic life is that life in which the net annual worth (cost) is a minimum. This life may not be the same as the useful or the actual life of the unit. If annualized costs are increasing from time zero, the economic life is always one year.

A challenger and defender should always be compared on the basis of equivalent annualized costs, as their lives will almost always be different.

Remember the question concerns replacement now, or keeping for one more year. If replacement is being considered now and the decision is to keep the existing unit, then presumably the situation will be reanalyzed after one more year.

PROBLEMS FOR CHAPTER 21, REPLACEMENT ANALYSIS

Problem Group A

Compare present worths, $PTZ \neq ETZ$.

A.1 Plumbing fixtures are being designed for a new building. Which of the two types has the lowest cost present worth if $i = 8$ percent?

	TYPE A	TYPE B
Estimated life	10 yr	20 yr
Cost new	$50,000	$120,000
Maintenance, yr 1–5	$3,000/yr	$1,000/yr
yr 6–10	$5,000/yr	$2,000/yr
yr 11–20	Repeat 1–5, 6–10	$4,000/yr
Replacement cost in 10 yr	$70,000	N/A

Problem Group B

Compare alternatives: Involves gradients, sunk cost.

B.1 An existing judicial center cost $400,000 to build just 38 years ago. A breakdown in the heating and air conditioning system occurred just last year and repairs costing $500,000 were made. Now the case load in the court rooms has created overcrowding and the judges and attorneys are asking for more room.

Two possibilities are under consideration. Renovate the existing building for $1,000,000 and purchase an existing office building for an annex across the street for $1,500,000 capital and remodeling costs. O&M costs on the existing building are $200,000 per year and rising at the rate of 10 percent per year. O&M costs on the annex would be about $50,000 and rise at the same rate. Twenty years from now another $2,000,000 worth of work will be needed.

The other alternative is to tear down the existing building, and build a new one at the same location. This is estimated to cost $5,000,000. The O&M costs initially are estimated at $300,000 per year but due to advanced design the annual increase would be limited to $25,000 per year. The lives of both alternatives may be estimated at 40 more years. (Actually these massive institutional structures will last for hundreds of years but the influence on present worth of any value remaining after 40 years is usually negligible at realistic interest rates. In addition, estimates of value of a property 40 years from now is subject to sizable error.)

When comparing the two alternatives there is a considerable intangible value attached to each. The existing structure is steeped in local history and is regarded as a city landmark. The newer structure would be more comfortable, efficient, and easier to maintain. The morale of the occupants would undoubtedly rise in the new building and it would go far in improving the image of downtown. On the balance, the new building is estimated to have about the same effect as $1,000,000 worth of public relations type of civic promotion. So the new building may be credited with an extra $1,000,000 worth of present worth.

Financing of all work will be done from funds which either draw 6 percent when not otherwise being used or which cost 6 percent to borrow. Which is the more desirable alternative?

B.2 Several designs for the exterior curtain walls of an office building have been submitted with the following estimates of cost and insulation value.

	ALTERNATIVE A	ALTERNATIVE B	ALTERNATIVE C
First cost	$90,000	$165,000	$200,000
Estimated life of structure	40 yr	40 yr	40 yr
Annual maintenance costs	$12,000	$10,000	$14,000
Insulation factor, $u =$	0.20	0.10	0.05
Salvage value	0	$10,000	$50,000

The heating and cooling costs are estimated as directly proportional to the u value. The annual costs for this building are calculated as $200,000 \times u$.

a. With $i = 9$ percent, which is the lowest cost alternative.

b. If heating and cooling costs are estimated to increase between 5 and 12 percent per year over the life of the structure, does this make a difference in the selection? Why?

B.3 A contractor has a continuing contract to excavate a large amount of earth for $1.50/yd^3. He has a dragline that is currently excavating an average of 120 yd^3/hr. The machine is on the job and operated for a total of 2,080 hours per year, but due to age the machine is down 20 percent of this time. Direct costs for O&M are currently $48 per hour (including the operator's wage) for the full 2,080 hours per year.

In addition the contractor calculates that $0.30/yd^3 is needed to cover supervision, overhead and profit. A new machine is available for $200,000 that will excavate 150 yd^3/hr with 95 percent availability. The O&M costs on it are estimated as $58 per hour including operator's wage. It is expected to last 12 years and have a salvage value of $25,000 at that time. Assume $i = 10$ percent.

a. Assume the present dragline could either (i) continue for another 12 years at the same reliability and hourly costs, with no salvage value, or (ii) in the alternative, the present dragline could be sold. What amount of sale price must the present dragline bring in order to be at a break-even point between challenger and defender?

b. Assume the present dragline broke down last month and required $20,000 worth of repairs to become operational again. What is the new break-even point sale price?

c. Assume availability of both units will decrease by 40 hours per year each year. What difference in break-even point sale price does this make?

Chapter 22
Double Arithmetic Gradient

DERIVATION OF THE DOUBLE ARITHMETIC GRADIENT

Change is accepted as a normal condition of modern life, and calculations of future incomes and costs must reflect this condition. In previous chapters simple periodic changes in payments were accounted for by either arithmetic gradients or geometric gradients. Where a gradient payment is itself subject to a gradient, the solution becomes more difficult unless new equations and tables of factors are developed. For instance, where preliminary project studies show anticipated periodic increases in customer revenue, the arithmetic gradient can accommodate a gradient increase in the number of customers, or in the revenue per customer, but not both. To calculate the value of gradient increases in both of these variables a double gradient is required.

The general form of an equation for double gradient may be developed using the following notations and assumptions.

G_1 represents one of the two gradients, say an increasing number of customers per year.

G_2 represents the second of the two gradients, say an increase in the payment per customer.

Deposits are made at the end of each year (period) except the first year (period) into an account bearing interest compounded at rate i. If the number of customers begins with zero the first year and increases to G_1 the second year, to $2G_1$ the third year, and so on, then the number of customers in the nth year will be $(n-1)G_1$ (see Chapter 5).

By the same reasoning, the payment per customer is zero the first year, increasing to G_2 the second year, to $2G_2$ the third year, and so on. Therefore, the payment per customer deposited at the end of the nth year is $(n-1)G_2$. The total amount deposited the first year is zero. The total amount deposited the second year is the product of G_1 customers times G_2 payment per customer, or simply G_1G_2. At the end of the third year the total amount deposited is $2G_1$ customers times $2G_2$ payments per customer, or $4G_1G_2$. The deposit at the end of the nth year will amount to the product of $(n-1)^2G_1G_2$. If each of these deposits bear interest compounded at the end of each year, the accumulation of principal and interest at the end of the nth year may be summed as the future value, F, of a series. This series can be examined one year at a time as shown below, where F_n is the future value of the series at the end of year n.

$$n=1, \quad F_1=0$$

In the arithmetic gradient series there is no deposit at the end of year 1.

$$n=2, \quad F_2=G_1G_2$$

The first deposit occurs at the end of year 2, with no time for interest to accumulate.

$$n=3, \quad F_3=G_1G_2(1+i)+2^2G_1G_2$$

At the end of year 3, the second deposit occurs and interest for one period accumulates on the first deposit.

The general term for the future value of the series in the nth year is

$$n=n, \quad F_n=G_1G_2(1+i)^{n-2}+2^2G_1G_2(1+i)^{n-3}+\cdots+(n-1)^2G_1G_2$$

The equation is made slightly less cumbersome by dividing both sides by G_1G_2, as follows.

$$F/G_1G_2=(1+i)^{n-2}+2^2(1+i)^{n-3}+3^2(1+i)^{n-4}$$
$$+\cdots+(n-1)^2 \tag{22.1}$$

or in factor form

$$F=G_1G_2(F/GG,i\%,n) \tag{22.2}$$

This equation was programmed into a computer which produced the tables of F/GG, P/GG and A/GG factors found at the end of this chapter. These tables greatly reduce the time and effort required for solving problems involving double gradients. An example of a typical problem application follows.

Example 22.1

A transportation authority (TA) has just let a contract for the construction of a small toll bridge to serve an area just opening up for development. To finance the bridge the TA borrowed $1,100,000 which must be repaid seven years from now in a lump sum with interest compounding annually at 7 percent on the loan balance. The bridge is expected to open for traffic one year from now, and during the first year of operation is expected to carry an average of 1,000 vehicles per day. Each year the TA expects the traffic count to increase by an additional 1,000 vehicles per day, so that 2,000 vehicles per day are expected the second year, 3,000 the third year, and so on. The TA plans to set aside the entire toll receipts from the bridge to repay the $1,100,000 borrowed to construct the bridge. The receipts will be deposited at the end of each year into a fund bearing interest compounding annually at 7 percent. To encourage traffic to use the toll bridge the TA proposed to set the tolls for the first year of operation at $0.05 per vehicle, and then raise them by an additional $0.05 per vehicle each year thereafter. (The toll will be $0.10/vehicle for the second year, $0.15 for the third year, and so on.) You are requested to determine if the amount accumulated in the account seven years from now (at the end of the sixth year of operation) will be sufficient to repay the loan if the estimates of traffic count and revenue prove correct.

SOLUTION

The two gradients are determined as

$$\text{traffic gradient, } G_1 = 1,000 \text{ vehicles/day/yr} \times 365 \text{ days/yr}$$

$$= 365,000 \text{ vehicles/yr/yr}$$

$$\text{toll/vehicle gradient, } G_2 = \$0.05/\text{vehicle/yr/yr}$$

Then $i = 7$ percent and $n = 6$ years of operation $+ 1$ year of construction $= 7$ years. The amount of principal and interest accumulating after 7 years is found as

$$F = G_1 G_2 (F/GG, 7\%, 7) = 365,000 \times 0.05 \times 98.7430$$

$$F = \$1,802,000 \text{ principal and interest accumulated at the end of 7 yr}$$

The actual cash flow occurs as tabulated below. The first deposit is made at the end of year 2 (end of the first year of operation). Each individual year end deposit accumulated interest in accordance with $(1+i)^n$ for however many years that deposit is in the account. The following table shows how each deposit accumulated to the total.

DEPOSIT CREDITED AT EOY	AMOUNT DEPOSITED TOLL$\times$VEHICLES/YR		PRINCIPAL +INTEREST FACTOR		ACCUMULATED FUTURE SUM
1	0			=	
2	$0.05\times365,000$	$\times$	$(1.07)^5$	=	\$ 25,600
3	$0.10\times730,000$	$\times$	$(1.07)^4$	=	95,690
4	$0.15\times1,095,000$	$\times$	$(1.07)^3$	=	201,210
5	$0.20\times1,460,000$	$\times$	$(1.07)^2$	=	334,310
6	$0.25\times1,825,000$	$\times$	$(1.07)^1$	=	488,190
7	$0.30\times2,190,000$	$\times$	$(1.07)^0$	=	657,000
	Total amount accumulated after 7 yr			=	\$1,802,000

The amount of repayment required for the $1,100,000 loan after seven years at $i=7$ percent is calculated as $F = P(1+i)^n$, or

$$F = \$1,100,000(1.07)^7 = \$1,766,400$$

Therefore, the proposed tolls will be sufficient to repay the accumulated P&I of the loan.

DOUBLE ARITHMETIC GRADIENT PLUS PERIODIC UNIFORM SERIES

Many common problem situations require a solution involving both the double arithmetic gradient and the periodic uniform series. For instance, in calculating utility rates, the units of consumption may increase at a gradient amount each year at the same time the amount charged per unit rises. The following example illustrates a case of this type.

Example 22.2

Construction on a new municipal water treatment plant has just been completed and the plant is now going into operation. The plant cost $1,800,000 which was borrowed under a loan agreement stipulating repayment in one lump sum ten years from now, with interest compounding annually at $5\frac{1}{2}$ percent on the outstanding balance of the loan. The utilities board is responsible for setting water rates and paying off the borrowed money. They are considering a proposal to pay off the loan by adding a charge of $0.30/1,000 gal (kgal) to the existing rate schedule and depositing the proceeds into a trust fund bearing 6 percent compounded annually. To further boost the fund the rates will be increased by $0.04/kgal each year starting with the second year, and this money also will be added to the trust fund. You are asked to determine if there will be a sufficient balance in the trust fund at the end of ten years to repay the loan. The following consumption data are provided.

current consumption   240,000 kgal/yr
anticipated growth   500 new dwelling units/yr at 120 kgal/yr
                       for each dwelling unit=60,000 kgal/yr

SOLUTION

The income the first year is the product of the consumption in kgal/yr (A_1) times the rate/kgal (A_2), or

$$A_1 A_2 = 240,000 \text{ kgal/yr} \times \$0.30/\text{kgal} = \$72,000/\text{yr}$$

Also, two single arithmetic gradient increases provide additional income, calculated as follows.

1. The gradient increase in consumption (G_1) of 60,000 kgal/yr each year at the base rate of \$0.30/kgal yields an arithmetic gradient of $G_1 A_2 = 60,000 \text{ kgal/yr/yr} \times \$0.30/\text{kgal} = \$18,000/\text{yr/yr}$.
2. The gradient increase in rate (G_2) of \$0.04/kgal/yr on the base amount of consumption (A_1) of 240,000 kgal/yr yields a gradient of $G_2 A_1 = \$0.04/\text{kgal/yr} \times 240,000 \text{ kgal/yr} = \$9,600/\text{yr/yr}$.

Finally there is the double gradient increase resulting from the increase in consumption (G_1) of 60,000 kgal/yr/yr, multiplied by the increase in rate of (G_2) of \$0.04/kgal/yr. This double gradient amount is

$$G_1 G_2 = 60,000 \text{ kgal/yr} \times \$0.04/\text{kgal/yr} = \$2,400/\text{yr/yr}$$

The future amount in the trust fund after ten years with interest compounding at 6 percent annually is the sum of all four of the components, as follows.

$$F = A_1 A_2 (F/A, i, n) + (G_1 A_2 + G_2 A_1)(F/G, i, n) + G_1 G_2 (F/GG, i, n)$$

Substituting and solving yields

$$F = \$72,000 \, (F/A, 6\%, 10) + (\$18,000 + \$9,600) \, (F/A, 6\%, 10)$$
$$\qquad\quad 13.1808 \qquad\qquad\qquad\qquad\qquad\quad 53.0132$$
$$+ \$2,400 \, (F/GG, 6\%, 10)$$
$$\qquad\quad 320.1260$$
$$F = \$949,000 + \$1,463,200 + \$768,300 = \$3,180,500$$

The amount owed at the end of the ten-year period as a result of the interest compounding annually at $5\frac{1}{2}$ percent on the loan of \$1,800,000 amounts to $F = \$1,800,000(1 + 0.055)^{10} = \$3,074,700$. Therefore, since the proposed water rate will yield a total of $3,180,500 in the trust fund, the proposed rate is adequate to pay off the loan.

OTHER APPLICATIONS

The double gradient can be applied to the solution of a wide variety of common problems involving a gradient change in two related variables. Three general categories are mentioned as examples of the scope.

1. Problems involving quantity sales of units to customers. In these cases, G_1 represents the gradient change in the number of customers and G_2 represents the gradient in the cost or income per customer. This type problem may include road user costs, traffic signal, and intersection problems (where G_1 is the change in number of vehicles, G_2 change in cost per vehicle); sales to utility and other types of customers (G_1 change in number of customers, G_2 change in the payment per customer); income from rental units (G_1 change in occupancy rate, G_2 change in rental rate); a variety of service point problems (G_1 is the change in units passing the service point, and G_2 is the change in cost per service).
2. Problems involving units of production, where G_1 represents the gradient in the number of units produced and G_2 represents the gradient in the cost or income per unit. Problems involving variables of this type could include productivity of construction equipment in terms of cubic yards, square yards, hours, pumping capacity, and so forth (G_1 is change in production units per time period or availability "up time" with increasing age, G_2 is change in unit price).
3. Probability problems (including insurance problems), where G_1 represents the gradient change in the frequency of occurrence and G_2 represents the gradient change in the cost of each occurrence. (G_1 is change in number of accidents or other probability occurrence per year and G_2 is change in cost per accident.)

SUMMARY

Double arithmetic gradients can be utilized to approximate a number of situations where an arithmetic gradient is itself subject to a gradient. An equation relating the future value F for a double arithmetic gradient G_1 and G_2 is developed which, in factor form, is

$$F = G_1 G_2 (F/GG, i\%, n)$$

Values for the GG factors are given in Table 22.1. Example problems illustrate the utility of this factor.

Table 22.1 INTEREST TABLES FOR THE DOUBLE ARITHMETIC GRADIENT

n	i = 1% F/GG	P/GG	A/GG	i = 2% F/GG	P/GG	A/GG	i = 3% F/GG	P/GG	A/GG	n
1	0.00000	0.00000	0.00000	0.00000	0.00000	0.00000	0.00000	0.00000	0.00000	1
2	1.00000	0.98030	0.49751	1.00000	0.96117	0.49505	1.00000	0.94260	0.49251	2
3	5.01000	4.86266	1.65341	5.02000	4.73046	1.64031	5.03000	4.60316	1.62736	3
4	14.0601	13.5115	3.46274	14.1204	13.0451	3.42594	14.1809	12.5995	3.38962	4
5	30.2007	28.7349	5.92054	30.4028	27.5368	5.84215	30.6063	26.4013	5.76484	5
6	55.5027	52.2861	9.02187	56.0109	49.7360	8.87917	56.5245	47.3384	8.73855	6
7	92.0577	85.8639	12.7618	93.1311	81.0762	12.5272	94.2203	76.6097	12.2953	7
8	141.978	131.115	17.1354	143.994	122.397	16.7767	146.047	115.291	16.4239	8
9	207.398	189.632	22.1377	210.874	176.450	21.6178	214.428	164.341	21.1070	9
10	290.472	262.961	27.7639	295.091	242.898	27.0410	301.861	224.613	26.3315	10
11	393.377	352.593	34.0090	402.013	323.324	33.0366	410.917	296.855	32.0833	11
12	518.311	459.974	40.8682	531.053	418.732	39.5951	544.244	381.722	38.3486	12
13	667.494	586.502	48.3364	685.674	530.048	46.7070	704.572	479.779	45.1134	13
14	843.169	733.525	56.4090	868.388	658.120	54.3628	894.709	591.508	52.3640	14
15	1047.60	902.350	65.0809	1081.76	803.760	62.5530	1117.55	717.313	60.0869	15
16	1283.08	1094.23	74.3473	1328.39	967.651	71.2683	1376.08	857.526	68.2683	16
17	1551.91	1310.40	84.2035	1610.96	1150.49	80.4993	1673.36	1012.41	76.8951	17
18	1856.43	1552.00	94.6444	1932.18	1352.83	90.2368	2012.56	1182.17	85.9538	18
19	2198.99	1820.19	105.665	2294.82	1575.24	100.471	2396.94	1366.94	95.4313	19
20	2581.98	2116.05	117.261	2701.72	1818.18	111.194	2829.84	1566.82	105.315	20
21	3007.80	2440.62	129.428	3155.75	2082.09	122.395	3314.74	1781.84	115.591	21
22	3478.88	2794.92	142.160	3659.87	2367.34	134.066	3855.18	2011.99	126.247	22
23	3997.67	3179.91	155.453	4217.06	2674.28	146.198	4454.84	2257.23	137.271	23
24	4566.64	3596.53	169.301	4830.41	3003.17	158.781	5117.43	2517.46	148.650	24
25	5188.31	4045.68	183.701	5503.01	3354.26	171.806	5847.01	2792.56	160.371	25
30	9176.32	6808.12	263.802	9868.80	5448.28	243.265	10642.3	4384.50	223.694	30
36	16245.1	11354.2	377.122	17771.6	8712.03	341.798	19522.4	6735.86	308.528	36
40	22616.1	15190.2	462.626	25030.0	11335.9	414.391	27849.5	8537.46	369.351	40
48	40174.4	24918.6	656.202	45535.3	17601.1	573.828	52032.3	12591.7	498.353	48
60	81560.4	44894.9	998.662	95960.4	29247.0	841.376	114443.	19424.7	701.873	60
120	783778.	237481.	3407.16	>10**6	106427.	2346.51	>10**6	51508.2	1591.08	120
180	>10**6	533967.	6408.50	>10**6	174005.	3581.50	>10**6	67580.5	2037.38	180
240	>10**6	856001.	9425.30	>10**6	215115.	4339.75	>10**6	73086.6	2194.42	240
300	>10**6	>10**6	12116.4	>10**6	236056.	4733.57	>10**6	74658.1	2240.06	300
360	>10**6	>10**6	14331.7	>10**6	245651.	4917.15	>10**6	75061.0	2251.88	360
INF	INF	2010000	20100.0	INF	252500.	5050.00	INF	75185.2	2255.56	INF

Table 22.1 (Continued)

n	i=5% F/GG	i=5% P/GG	i=5% A/GG	i=6% F/GG	i=6% P/GG	i=6% A/GG	i=8% F/GG	i=8% P/GG	i=8% A/GG	n
1	0.00000	0.00000	0.00000	0.00000	0.00000	0.00000	0.00000	0.00000	0.00000	1
2	1.00000	0.90703	0.48780	1.00000	0.89000	0.48544	1.00000	0.85734	0.48077	2
3	5.05000	4.36238	1.60190	5.06000	4.24847	1.58940	5.08000	4.03267	1.56481	3
4	14.3025	11.7667	3.31835	14.3636	11.3773	3.28340	14.4864	10.6479	3.21483	4
5	31.0176	24.3031	5.61341	31.2254	23.3334	5.53928	31.6453	21.5373	5.39415	5
6	57.5685	42.9585	8.46358	58.0989	40.9575	8.32922	59.1769	37.2915	8.06673	6
7	96.4469	68.5430	11.8456	97.5849	64.8995	11.6258	99.9111	58.2972	11.1973	7
8	150.269	101.708	15.7365	152.440	95.6427	15.4019	156.904	84.7703	14.7513	8
9	221.783	142.963	20.1135	225.586	133.524	19.6310	233.456	116.786	18.6951	9
10	313.872	192.690	24.9543	320.122	178.754	24.2870	333.133	154.305	22.9960	10
11	429.565	251.158	30.2366	439.329	231.433	29.3441	459.783	197.193	27.6221	11
12	572.044	318.535	35.9389	586.689	291.566	34.7772	617.566	245.244	32.5427	12
13	744.646	394.902	42.0396	765.890	359.079	40.5616	810.971	298.192	37.7279	13
14	950.878	480.258	48.5176	980.843	433.828	46.6733	1044.85	355.730	43.1490	14
15	1194.42	574.537	55.3523	1235.59	515.612	53.0888	1324.44	417.518	48.7784	15
16	1479.14	677.613	62.5232	1534.84	604.182	59.7851	1655.39	483.193	54.5896	16
17	1809.10	789.305	70.0106	1882.93	699.252	66.7399	2043.82	552.382	60.6573	17
18	2188.56	909.390	77.7949	2284.90	800.501	73.9315	2496.33	624.704	66.6572	18
19	2621.98	1037.61	85.8569	2746.00	907.587	81.3387	3020.04	699.779	72.8663	19
20	3114.08	1173.66	94.1779	3271.75	1020.15	88.9412	3622.64	777.231	79.1627	20
21	3669.79	1317.24	102.740	3868.06	1137.81	96.7191	4312.45	856.693	85.5256	21
22	4294.28	1468.00	111.525	4541.14	1260.19	104.653	5098.45	937.811	91.9355	22
23	4992.99	1625.57	120.515	5297.51	1386.90	112.725	5990.32	1020.24	98.3741	23
24	5771.64	1789.60	129.694	6144.47	1516.55	120.917	6998.55	1103.67	104.824	24
25	6636.22	1959.69	139.045	7089.14	1651.76	129.212	8134.43	1187.77	111.269	25
30	12479.9	2887.56	187.840	13572.2	2363.05	171.673	16192.0	1609.12	142.934	30
36	23865.8	4120.60	249.027	26563.6	3260.44	222.997	33358.2	2089.04	178.289	36
40	35055.8	4979.53	290.198	39669.3	3856.74	256.325	51693.4	2379.49	199.545	40
48	69700.8	6701.18	370.699	81745.3	4986.33	318.615	115493.	2872.20	235.636	48
60	169939.	9097.75	480.618	211734.	6418.59	397.155	343544.	3392.79	274.131	60
120	>10**6	15252.4	764.814	>10**6	9246.46	555.298	>10**6	4040.89	323.303	120
180	>10**6	16388.0	813.923	>10**6	9518.94	571.152	>10**6	4062.05	324.964	180
240	>10**6	16399.8	819.448	>10**6	9536.11	572.167	>10**6	4062.49	324.999	240
300	>10**6	16399.1	819.955	>10**6	9536.99	572.220	>10**6	4062.50	325.000	300
360	>10**6	16399.9	819.997	>10**6	9537.04	572.222	>10**6	4062.50	325.000	360
INF	INF	16400.0	820.000	INF	9537.04	572.222	INF	4062.50	325.000	INF

	i=10%			i=15%			i=20%			
n	F/GG	P/GG	A/GG	F/GG	P/GG	A/GG	F/GG	P/GG	A_r/GG	n
1	0.00000	0.00000	0.00000	0.00000	0.00000	0.00000	0.00000	0.00000	0.00000	1
2	1.00000	0.82645	0.47619	1.00000	0.75614	0.46512	1.00000	0.69444	0.45455	2
3	5.10000	3.83171	1.54079	5.15000	3.38621	1.48308	5.20000	3.00926	1.42857	3
4	14.6100	9.97883	3.14803	14.9225	8.53199	2.98846	15.2400	7.34954	2.83905	4
5	32.0710	19.9136	5.25315	33.1609	16.4868	4.91827	34.2880	13.7796	4.60761	5
6	60.2781	34.0254	7.81249	63.1350	27.2950	7.21235	66.1456	22.1520	6.66124	6
7	102.306	52.4991	10.7836	108.605	40.8287	9.81361	115.375	32.1990	8.93276	7
8	161.537	75.3580	14.1254	173.896	56.8469	12.6683	187.450	43.5948	11.3612	8
9	241.690	102.500	17.7982	263.980	75.0397	15.7264	288.940	55.9984	13.8921	9
10	346.859	133.729	21.7638	384.578	95.0617	18.9412	427.728	69.0804	16.4772	10
11	481.545	168.779	25.9857	542.264	116.556	22.2702	613.273	82.5392	19.0751	11
12	650.700	207.333	30.4289	744.604	139.172	25.6745	856.928	96.1101	21.5502	12
13	859.770	249.045	35.0601	1000.29	162.576	29.1190	1172.31	109.569	24.1731	13
14	1114.75	293.548	39.8480	1319.34	186.460	32.5725	1575.78	122.732	26.5197	14
15	1422.22	340.468	44.7627	1713.24	210.548	36.0072	2086.93	135.453	28.9710	15
16	1789.44	389.435	49.7763	2195.23	234.592	39.3992	2729.32	147.623	31.2128	16
17	2224.39	440.083	54.8626	2780.51	258.381	42.7277	3531.18	159.162	33.3349	17
18	2735.83	492.062	59.9973	3486.59	281.734	45.9751	4526.42	170.017	35.3304	18
19	3333.41	545.039	65.1577	4333.57	304.500	49.1269	5755.70	180.158	37.1959	19
20	4027.75	598.699	70.3230	5344.61	326.557	52.1712	7267.84	189.575	38.9304	2.0
21	4830.52	652.752	75.4740	6546.30	347.809	55.0988	9121.41	198.269	40.5350	21
22	5754.58	706.927	80.5932	7968.45	368.184	57.9027	11386.7	206.125	42.0125	22
23	6814.04	760.979	85.6648	9648.63	387.628	60.5779	14148.0	213.564	43.3673	23
24	8024.44	814.686	90.6744	11624.9	406.108	63.1214	17506.6	220.218	44.6047	24
25	9402.88	867.848	95.6092	13944.7	423.606	65.5316	21584.0	226.256	45.7306	25
30	19543.7	1120.02	118.811	32875.6	496.523	75.6205	59003.5	248.565	49.9233	30
36	42656.6	1379.91	142.604	85086.4	555.569	83.8830	186366.	262.930	52.6604	36
40	68944.4	1523.32	155.774	155780.	581.564	87.5615	393912.	268.009	53.6383	40
48	163996.	1741.92	176.005	501725.	612.307	91.9583	>10**6	272.754	54.5594	48
50	589311.	1935.46	194.184	>10**6	630.201	94.9517	>10**6	274.622	54.9255	60
120	>10**6	2098.17	209.819	>10**6	637.031	95.5547	>10**6	275.000	55.0000	120
180	>10**6	2099.99	209.999	10**6	637.037	95.5556	>10**6	275.000	55.0000	180
240	>10**6	2100.00	210.000	10**6	637.037	95.5556	>10**6	275.000	55.0000	240
300	>10**6	2100.00	210.000	10**6	637.037	95.5556	>10**6	275.000	55.0000	300
360	>10**6	2100.00	210.000	10**6	637.037	95.5556	>10**6	275.000	55.0000	360
INF	INF	2100.00	210.000	INF	637.037	95.5556	INF	275.000	55.0000	INF

Table 22.1 (Continued)

n	i = 30% F/GG	P/GG	A/GG	i = 50% F/GG	P/GG	A/GG	n
1	0.00000	0.00000	0.00000	0.00000	0.00000	0.00000	1
2	1.00000	0.59172	0.43478	1.00000	0.44444	0.40000	2
3	5.30000	2.41238	1.32832	5.50000	1.62963	1.15789	3
4	15.8900	5.56353	2.56829	17.2500	3.40741	2.12308	4
5	36.6570	9.87280	4.05359	41.8750	5.51440	3.17536	5
6	72.6541	15.0522	5.69567	87.8125	7.70919	4.22556	6
7	130.450	20.7894	7.41918	167.719	9.81619	5.21321	7
8	218.585	26.7963	9.16205	300.578	11.7281	6.10214	8
9	348.161	32.8314	10.8749	514.867	13.3929	6.87528	9
10	533.609	38.7070	12.5203	853.301	14.7975	7.52934	10
11	793.692	44.2869	14.0712	1379.95	15.9536	8.07012	11
12	1152.80	49.4805	15.5098	2190.93	16.8862	8.50870	12
13	1642.64	54.2349	16.8260	3430.39	17.6261	8.85859	13
14	2304.43	58.5271	18.0157	5314.59	18.2050	9.13381	14
15	3191.76	62.3563	19.0796	8167.88	18.6526	9.34767	15
16	4374.29	65.7376	20.0222	12476.8	18.9952	9.51208	16
17	5942.58	68.6970	20.8501	18971.2	19.2550	9.63729	17
18	8014.35	71.2669	21.5719	28745.8	19.4506	9.73187	18
19	10742.7	73.4832	22.1968	43442.8	19.5967	9.80278	19
20	14326.5	75.3827	22.7344	65525.1	19.7053	9.85561	20
21	19024.4	77.0017	23.1944	98687.7	19.7855	9.89472	21
22	25172.7	78.3747	23.5359	148473.	19.8444	9.92354	22
23	33208.5	79.5339	23.9175	223193.	19.8875	9.94466	23
24	43700.1	80.5085	24.1971	335318.	19.9190	9.96008	24
25	57386.1	81.3248	24.4321	503553.	19.9418	9.97129	25
30	219433.	83.7532	25.1355	>10**6	19.9893	9.99468	30
36	>10**6	84.7736	25.4341	>10**6	19.9987	9.99934	36
40	>10**6	85.0106	25.5039	>10**6	19.9997	9.99984	40
48	>10**6	85.1552	25.5466	>10**6	20.0000	9.99999	48
60	>10**6	85.1832	25.5550	>10**6	20.0000	10.0000	60
120	>10**6	85.1852	25.5556	>10**6	20.0000	10.0000	120
180	>10**6	85.1852	25.5556	>10**6	20.0000	10.0000	180
240	>10**6	85.1852	25.5556	>10**6	20.0000	10.0000	240
300	>10**6	85.1852	25.5556	>10**6	20.0000	10.0000	300
360	>10**6	85.1852	25.5556	>10**6	20.0000	10.0000	360
INF	INF	85.1852	25.5556	INF	20.0000	10.0000	INF

PROBLEMS FOR CHAPTER 22, DOUBLE ARITHMETIC GRADIENT

Problem Group A

Compare alternatives involving double arithmetic gradients.

A.1 Two alternate alignments are under consideration for the location of a proposed highway. The characteristics of each are listed in the table. The motoring public is assumed to borrow and lend at an average $i = 10$ percent. Both routes will cost the same to construct and maintain. You are asked to find the route with the least user cost over the ten-year period.

	A	B
Traffic count the first year	10,000 vehicles/day	15,000 vehicles/day
Gradient increase in each subsequent year	1,000 vehicles/day/yr	0
Operating cost per vehicle traversing the entire route the first year	$1.00/vehicle	$1.25/vehicle
Gradient increase in each subsequent year	$0.10/vehicle/yr	$0.10/vehicle/yr

A.2 A contractor buys a new scraper for $95,000, expects to keep it for five years and then trade it in on a new one. He asks you to determine if the extra amount he is charging per cubic yard of excavation is sufficient to fund the purchase of a new scraper at the end of the next five-year period. The following data are provided as a basis for your calculations.

Heavy equipment prices for new equipment are rising at the rate of 5.5 percent compounded annually, so the replacement five years from now will cost more than the present model. During the first year, the scraper moves dirt at an average rate of 300 yd^3/hr, but with age and loss of efficiency, the rate declines by 30 yd^3/hr each year. The contractor's equipment works an average of 1,800 hr/yr.

The contractor is charging $0.10/yd^3 this year to cover the replacement cost, but expects to raise the amount by $0.02/yd^3 each year starting one year from now. He deposits this extra $0.10/yd^3 charge into an account bearing 9 percent interest compounded annually. The existing machine can be traded in on the new machine for an estimated $20,000 trade-in allowance at the end of the five-year period.

A.3 Find the equivalent annual cost of a bulldozer over a five-year life span using the following data. Use $i = 9$ percent.

cost new	$80,000
resale price after 5 yr	$35,000
cost per hour to operate and maintain the first year	$16.00/hr
gradient in hourly costs each year	$1.10/hr/yr
number of hours used the first year	2,000 hr/yr
gradient in hours of use per year	-100 hr/yr

(Machine is used 1,900 hr the second year, 1,800 hr the third year, and so on.)

A.4 At the present time a certain downstream area floods when it is subject to a ten-year storm. (The area has a 1 in 10 probability, or $p = .10$, of flooding this year.) With increasing development upstream, this probability will increase by $p = .01$ each year for the next 20 years (next year $p = .11$, the third year $p = .12$, etc.). If the area floods this year, the cost of flood damage is estimated at $100,000. Due to the increasing value of the property in the area, if flooding occurs in any future year, the damage is expected to increase by $10,000 per year each year after this year. A flood prevention structure can be built to protect the area for the next 20 years at an estimated cost of $200,000. Is the present worth of the probable flood damage greater or less than the cost of the flood prevention structure? Which alternative do you recommend and why? Use $i = 8$ percent.

A.5 Assume a friend purchased a car four years ago for $9,600. Expenses have been as listed in the table below. At what minimum price can she sell the car now if her goal is to keep mileage costs under $0.28/mi. Use $i = 10$ percent.

	YEAR			
	1	2	3	
Gas (cents/gal)	120	130	140	150
Tires (cents/mi)	1.6	1.7	1.8	1.9
Repairs and maintenance (cents/mi)	2.2	2.6	3.0	3.4
License and insurance (cents/mi)	4.0	4.2	4.4	4.6
mi/yr driven	20,000	18,000	16,000	14,000

The car gets 20 miles to the gallon. Make the simplifying assumption that all car costs are paid at the end of the year in which incurred.

Chapter 23
Inflation

KEY EXPRESSIONS IN THIS CHAPTER

i_i = Inflation rate.

i_c = Combined interest-inflation rate. Rates calculated for loans or investments including the combined effects of interest charges and inflation. This rate is not discounted for the effects of inflation on buying power. Also known as the nominal market interest rate, apparent rate, or published rate.

i_r = Real interest rate after inflation, or real increase in buying power, periodically compounded.

NATURE OF INFLATION

Throughout this text money, expressed in dollars, serves as a convenient, easily understood, readily accepted measure of the value, or worth of goods and services. The concept of comparing alternatives on the basis of *equivalent* worth is valid as long as the basic value of the dollar remains constant. Unfortunately, the basic value of the dollar, expressed in terms of the goods and services it can purchase, often changes with time. Inflation occurs when the price of goods and services increases (inflates) as the purchasing power of money declines. For example, if a given loaf of bread costs $1.00 now and $1.12 a year later, the price of the bread has been *inflated* by $0.12 in one year. If there has been no corresponding increase in the size of the loaf or quality of the product, then money is simply not worth as much (in terms of bread) and the 12 percent price increase is attributed to inflation.

Inflation is an important ingredient in cost analysis and has been observed throughout much of the commercial history of the civilized world. (For instance, in 301 A.D. the Emperor Diocletian claimed that inflation threatened to destroy the economy of the Roman Empire.) Inflation is often treated as affecting all goods and services equally, but actually the rate of price increase often varies greatly from one item to another. For instance gasoline may double in price during the same time period that the price of bread increases by a much smaller amount.

INFLATION AND PRICE INDEXES

Price trends for groups of items of special interest are often condensed into an index format. Thus the Consumer Price Index (CPI) is one widely watched index of the cost of a cross section of typical consumer items and is often referred to as the "cost of living" index. In construction a number of indexes are used to measure cost trends, including the Engineering News Record Construction Cost Index and many others. Rises in these indexes indicate rises in prices and the corresponding decrease in purchasing power of money.

Over the past several decades the costs of goods and services have grown at an increasing rate, as measured by the growth in the CPI. The growth in the CPI averaged between 1 and 2 percent per year for the period between 1955 and 1965. Between 1965 and 1975 the growth in the CPI steadily increased from around 1 percent to approximately 7 percent. Since 1975 the growth rate has been even higher.

HOW INFLATION WORKS

An example of how inflation works is shown in the following example. Assume that current wages and prices for a given worker, time, and place are

> pay for work $10/hr
> price for bread $1/loaf

Therefore, this worker can buy ten loaves of bread with one hour's pay.

Assume inflation occurs at a rate of 10 percent per year ($i_i = 10\%$). Then one year later, if both of these items have kept pace with inflation,

> pay for work $11/hr
> price for bread $1.10/loaf

Therefore, despite the pay increase the worker can still buy only ten loaves with one hour's pay. A dollar is simply worth about 10 percent less than one year previously. If all costs and incomes keep pace exactly with inflation, then no one's purchasing power is decreased (or increased). However, many people are on fixed incomes and are not able to raise their income to compensate for inflation. Even though the nominal amount of their income or wealth is unchanged, their buying power declines together with their standard of living in many instances. Others find themselves in higher tax brackets with less buying power, while some manage to profit from inflation as property prices increase while mortgage payments remain constant.

Investors Seek a Gain in Purchasing Power

Normally investors are not satisfied just to keep pace with inflation, as this nets no real gain on an investment. To be successful, an investment must result in a net gain in buying power over and above the increase required to keep up with inflation.

For example, if an investor desires a real return of 4 percent increase in buying power after inflation ($i_r = 4\%$) and inflation is at 10 percent ($i_i = 10\%$) then for an investment of \$1,000 for a duration of one year, the cash flow diagram is as shown in Figure 23.1. To regain the same buying power that the \$1,000 had at the beginning of the year (with no real increase in purchasing power), F must equal $P(1+i_i)^n$ or,

$$F = \$1,000 \times 1.10 = \$1,100$$

This \$1,100 at EOY 1 will buy only the same 1,000 loaves of bread or 100 hours of work that \$1,000 would buy at EOY 0. If the investor wants 4 percent increase in buying power then he or she needs to invest at a rate which will yield $\$1,100(1+i_r)^n$ or $\$1,100(1.04) = \$1,144$. Expressed as an equation the investment must yield an FW of

$$F = P(1+i_i)^n(1+i_r)^n$$

so that

$$F = \$1,000(1.10)(1.04) = \$1,144$$

Expressed in terms of interest, the combined interest-inflation rate to account for real return as well as inflation must be

combined interest-inflation $i_c = (1+i_i)(1+i_r) - 1$

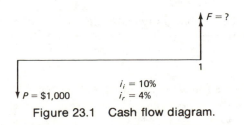

$F = ?$

1

$i_i = 10\%$
$i_r = 4\%$

$P = \$1,000$

Figure 23.1   Cash flow diagram.

so that

$$i_c = (1.10)(1.04) - 1 = 0.144 \quad \text{or} \quad 14.4\%$$

If the investment is for five years, the same procedure is used to find F.

$$F_1 = P(1 + i_i)^n \times (1 + i_r)^n$$

$$F_1 = \$1,000(1.10^5 \times 1.04^5)$$

$$F_1 = \$1,000(1.10 \times 1.04)^5$$

$$F_1 = \$1,000 \times 1.144^5 = \$1,959$$

Thus, a return of $1,959 on every $1,000 invested would provide a real increase in buying power of 4 percent per year (before taxes) compounded annually over the five-year period, with inflation at 10 percent.

REAL INTEREST RATE

It is evident from the foregoing that in order to net a real increase in buying power at a rate i_r, that the combined interest-inflation rate of return, i_c, must exceed the inflation rate i_r, and can be calculated as combined interest-inflation, i_c.

$$i_c = (1 + i_i)(1 + i_r) - 1 \tag{23.1}$$

Conversely, the rate of return after inflation (ROR i_r) can be found as

$$\text{real gain in buying power, } i_r = \frac{1 + i_c}{1 + i_i} - 1 \tag{23.2}$$

The next example illustrates the use of the equations in determining the real after-inflation rate of return, i_r, compared to the combined interest-inflation rate of return, i_c.

Example 23.1 (*Given P, F, i_i, n, find i_c, i_r*)

Given

$$P = \$1,000$$
$$F = \$2,000$$
$$n = 4 \text{ yr}$$
$$i_i = 10\%$$

Find (a) ROR i_c, the combined interest-inflation rate of return; (b) ROR i_r, the real after-inflation rate of return.

SOLUTION

a. The ROR i_c is found in the same way as the ROR found in earlier chapters, as follows.

$$F/P = (1 + i_c)^n$$

Thus,

$$2,000/1,000 = (1+i_c)^4$$

and

$$i_c = 2^{0.25} - 1 = 0.1892 \quad \text{or} \quad \underline{18.9\%}$$

b. The real *ROR* i_r is found as

$$i_r = \frac{(1+i_c)}{(1+i_i)} - 1$$

Thus,

$$i_r = \frac{1.1892}{1.1} - 1 = 0.0811 \quad \text{or} \quad \underline{8.1\%} \quad \begin{array}{l} \text{real interest gain in} \\ \text{buying power} \end{array}$$

Thus, the real gain in buying power from this investment equals 8.1 percent compounded per year over the four-year period. Using the $1 loaf of bread and $10 per hour labor wage to illustrate, for each $100 invested, (equal to 100 loaves or 10 hours of labor) the purchasing power of the balance in the account increases enough at the end of the first year to buy an additional 8.1 loaves of bread or 0.81 hours of labor, even though the price of bread and labor rose 10 percent during the year. These increases in purchasing power are compounded annually at 8.1 percent throughout the four-year investment period.

INFLATION AND SERIES PAYMENTS

Many series payments in an inflationary environment can be classified in one of two general categories.

1. *Equal payment series.* The amounts remain constant but the buying power of each payment shrinks due to inflation, for example, bonds yield a series of future fixed-interest payments, but with inflation each payment will buy less.
2. *Series of increasing payments with constant buying power.* Each payment in the series is larger than the previous one by an amount sufficient to maintain constant buying power, for example, an investment in a rental building with rents geared to rise with inflation. Another type example is any project incurring a commitment for a future series of maintenance person-hours per year relatively fixed in number such as a highway, utility, or similar project. The costs per person-hour increase as wages rise each year with inflation, but the number of person-hours remains relatively constant.

Series of Fixed Payments With Declining Buying Power

An example follows illustrating how to find the real rate of return (rate of compound increase in buying power) for a given P, A, F, and i_i.

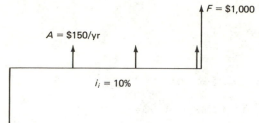

Figure 23.2   Cash flow diagram for Example 23.2

Example 23.2   (*Given P, A, F, n, i_i, find i_c, i_r*)

Given

$P = -\$1{,}000$
$A = +\$150/\text{yr}$
$F = +\$1{,}000$
$n = 3$
$i_i = 10\%$

Find i_c, combined interest-inflation rate of return and i_r, real rate of return. See Figure 23.2.

SOLUTION
Solve for *ROR* i_c using the present worth method (or any other valid method).

$$PW = 0 = -\$1{,}000 + 150(P/A, i_c, 3) + 1{,}000(P/F, i_c, 3)$$

Solving the combined interest-inflation rate, $i_c = 15$ percent. To determine the real interest rate, i_r, assume as before that

pay for work $= \$10/\text{hr}$ at *PTZ*
price for bread $= \$1/\text{loaf}$ at *PTZ*

If the 10 percent inflation raises the wages for work as well as the cost of bread then at EOY 1 work earns \$11 per hour, which still only buys 10 loaves of bread since bread is \$1.10 per loaf.

The interest payment of \$150 at EOY 1 is worth only $\$150/(\$11/\text{hr}) = 13.64$ hours of work instead of 15 hours. Thus from the initial investment of \$1,000 or 100 hours worth of work, the return the first year is only 13.64 hours worth. In subsequent years the table below shows the value of the interest payment in terms of hours of work. Assume inflation at 10 percent per year.

HOURLY WAGE	EOY	INVESTMENT	RETURN
\$10/hr	0	\$1,000 or 100 hr	
\$11/hr	1		\$150 or 13.64 hr
\$12.10/hr	2		\$150 or 12.40 hr
\$13.31/hr	3		\$150 or 11.27 hr
\$13.31/hr	3		\$1,000 or 75.13 hr

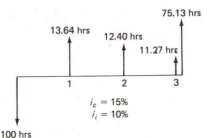

Figure 23.3   Cash flow diagram in terms of hours of work.

The "cash flow" diagram for this real investment and return is shown in Figure 23.3. Based on the buying power of the investment and return, the real rate of return i_r, may be calculated as follows.

EOY		4%	5%
0	$P_1 =$	− 100.00 hr	− 100.00 hr
1	$P_2 = 13.64 (P/F, i, 1)$	+ 13.12	
	4% 0.96154		
	5% 0.95239		+ 12.99
2	$P_3 = 12.40 (P/F, i, 2)$	+ 11.46	
	4% 0.92456		
	5% 0.80703		+ 11.25
3	$P_4 = 11.27 (P/F, i, 3)$	+ 10.02	
	4% 0.88900		
	5% 0.86384		+ 9.74
3	$P_5 = 75.13 (P/F, i, 3)$	+ 66.79	+ 64.90
		+ 1.39 hr	− 1.12 hr

$$i_r = 4\% + \frac{1.39}{1.39 + 1.12} \times 1\% = 4.55\%$$

Thus the real interest rate of return, $i_r = 4.55$ percent, or the investment earned additional purchasing power at the rate of 4.55 percent compounded annually over and above the inflation rate of 10 percent.

A faster route to the same conclusion involves use of the i_r equation, as follows. Since $i_c = 15$ percent and $i_i = 10$ percent then

$$i_r = \frac{1 + i_c}{1 + i_i} - 1 = \frac{1.15}{1.10} - 1 = 0.0455 \quad \text{or} \quad \underline{\underline{4.55\%}}$$

Thus the same resulting ROR $i_r = 4.55\%$ is determined more easily.

TAXES AND INFLATION

In an environment of inflation, taxes are treated as just another expense as in previous chapters. For example, assume an investor is in the 30 percent tax bracket and therefore is allowed to keep 70 percent of each $150 interest payment (or $0.7 \times \$150 = \$105$ after-tax income) as shown in the cash flow diagram (Figure 23.4). The after-tax real interest rate of return after

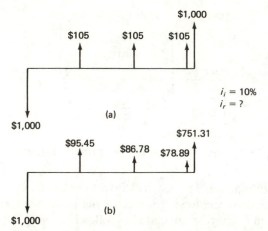

Figure 23.4 (a) Actual cash flow 30% tax bracket. (b) Real after-tax purchasing power in terms of *PTZ* dollars. Purchasing power of interest payments $= (150 - 45)/(1 + i_i)^n$.

inflation (at $i_i = 10\%$) can then be calculated by the usual method for finding the *ROR* as follows.

	1%	0%
$P_1 = \$1,000$	$= -1,000.00$	$-1,000.00$
$P_2 = 0.7 \times \$150 \, (P/F, i, 1) \, / \, (F/P, i_i, 1)$ $\qquad 1\% \, 0.9901 \qquad\qquad 1.100$	$= + \quad 94.51 +$	95.45
$P_3 = 0.7 \times \$150(P/F, i, 2) \, / \, (F/P, i_i, 2)$ $\qquad 1\% \, 0.9803 \qquad\qquad 1.210$	$= + \quad 85.07 +$	86.78
$P_4 = (0.7 \times \$150 + \$1,000)(P/F, i, 3)/(F/P, i_i, 3)$ $\qquad 1\% \, 0.9706 \qquad\qquad\qquad 1.331$	$= + \quad 805.79 +$	830.20
$P_{TOT} =$	$-14.63 +$	12.43

$$i_r = \frac{12.43}{12.43 + 14.63} \times 1\% = 0.46 = \underline{\underline{0.5\%}} \text{ real rate of return}$$

Thus the real *after-tax* purchasing power of the investment increases at only 0.5 percent compounded annually.

Alternate Method

The after-tax real rate of return can be obtained a little more expeditiously by first finding the after-tax *ROR* i_c and then deriving the real after-tax *ROR* i_r, as follows.

Since the $1,000 investment in this example is returned nominally intact at EOY 3, then the after-tax combined interest-inflation rate of return can be calculated simply as

$$i_c = \frac{150 \times 0.7}{1,000} = 10.5\%$$

The after-tax real rate, i_r, is determined as

$$i_r = \frac{1+i_c}{1+i_i} - 1$$

and

$$i_r = \frac{1.105}{1.100} - 1 = 0.0045 \quad \text{or} \quad 0.5\%$$

Conclusion. Use of the i_r equation produces the results more rapidly once a clear mental image of the process is obtained.

Example of a Series of Payments with Constant Buying Power and Nominal Increases in Payment Amounts with Inflation

Many investors in engineering projects desire a fixed level of income in terms of buying power, consequently the nominal income must rise with inflation. Examples of such projects include rental buildings, production facilities such as electric generating plants, and many others.

Example 23.3

Consider again the previous example. With the inflation rate equal to zero, the anticipated income is shown in the cash flow diagram (Figure 23.5) yielding $ROR\ i_c = 15$ percent. Now, assume that inflation occurs at a rate of 10 percent ($i_i = 10\%$) and that the cash flows are going to increase proportionately to inflation much as an investor in a rental building might plan to increase the rental income with inflation.

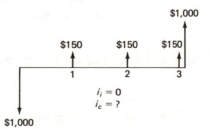

Figure 23.5 Cash flow diagram A, with zero inflation ($i_i = 0$). Buying power remains constant.

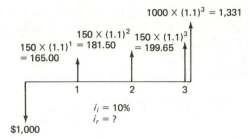

Figure 23.6 Cash flow diagram B, with inflation at rate $i=10\%$. Rents and resale value increase with inflation.

Now the cash flow diagram A (Figure 23.5) represents the buying-power value of the nominal cash flows in the cash flow diagram B (Figure 23.6).

Given that the cash flow diagram B is the actual cash flow, the combined interest-inflation rate of return, i_c, can be found together with the real rate of return, i_r, as follows.

SOLUTION
Since the third i value (of the three i's: i_c, i_i, i_r) can be found if any two are already known, and one is already known ($i_i=10\%$), then the easiest-to-find of the other pair should be sought. This is obviously i_r, since i_r involves the simplest series.

The i_r value is obtained directly from cash flow diagram A (Figure 23.5) showing the buying power of the investment and return. (If the resale value were different from the initial cost, the usual trial and error system for determining ROR i_r would be used.) Since the resale value is the same as the initial cost, then

$$i_r = \frac{150}{1,000} = 15\%$$

Then the combined interest-inflation rate of return, i_c, can be derived as

$$i_c = (1+i_r)(1+i_i) - 1 = (1.15 \times 1.10) - 1 = 0.265 = 26.5\%$$

Conclusion. When inflation is at $i_i=10$ percent, in order to realize a real rate of increase in buying power of $i_r=15$ percent, a combined interest-inflation rate of $i_c=26.5$ percent is required.

COMPARISON OF ALTERNATIVES IN AN ENVIRONMENT OF INFLATION

Often when selecting among alternatives, one alternative features a high cost and low maintenance while a competing alternative has the opposite feature. To illustrate the effect that inflation has on this choice the following example is presented. Both alternatives in this example have the same total payment cash flow of $1,200,000 over the 20-year period, in order to better illustrate the effects of the difference in timing of payments and interest rates.

Example 23.4 (*Comparison of two alternative proposals, with the same total cash flow but different timing*)

Assume a proposed road construction project can be paved either with material A or material B. Material A features a low first cost and high maintenance, while material B has a high first cost but low maintenance. If there were zero inflation the cash flow diagrams for the two materials would appear as follows in Figures 23.7 and 23.8. In other words, these are the anticipated payments in terms of dollars with a constant purchasing power. ($1,000 is abbreviated as $k below.)

Material A
construction cost   = $600k
O & M 20 yr × $30k/yr=   600k
total cash flow   = $1,200k

Material B
construction cost   = $1,000k
O & M 20 yr × $10k/yr=   200k
total cash flow   = $1,200k

SOLUTION
To find the more economical alternative under these conditions, an *IROR* can be calculated on the incremental investment of $400k required if

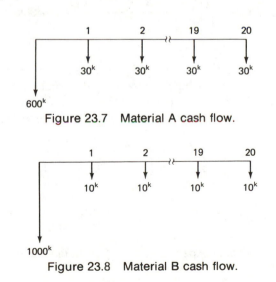

Figure 23.7   Material A cash flow.

Figure 23.8   Material B cash flow.

material B is selected rather than the lower cost material A. The value of i_c at which the $NPW = 0$ is found as follows.

$$P = -\text{material A} + \text{material B} \quad i_c = 0\%$$

$$\text{construction, } P_1 = +\$600k - \$1,000k \qquad = -\$400k$$

$$\text{O\&M, } P_2 = (+30k - 10k)(P/A, i, 20) = +\$400k$$

$$0$$

Only one trial rate, i_c, is necessary since $NPW = 0$ when $i_c = 0$.

Conclusion. If inflation is zero ($i_i = 0$) and the cash flows of two alternatives total the same, then the alternative with the lowest first cost has the lowest present worth for all positive interest rates ($i_c > 0$). (The lowest first cost alternative will incur the least amount borrowed at interest for the shortest time.) Thus for $MARR > 0$, select material A in this case.

Now assume inflation is anticipated at a sustained rate of 10 percent ($i_i = 10\%$). Then the actual cash flow diagram in terms of current (decreasing purchasing power) dollars is shown in Figures 23.9 and 23.10. The increasing dollar amounts shown are needed to buy a constant level of service for operating and maintaining the road.

To find the *IROR* for this case the geometric gradient is used since the dollar amounts increase at the inflation rate of 10 percent per year ($i_i = 10\%$).

The combined interest-inflation rate, i_c, at which $\Delta P_{TOT} = 0$ is found as follows.

$$i_c = \quad 10\%$$

$$\text{construction cost, } \Delta P_1 = (+\$600k - \$1,000k) = \$-400k$$

$$\text{O\&M costs, } \Delta P_2 = \frac{cn}{1+r} = \frac{(+33-11)\times 20}{1.1} = \quad +400k$$

$$0$$

Conclusion. When inflation is at 10 percent ($i_i = 10\%$) then for equal total real cash flows (equal buying power of the total cash flow for each alternative) the *IROR* $i_c = 10$ percent. Therefore, if the *MARR* $i_c > 10$ percent for the combined interest-inflation rate value (infers that $i_r > 0$) then

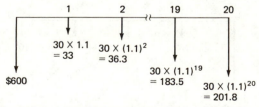

Figure 23.9   Inflated expenditures required to maintain material A.

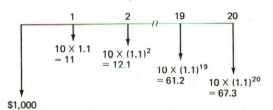

Figure 23.10   Inflated expenditures required to maintain material B.

use material A. And if the *MARR* $i_c < 10\%$ for the combined interest-inflation rate (infers that $i_r < 0$) then use material B.

The general conclusion is: Whenever the total real buying power cash flows of the two alternatives are equal then:

1. For *MARR* $i_c > i_i$ the alternative with the lowest initial cost is more economical.
2. Conversely, for *MARR* $i_c < i_i$ the alternative with the higher initial cost is more economical.

PROJECTS WITH FINANCING

Some roads are financed with borrowed money (often the borrowed money is obtained by selling road bonds) while others are paid out of current tax revenue. The current tax revenues should also be considered as borrowed funds since most of the taxpayers owe debts of their own, and had the taxes not been levied the taxpayers could use these funds to repay debts. Therefore, interest should be charged against the funds used in construction projects even when channeled through current tax revenues. The next example illustrates the point.

Example 23.5   (*Shows effect of financing*)

Find the more economical selection if the project outlined in Example 23.4 is now financed by road bonds at 10%.

SOLUTION
The cash flow diagrams now show a receipt of the borrowed construction funds in addition to annual amortization payments of principal plus interest over the 20-year period. With road bonds at $i = 10\%$, the annual repayment amount, A, is

$$A = P(A/P, 10\%, 20) = P \times 0.1175$$

$$A_A = \$600k \times 0.1175 = \$70.5k$$

$$A_B = \$1{,}000 \times 0.1175 = \$117.5k$$

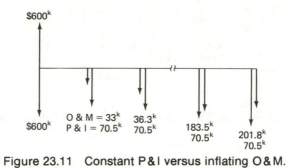

Figure 23.11   Constant P & I versus inflating O & M.

These repayments are combined with the O&M payments of Example 23.4 corrected for inflation and shown in Figures 23.11 and 23.12.

Even though the owner's net investment in the project is zero for both alternatives (so that there is no increment of investment and thus no real *IROR*) the interest rate at which the present worths equal each other can be calculated.

Assume

$$i_i = 10\%$$
$$i_{bonds} = 10\%$$
$$n = 20 \text{ yr}$$

Then for the geometric gradient (for O&M costs)

$$r = i_i = 0.10$$

Since i_c is not yet known, a trial value of $i_c = 0.10$ is assumed. Then

$$O\&M = \Delta P_{-A+B} = \frac{cn}{1+r} = \frac{(33-11)\times20}{1.1} = \$400k$$

$$P\&I = \Delta P_{-A+B} = (+70.5 - 117.5)(P/A, i, 20) = \underline{\$400}$$
$$8.1536$$

$$\Delta P_{TOT} \qquad\qquad = \quad 0$$

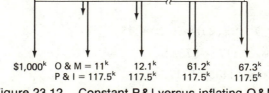

Figure 23.12   Constant P & I versus inflating O & M.

Conclusion

1. If bond money can be borrowed at the same rate as the inflation rate (if $i_{bonds} = i_i$), and if the combined interest-inflation rate is greater than the inflation rate (if $i_c > i_i$) then the alternative with the lower first cost is more economical (material A is preferred). (Money earns more if left with the taxpayer longer and this A series total begins with a smaller amount and gradually increases to exceed the totals for the B series.) But conversely if the combined interest-inflation rate is lower than the inflation rate (if $i_c < i_i$) then the higher first cost is preferred (material B is preferred). (Money earns more if the taxpayer invests it through taxes to pay for roads.)

2. If the real rate of return is positive, ($i_r > 0$ which implies that $i_c > i_i$), and bond funds are available at less than the market rate ($i_{bonds} < i_i$) (since bonds are usually low-risk investments, the interest rate on bonds is often lower than average), then the lower first cost alternative (material A) is the better selection (less money paying less interest for less time, and purchasing power is rising).

SUMMARY

Inflation plays an important role in today's economic planning. To select the most economical from among alternatives, the inflationary loss in purchasing power of future income and costs must be accounted for and real incremental rates of return calculated to determine the true anticipated benefits of proposed projects.

PROBLEMS FOR CHAPTER 23, INFLATION

Problem Group A

Find the amounts to invest in order to obtain a given objective.

A.1 A young engineer on her twenty-third birthday decides to start saving (investing) toward building up a retirement fund. She feels that $300,000 worth of buying power in terms of today's dollars will be adequate to see her through her sunset years starting at her sixty-third birthday.

a. If the inflation rate is estimated at 6 percent for the intervening period, how much must the investment be worth on her sixty-third birthday?

b. If her rate of return from her savings (investment) is estimated at 15 percent and she expects to make annual end of year deposits (with the first deposit on her twenty-third birthday and the last on her sixty-third) and each subsequent deposit will be 6 percent more than the previous one, how much should her first deposit amount to?

c. What will be the amount of her last deposit?

Problem Group B

Find the *ROR*.

B.1 A client expects to deposit $1,000 per year for ten years (first deposit at EOY 1, last deposit at EOY 10) into an account that pays interest compounded at 8 percent per year. If inflation is estimated at 5 percent per year find the real interest rate, i_r, in terms of real compound increase in buying power.

B.2 A home buyer purchases a house for $70,000. He pays $10,000 down and gets a $60,000 mortgage at 10 percent and 30 years payout period. If inflation occurs at 6 percent, what is the real interest rate, i_r?

B.3 Find the real rate of return, i_r, on the investments illustrated by each of the following cash flow diagrams.
  a. Given: $i_i = 8$ percent.
  b. Given: $i_i = 8$ percent.
  c. Given: $i_i = 10$ percent.

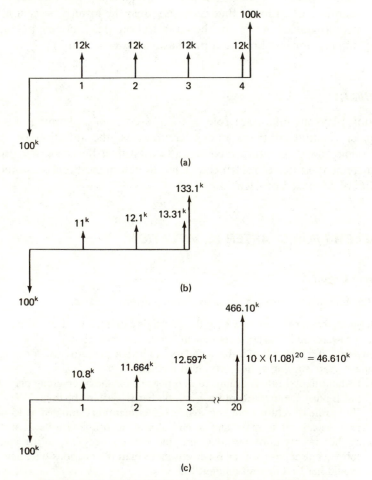

Problem B.3   (a), (b), and (c) Cash flow diagrams.

Problem Group C

Find the after-tax real rate of return i_c.

C.1 An investor is considering the purchase of a bond with a maturity amount of
$5,000 and date ten years from now. The coupons on the bond provide for $450
interest paid annually at the end of each year until maturity. Assume $i_i = 6$
percent and the investor is in the 30 percent tax bracket for ordinary income.
Capital gain tax is paid on 40 percent of the gain at the ordinary rate. If the
bond can be purchased now for $4,000 including commission, find the real rate
of return, i_r.

C.2 Find the real after-tax *ROR* for the before-tax income illustrated in the cash
flow diagram below. Given: 45 percent tax bracket and $i_i = 8$ percent.

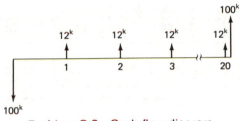

Problem C.2   Cash flow diagram.

C.3 Find the real after-tax *IROR* for the before-tax costs and income listed below.
Given: 42 percent tax bracket; $i_i = 8$ percent; investment tax credit allowed 10
percent at EOY 1; depreciation by *SL*.

	A	B
Cost new	100k	150k
Salvage value @ EOY 10	20k	30k
O&M costs	55k/yr	45k/yr
Grad. O&M costs—both increase 8%/yr compounded		

Appendix A   INTEREST TABLES

	$i = 0.5\%$			$i = 0.5\%$			$i = 0.5\%$			
	PRESENT SUM, P			UNIFORM SERIES, A			FUTURE SUM, F			
n	P/F	P/A	P/G	A/F	A/P	A/G	F/P	F/A	F/G	n
1	0.99502	0.9950	0.0000	1.00000	1.00500	0.0000	1.0050	1.0000	0.0000	1
2	0.99007	1.9851	0.9900	0.49875	0.50375	0.4987	1.0100	2.0050	1.0000	2
3	0.98515	2.9702	2.9603	0.33167	0.33667	0.9966	1.0150	3.0150	3.0050	3
4	0.98025	3.9505	5.9011	0.24813	0.25313	1.4937	1.0201	4.0301	6.0200	4
5	0.97537	4.9258	9.8026	0.19801	0.20301	1.9900	1.0252	5.0502	10.050	5
6	0.97052	5.8963	14.655	0.16460	0.16960	2.4854	1.0303	6.0755	15.100	6
7	0.96569	6.8620	20.449	0.14073	0.14573	2.9800	1.0355	7.1058	21.175	7
8	0.96089	7.8229	27.175	0.12283	0.12783	3.4738	1.0407	8.1414	28.281	8
9	0.95610	8.7790	34.824	0.10891	0.11391	3.9667	1.0459	9.1821	36.423	9
10	0.95135	9.7304	43.386	0.09777	0.10277	4.4588	1.0511	10.228	45.505	10
11	0.94661	10.677	52.852	0.08866	0.09366	4.9501	1.0564	11.279	55.833	11
12	0.94191	11.618	63.213	0.08107	0.08607	5.4405	1.0616	12.335	67.112	12
13	0.93722	12.556	74.460	0.07464	0.07964	5.9301	1.0669	13.397	79.448	13
14	0.93256	13.488	86.583	0.06914	0.07414	6.4189	1.0723	14.464	92.845	14
15	0.92792	14.416	99.574	0.06436	0.06936	6.9069	1.0776	15.536	107.31	15
16	0.92330	15.339	113.42	0.06019	0.06519	7.3940	1.0830	16.614	122.84	16
17	0.91871	16.258	128.12	0.05651	0.06151	7.8803	1.0884	17.697	139.46	17
18	0.91414	17.172	143.66	0.05323	0.05823	8.3657	1.0939	18.785	157.15	18
19	0.90959	18.082	150.03	0.05030	0.05530	8.8504	1.0994	19.879	175.94	19
20	0.90506	18.987	177.23	0.04767	0.05267	9.3341	1.1049	20.979	195.82	20
21	0.90056	19.888	195.24	0.04528	0.05028	9.8171	1.1104	22.084	216.80	21
22	0.89608	20.784	214.06	0.04311	0.04811	10.299	1.1159	23.194	238.88	22
23	0.89162	21.675	233.67	0.04113	0.04613	10.780	1.1215	24.310	262.08	23
24	0.88719	22.562	254.08	0.03932	0.04432	11.261	1.1271	25.432	286.39	24
25	0.88277	23.445	275.26	0.03765	0.04265	11.740	1.1328	26.559	311.82	25
26	0.87838	24.324	297.22	0.03611	0.04111	12.219	1.1384	27.691	338.38	26
27	0.87401	25.198	319.95	0.03459	0.03959	12.697	1.1441	28.830	366.07	27
28	0.86966	26.067	343.43	0.03336	0.03836	13.174	1.1498	29.974	394.90	28
29	0.86533	26.933	367.66	0.03213	0.03713	13.651	1.1556	31.124	424.87	29
30	0.86103	27.794	392.63	0.03098	0.03598	14.126	1.1614	32.280	456.00	30
32	0.85248	29.503	444.76	0.02889	0.03389	15.075	1.1730	34.608	521.72	32
34	0.84402	31.195	499.75	0.02706	0.03206	16.020	1.1848	36.960	592.11	34
36	0.83564	32.871	557.56	0.02542	0.03042	16.962	1.1966	39.336	667.22	36
48	0.78710	42.580	959.91	0.01849	0.02349	22.543	1.2704	54.097	1219.5	48
60	0.74137	51.725	1448.6	0.01433	0.01933	28.006	1.3488	69.770	1954.0	60
120	0.54963	90.073	4823.5	0.00610	0.01110	53.550	1.8194	163.87	8775.8	120
180	0.40748	118.50	9031.3	0.00344	0.00844	76.211	2.4540	290.81	22163.	180
240	0.30210	139.58	13415.	0.00216	0.00716	96.113	3.3102	462.04	44408.	240
300	0.22397	155.20	17603.	0.00144	0.00644	113.41	4.4649	692.99	78598.	300
360	0.16604	166.79	21403.	0.00100	0.00600	128.32	6.0225	1004.5	128903.	360
INF	0.00000	200.00	43000.	0.0000	0.00500	200.00	INF	INF	INF	INF

i = 0.75%

	PRESENT SUM, P			UNIFORM SERIES, A			FUTURE SUM, F			
n	P/F	P/A	P/G	A/F	A/P	A/G	F/P	F/A	F/G	n
1	0.99256	0.9925	0.0000	1.00000	1.00750	0.0000	1.0075	1.0000	0.0000	1
2	0.98517	1.9777	0.9851	0.49813	0.50563	0.4981	1.0150	2.0075	1.0000	2
3	0.97783	2.9555	2.9408	0.33085	0.33835	0.9950	1.0226	3.0225	3.0075	3
4	0.97055	3.9261	5.8525	0.24721	0.25471	1.4906	1.0303	4.0452	6.0300	4
5	0.96333	4.8894	9.7058	0.19702	0.20452	1.9850	1.0380	5.0755	10.075	5
6	0.95616	5.8456	14.486	0.16357	0.17107	2.4782	1.0458	6.1136	15.150	6
7	0.94904	6.7946	20.180	0.13967	0.14717	2.9701	1.0537	7.1594	21.264	7
8	0.94198	7.7366	26.774	0.12176	0.12926	3.4607	1.0616	8.2131	28.424	8
9	0.93496	8.6715	34.254	0.10782	0.11532	3.9501	1.0695	9.2747	36.637	9
10	0.92800	9.5995	42.606	0.09667	0.10417	4.4383	1.0775	10.344	45.911	10
11	0.92109	10.520	51.817	0.08755	0.09505	4.9252	1.0856	11.421	56.256	11
12	0.91424	11.434	61.874	0.07995	0.08745	5.4109	1.0938	12.507	67.678	12
13	0.90743	12.342	72.763	0.07352	0.08102	5.8954	1.1020	13.601	80.185	13
14	0.90068	13.243	84.472	0.06801	0.07551	6.3786	1.1102	14.703	93.787	14
15	0.89397	14.137	96.987	0.06324	0.07074	6.8605	1.1186	15.813	108.49	15
16	0.88732	15.024	110.29	0.05906	0.06656	7.3412	1.1269	16.932	124.30	16
17	0.88071	15.905	124.38	0.05537	0.06287	7.8207	1.1354	18.059	141.23	17
18	0.87416	16.779	139.24	0.05210	0.05960	8.2989	1.1439	19.194	159.29	18
19	0.86765	17.646	154.86	0.04917	0.05667	8.7759	1.1525	20.338	178.49	19
20	0.86119	18.508	171.23	0.04653	0.05403	9.2516	1.1611	21.491	198.82	20
21	0.85478	19.362	188.32	0.04415	0.05165	9.7261	1.1698	22.652	220.32	21
22	0.84842	20.211	206.14	0.04198	0.04948	10.199	1.1786	23.822	242.97	22
23	0.84210	21.053	224.66	0.04000	0.04750	10.671	1.1875	25.001	266.79	23
24	0.83583	21.889	243.89	0.03818	0.04568	11.142	1.1964	26.188	291.79	24
25	0.82961	22.718	263.80	0.03652	0.04402	11.611	1.2053	27.384	317.98	25
26	0.82343	23.542	284.38	0.03498	0.04248	12.080	1.2144	28.590	345.36	26
27	0.81730	24.359	305.63	0.03359	0.04105	12.547	1.2235	29.804	373.96	27
28	0.81122	25.170	327.54	0.03223	0.03973	13.012	1.2327	31.028	403.76	28
29	0.80518	25.975	350.08	0.03100	0.03850	13.477	1.2419	32.260	434.79	29
30	0.79919	26.775	373.26	0.02985	0.03735	13.940	1.2512	33.502	467.05	30
32	0.78733	28.355	421.46	0.02777	0.03527	14.863	1.2701	36.014	535.31	32
34	0.77565	29.912	472.07	0.02593	0.03343	15.781	1.2892	38.564	608.61	34
36	0.76415	31.446	524.99	0.02430	0.03180	16.694	1.3086	41.152	687.02	36
48	0.69860	40.184	886.84	0.01739	0.02489	22.069	1.4314	57.520	1269.4	48
60	0.63870	48.173	1313.5	0.01326	0.02076	27.266	1.5656	75.424	2056.5	60
120	0.40794	78.941	3998.5	0.00517	0.01267	50.652	2.4513	193.51	9801.9	120
180	0.26055	98.593	6892.6	0.00264	0.01014	69.909	3.8380	378.40	26454.	180
240	0.16641	111.16	9494.1	0.00150	0.00900	85.421	6.0091	667.88	57051.	240
300	0.10629	119.16	11636.	0.00089	0.00839	97.654	9.4084	1121.1	109483.	300
360	0.06789	124.28	13312.	0.00055	0.00805	107.11	14.730	1830.7	196099.	360
INF	0.00000	133.33	17777.	0.0000	0.00750	133.33	INF	INF	INF	INF

Appendix A (Continued)

	i = 1%			i = 1%			i = 1%			
	PRESENT SUM, P			UNIFORM SERIES, A			FUTURE SUM, F			
n	P/F	P/A	P/G	A/F	A/P	A/G	F/P	F/A	F/G	n
1	0.99010	0.9901	0.0000	1.00000	1.01000	0.0000	1.0100	1.0000	0.0000	1
2	0.98030	1.9704	0.9803	0.49751	0.50751	0.4975	1.0201	2.0100	1.0000	2
3	0.97059	2.9409	2.9214	0.33002	0.34002	0.9933	1.0303	3.0301	3.0100	3
4	0.96098	3.9019	5.8044	0.24628	0.25628	1.4875	1.0406	4.0604	6.0401	4
5	0.95147	4.8534	9.6102	0.19604	0.20604	1.9801	1.0510	5.1010	10.100	5
6	0.94205	5.7954	14.320	0.16255	0.17255	2.4709	1.0615	6.1520	15.201	6
7	0.93272	6.7281	19.916	0.13863	0.14863	2.9602	1.0721	7.2135	21.353	7
8	0.92348	7.6516	25.381	0.12069	0.13069	3.4477	1.0828	8.2856	28.567	8
9	0.91434	8.5660	33.695	0.10674	0.11674	3.9336	1.0936	9.3685	36.852	9
10	0.90529	9.4713	41.843	0.09558	0.10558	4.4179	1.1046	10.462	46.221	10
11	0.89632	10.367	50.806	0.08645	0.09645	4.9005	1.1156	11.566	56.683	11
12	0.88745	11.255	60.568	0.07885	0.08885	5.3814	1.1268	12.682	68.250	12
13	0.87866	12.133	71.112	0.07241	0.08241	5.8607	1.1380	13.809	80.932	13
14	0.86996	13.003	82.422	0.06690	0.07690	6.3383	1.1494	14.947	94.742	14
15	0.86135	13.865	94.481	0.06212	0.07212	6.8143	1.1609	16.096	109.69	15
16	0.85282	14.717	107.27	0.05794	0.06794	7.2886	1.1725	17.257	125.78	16
17	0.84438	15.562	120.78	0.05426	0.06426	7.7613	1.1843	18.430	143.04	17
18	0.83602	16.398	134.99	0.05098	0.06098	8.2323	1.1961	19.614	161.47	18
19	0.82774	17.226	149.89	0.04805	0.05805	8.7016	1.2081	20.810	181.09	19
20	0.81954	18.045	165.46	0.04542	0.05542	9.1693	1.2201	22.019	201.90	20
21	0.81143	18.857	181.69	0.04303	0.05303	9.6354	1.2323	23.239	223.91	21
22	0.80340	19.660	198.56	0.04086	0.05086	10.099	1.2447	24.471	247.15	22
23	0.79544	20.455	216.06	0.03889	0.04889	10.562	1.2571	25.716	271.63	23
24	0.78757	21.243	234.18	0.03707	0.04707	11.023	1.2697	26.973	297.34	24
25	0.77977	22.023	252.89	0.03541	0.04541	11.483	1.2824	28.243	324.32	25
26	0.77205	22.795	272.19	0.03387	0.04387	11.940	1.2952	29.525	352.56	26
27	0.76440	23.559	292.07	0.03245	0.04245	12.397	1.3082	30.820	382.08	27
28	0.75684	24.316	312.50	0.03112	0.04112	12.851	1.3212	32.129	412.91	28
29	0.74934	25.065	333.48	0.02990	0.03990	13.304	1.3345	33.450	445.03	29
30	0.74192	25.807	355.00	0.02875	0.03875	13.755	1.3478	34.784	478.48	30
32	0.72730	27.269	399.58	0.02667	0.03667	14.653	1.3749	37.494	549.40	32
34	0.71297	28.702	446.15	0.02484	0.03484	15.544	1.4025	40.257	625.77	34
36	0.69892	30.107	494.62	0.02321	0.03321	16.428	1.4307	43.076	707.68	36
48	0.62026	37.974	820.14	0.01533	0.02633	21.597	1.6122	61.222	1322.2	48
60	0.55045	44.955	1192.8	0.01224	0.02224	26.533	1.8167	81.669	2166.9	60
120	0.30299	69.700	3334.1	0.00435	0.01435	47.834	3.3003	230.03	11003.	120
180	0.16678	83.321	5330.0	0.00200	0.01200	63.969	5.9958	499.58	31958.	180
240	0.09181	90.819	6878.6	0.00101	0.01101	75.739	10.892	989.25	74925.	240
300	0.05053	94.946	7978.6	0.00053	0.01053	84.032	19.788	1878.8	157885.	300
360	0.02782	97.218	8720.4	0.00029	0.01029	89.699	35.949	3494.9	313496.	360
INF	0.00000	100.00	10000.	0.0000	0.01000	100.00	INF	INF	INF	INF

i = 1.5%

	PRESENT SUM, P			UNIFORM SERIES, A			FUTURE SUM, F			
n	P/F	P/A	P/G	A/F	A/P	A/G	F/P	F/A	F/G	n
1	0.98522	0.9852	0.0000	1.00000	1.01500	0.0000	1.0150	1.0000	0.0000	1
2	0.97066	1.9558	0.9706	0.49628	0.51128	0.4962	1.0302	2.0150	1.0000	2
3	0.95632	2.9122	2.8833	0.32838	0.34338	0.9900	1.0456	3.0452	3.0150	3
4	0.94218	3.8543	5.7098	0.24444	0.25944	1.4813	1.0613	4.0909	6.0602	4
5	0.92826	4.7826	9.4228	0.19409	0.20909	1.9702	1.0772	5.1522	10.151	5
6	0.91454	5.6971	13.995	0.16053	0.17553	2.4565	1.0934	6.2295	15.303	6
7	0.90103	6.5982	19.401	0.13656	0.15156	2.9404	1.1098	7.3229	21.532	7
8	0.88771	7.4859	25.615	0.11858	0.13358	3.4218	1.1264	8.4328	28.855	8
9	0.87459	8.3605	32.612	0.10461	0.11961	3.9007	1.1433	9.5593	37.288	9
10	0.86167	9.2221	40.367	0.09343	0.10843	4.3772	1.1605	10.702	46.848	10
11	0.84893	10.071	48.856	0.08429	0.09929	4.8511	1.1779	11.863	57.550	11
12	0.83639	10.907	58.057	0.07668	0.09168	5.3226	1.1956	13.041	69.414	12
13	0.82403	11.731	67.945	0.07024	0.08524	5.7916	1.2135	14.236	82.455	13
14	0.81185	12.543	78.499	0.06472	0.07972	6.2582	1.2317	15.450	96.692	14
15	0.79985	13.343	89.697	0.05994	0.07494	6.7223	1.2502	16.682	112.14	15
16	0.78803	14.131	101.51	0.05577	0.07077	7.1839	1.2689	17.932	128.82	16
17	0.77639	14.907	113.94	0.05208	0.06708	7.6430	1.2880	19.201	146.75	17
18	0.76491	15.672	126.09	0.04881	0.06381	8.0997	1.3073	20.489	165.95	18
19	0.75361	16.426	140.50	0.04588	0.06088	8.5539	1.3269	21.796	186.44	19
20	0.74247	17.168	154.61	0.04325	0.05825	9.0056	1.3468	23.123	208.24	20
21	0.73150	17.900	169.24	0.04087	0.05587	9.4549	1.3670	24.470	231.36	21
22	0.72069	18.620	184.38	0.03870	0.05370	9.9018	1.3875	25.837	255.83	22
23	0.71004	19.330	200.00	0.03673	0.05173	10.346	1.4083	27.225	281.67	23
24	0.69954	20.030	216.09	0.03492	0.04992	10.788	1.4295	28.633	308.90	24
25	0.68921	20.719	232.63	0.03326	0.04826	11.227	1.4509	30.063	337.53	25
26	0.67902	21.398	249.60	0.03173	0.04673	11.664	1.4727	31.514	367.59	26
27	0.66899	22.067	267.00	0.03032	0.04532	12.099	1.4948	32.986	399.11	27
28	0.65910	22.726	284.79	0.02900	0.04400	12.531	1.5172	34.481	432.09	28
29	0.64936	23.376	302.97	0.02778	0.04278	12.961	1.5399	35.998	466.58	29
30	0.63976	24.015	321.53	0.02664	0.04164	13.388	1.5630	37.538	502.57	30
32	0.62099	25.267	359.69	0.02458	0.03958	14.235	1.6103	40.688	579.21	32
34	0.60277	26.481	399.16	0.02276	0.03776	15.073	1.6590	43.933	662.20	34
36	0.58500	27.660	439.83	0.02115	0.03615	15.900	1.7091	47.276	751.73	36
48	0.48936	34.042	703.54	0.01437	0.02937	20.666	2.0434	69.565	1437.6	48
60	0.40930	39.380	988.16	0.01039	0.02539	25.093	2.4432	96.214	2414.3	60
120	0.16752	55.498	2359.7	0.00302	0.01802	42.518	5.9693	331.28	14085.	120
180	0.06857	62.095	3376.9	0.00110	0.01610	53.416	14.584	905.62	48375.	180
240	0.02806	64.795	3870.6	0.00043	0.01543	59.736	35.632	2308.8	137924.	240
300	0.01149	65.900	4163.6	0.00017	0.01517	63.180	87.058	5737.2	362484.	300
360	0.00470	66.353	4310.7	0.00007	0.01507	64.966	212.70	14113.	916906.	360
INF	0.00000	66.666	4444.4	0.0000	0.01500	66.666	INF	INF	INF	INF

Appendix A (Continued)

| | i = 2% | | | i = 2% | | | i = 2% | | | |
| | PRESENT SUM, P | | | UNIFORM SERIES, A | | | FUTURE SUM, F | | | |
n	P/F	P/A	P/G	A/F	A/P	A/G	F/P	F/A	F/G	n
1	0.98039	0.9803	0.0000	1.00000	1.02000	0.0000	1.0200	1.0000	0.0000	1
2	0.96117	1.9415	0.9611	0.49505	0.51505	0.4950	1.0404	2.0200	1.0000	2
3	0.94232	2.8838	2.8458	0.32675	0.34675	0.9868	1.0612	3.0604	3.0200	3
4	0.92385	3.8077	5.6173	0.24262	0.26262	1.4752	1.0824	4.1216	6.0804	4
5	0.90573	4.7134	9.2402	0.19216	0.21216	1.9604	1.1040	5.2040	10.202	5
6	0.88797	5.6014	13.680	0.15853	0.17853	2.4422	1.1261	6.3081	15.406	6
7	0.87056	6.4719	18.903	0.13451	0.15451	2.9208	1.1486	7.4342	21.714	7
8	0.85349	7.3254	24.877	0.11651	0.13651	3.3960	1.1716	8.5829	29.148	8
9	0.83676	8.1622	31.572	0.10252	0.12252	3.8680	1.1950	9.7546	37.731	9
10	0.82035	8.9825	38.955	0.09133	0.11133	4.3367	1.2189	10.949	47.486	10
11	0.80426	9.7868	46.997	0.08218	0.10218	4.8021	1.2433	12.168	58.435	11
12	0.78849	10.575	55.671	0.07456	0.09456	5.2642	1.2682	13.412	70.604	12
13	0.77303	11.348	64.947	0.06812	0.08812	5.7230	1.2936	14.680	84.016	13
14	0.75788	12.106	74.799	0.06260	0.08260	6.1786	1.3194	15.973	98.696	14
15	0.74301	12.849	85.202	0.05783	0.07783	6.6309	1.3458	17.293	114.67	15
16	0.72845	13.577	96.128	0.05365	0.07365	7.0799	1.3727	18.639	131.96	16
17	0.71416	14.291	107.55	0.04997	0.06997	7.5256	1.4002	20.012	150.60	17
18	0.70016	14.992	119.45	0.04670	0.06670	7.9681	1.4282	21.412	170.61	18
19	0.68643	15.678	131.81	0.04378	0.06378	8.4073	1.4568	22.840	192.02	19
20	0.67297	16.351	144.60	0.04116	0.06116	8.8432	1.4859	24.297	214.86	20
21	0.65978	17.011	157.79	0.03878	0.05878	9.2759	1.5156	25.783	239.16	21
22	0.64684	17.658	171.37	0.03663	0.05663	9.7054	1.5459	27.299	264.94	22
23	0.63416	18.292	185.33	0.03467	0.05467	10.131	1.5769	28.845	292.24	23
24	0.62172	18.913	199.63	0.03287	0.05287	10.554	1.6084	30.421	321.09	24
25	0.60953	19.523	214.25	0.03122	0.05122	10.974	1.6406	32.030	351.51	25
26	0.59758	20.121	229.19	0.02970	0.04970	11.391	1.6734	33.670	383.54	26
27	0.58586	20.706	244.43	0.02829	0.04829	11.804	1.7068	35.344	417.21	27
28	0.57437	21.281	259.93	0.02699	0.04699	12.214	1.7410	37.051	452.56	28
29	0.56311	21.844	275.70	0.02578	0.04578	12.621	1.7758	38.792	489.61	29
30	0.55207	22.396	291.71	0.02465	0.04465	13.025	1.8113	40.568	528.40	30
32	0.53063	23.468	324.40	0.02261	0.04261	13.823	1.8845	44.227	611.35	32
34	0.51003	24.498	357.88	0.02082	0.04082	14.608	1.9606	48.033	701.69	34
36	0.49022	25.488	392.04	0.01923	0.03923	15.380	2.0398	51.994	799.71	36
48	0.38654	30.673	605.96	0.01260	0.03260	19.755	2.5870	79.353	1567.6	48
60	0.30478	34.760	823.69	0.00877	0.02877	23.696	3.2810	114.05	2702.5	60
120	0.09289	45.355	1710.4	0.00205	0.02205	37.711	10.765	488.25	18412.	120
180	0.02831	48.584	2174.4	0.00058	0.02058	44.755	35.320	1716.0	76802.	180
240	0.00863	49.568	2374.8	0.00017	0.02017	47.911	115.88	5744.4	275222.	240
300	0.00263	49.868	2453.9	0.00005	0.02005	49.208	380.23	18961.	9330086.	300
360	0.00080	49.959	2483.5	0.00002	0.02002	49.711	1247.5	62328.	>10**6	360
INF	0.00000	50.000	2500.0	0.0000	0.02000	50.000	INF	INF	INF	INF

494

| | i = 2.5% | | | i = 2.5% | | | i = 2.5% | | | |
| | PRESENT SUM, P | | | UNIFORM SERIES, A | | | FUTURE SUM, F | | | |
n	P/F	P/A	P/G	A/F	A/P	A/G	F/P	F/A	F/G	n
1	0.97561	0.9756	0.0000	1.00000	1.02500	0.0000	1.0250	1.0000	0.0000	1
2	0.95181	1.9274	0.9518	0.49383	0.51883	0.4938	1.0506	2.0250	1.0000	2
3	0.92860	2.8560	2.8090	0.32514	0.35014	0.9835	1.0768	3.0756	3.0250	3
4	0.90595	3.7619	5.5268	0.24082	0.26582	1.4691	1.1038	4.1525	6.1006	4
5	0.88385	4.6458	9.0622	0.19025	0.21525	1.9506	1.1314	5.2563	10.253	5
6	0.86230	5.5081	13.373	0.15655	0.18155	2.4280	1.1596	6.3877	15.509	6
7	0.84127	6.3493	18.421	0.13250	0.15750	2.9012	1.1886	7.5474	21.897	7
8	0.82075	7.1701	24.166	0.11447	0.13947	3.3704	1.2184	8.7361	29.444	8
9	0.80073	7.9708	30.572	0.10046	0.12546	3.8355	1.2488	9.9545	38.180	9
10	0.78120	8.7520	37.603	0.08926	0.11426	4.2964	1.2800	11.203	48.135	10
11	0.76214	9.5142	45.224	0.08011	0.10511	4.7533	1.3120	12.483	59.338	11
12	0.74356	10.257	53.403	0.07249	0.09749	5.2061	1.3448	13.795	71.822	12
13	0.72542	10.983	62.108	0.06605	0.09105	5.6549	1.3785	15.140	85.617	13
14	0.70773	11.690	71.309	0.06054	0.08554	6.0995	1.4129	16.519	100.75	14
15	0.69047	12.381	80.975	0.05577	0.08077	6.5401	1.4483	17.931	117.27	15
16	0.67362	13.055	91.080	0.05150	0.07660	6.9766	1.4845	19.380	135.20	16
17	0.65720	13.712	101.59	0.04793	0.07293	7.4091	1.5216	20.864	154.58	17
18	0.64117	14.353	112.45	0.04467	0.06967	7.8375	1.5596	22.386	175.45	18
19	0.62553	14.978	123.75	0.04176	0.06676	8.2619	1.5986	23.946	197.84	19
20	0.61027	15.589	135.35	0.03915	0.06415	8.6823	1.6386	25.544	221.78	20
21	0.59539	16.184	147.25	0.03679	0.06179	9.0986	1.6795	27.183	247.33	21
22	0.58086	16.765	159.45	0.03465	0.05965	9.5109	1.7215	28.862	274.51	22
23	0.56670	17.332	171.92	0.03270	0.05770	9.9193	1.7646	30.584	303.37	23
24	0.55288	17.885	184.63	0.03091	0.05591	10.323	1.8087	32.349	333.96	24
25	0.53939	18.424	197.58	0.02928	0.05428	10.724	1.8539	34.157	366.31	25
30	0.47674	20.930	265.12	0.02278	0.04778	12.666	2.0975	43.902	556.10	30
35	0.42137	23.145	335.88	0.01821	0.04321	14.512	2.3732	54.928	797.12	35
36	0.41109	23.556	350.27	0.01745	0.04245	14.869	2.4325	57.301	852.05	36
40	0.37243	25.102	408.22	0.01484	0.03984	16.262	2.6850	67.402	1096.1	40
48	0.30567	27.773	524.03	0.01101	0.03601	18.868	3.2714	90.859	1714.3	48
50	0.29094	28.362	552.60	0.01026	0.03526	19.483	3.4371	97.484	1899.3	50
60	0.22728	30.908	690.86	0.00735	0.03235	22.351	4.3997	135.99	3039.6	60
100	0.08465	36.614	1125.9	0.00231	0.02731	30.752	11.813	432.54	13301.	100
120	0.05166	37.933	1269.3	0.00136	0.02636	33.463	19.358	734.32	24573.	120
180	0.01174	39.530	1496.6	0.00030	0.02530	37.861	85.171	3366.8	127475.	180
240	0.00267	39.893	1570.1	0.00007	0.02507	39.357	374.73	14949.	588381.	240
300	0.00061	39.975	1591.7	0.00002	0.02502	39.817	1668.7	65910.	>10**6	300
360	0.00014	39.994	1597.7	0.00000	0.02500	39.950	7254.2	290129	>10**6	360
INF	0.00000	40.000	1600.0	0.0000	0.02500	40.000	INF	INF	INF	INF

Appendix A (*Continued*)

	$i=3\%$			$i=3\%$			$i=3\%$			
	PRESENT SUM, P			UNIFORM SERIES, A			FUTURE SUM, F			
n	P/F	P/A	P/G	A/F	A/P	A/G	F/P	F/A	F/G	n
1	0.97087	0.9708	0.0000	1.00000	1.03000	0.0000	1.0300	1.0000	0.0000	1
2	0.94260	1.9134	0.9426	0.49261	0.52261	0.4926	1.0609	2.0300	1.0000	2
3	0.91514	2.8286	2.7728	0.32353	0.35353	0.9803	1.0927	3.0909	3.0300	3
4	0.88849	3.7171	5.4383	0.23903	0.26903	1.4630	1.1255	4.1836	6.1209	4
5	0.86261	4.5797	8.8887	0.18835	0.21835	1.9409	1.1592	5.3091	10.304	5
6	0.83748	5.4171	13.076	0.15460	0.18460	2.4138	1.1940	6.4684	15.613	6
7	0.81309	6.2302	17.954	0.13051	0.16051	2.8818	1.2298	7.6624	22.082	7
8	0.78941	7.0196	23.480	0.11246	0.14246	3.3449	1.2667	8.8923	29.744	8
9	0.76642	7.7861	29.611	0.09843	0.12843	3.8031	1.3047	10.159	38.636	9
10	0.74409	8.5302	36.308	0.08723	0.11723	4.2565	1.3439	11.463	48.796	10
11	0.72242	9.2526	43.533	0.07808	0.10808	4.7049	1.3842	12.807	60.259	11
12	0.70138	9.9540	51.248	0.07046	0.10046	5.1485	1.4257	14.192	73.067	12
13	0.68095	10.635	59.419	0.06403	0.09403	5.5872	1.4685	15.617	87.259	13
14	0.66112	11.296	68.014	0.05853	0.08853	6.0210	1.5125	17.086	102.87	14
15	0.64186	11.937	77.000	0.05377	0.08377	6.4500	1.5579	18.598	119.96	15
16	0.62317	12.561	86.347	0.04961	0.07961	6.8742	1.6047	20.156	138.56	16
17	0.60502	13.166	96.028	0.04595	0.07595	7.2935	1.6528	21.761	158.72	17
18	0.58739	13.753	106.021	0.04271	0.07271	7.7081	1.7024	23.414	180.48	18
19	0.57029	14.323	116.279	0.03981	0.06981	8.1178	1.7535	25.116	203.89	19
20	0.55368	14.877	126.79	0.03722	0.06722	8.5228	1.8061	26.870	229.01	20
21	0.53755	15.415	137.55	0.03487	0.06487	8.9230	1.8602	28.676	255.88	21
22	0.52189	15.936	148.50	0.03275	0.06275	9.3185	1.9161	30.536	284.55	22
23	0.50669	16.443	159.65	0.03081	0.06081	9.7093	1.9735	32.452	315.09	23
24	0.49193	16.935	170.97	0.02905	0.05905	10.095	2.0327	34.426	347.54	24
25	0.47761	17.413	182.43	0.02743	0.05743	10.476	2.0937	36.459	381.97	25
26	0.46369	17.876	194.02	0.02594	0.05594	10.853	2.1565	38.553	418.43	26
28	0.43708	18.764	217.53	0.02329	0.05329	11.593	2.2879	42.930	497.69	28
30	0.41199	19.600	241.36	0.02102	0.05102	12.314	2.4272	47.575	585.84	30
35	0.35538	21.487	301.62	0.01654	0.04654	14.037	2.8138	60.462	848.73	35
36	0.34503	21.832	313.70	0.01580	0.04580	14.368	2.8982	63.275	909.19	36
40	0.30656	23.114	351.75	0.01326	0.04326	15.650	3.2620	75.401	1180.0	40
45	0.26444	24.518	420.63	0.01079	0.04079	17.155	3.7816	92.719	1590.6	45
48	0.24200	25.266	455.02	0.00958	0.03958	18.008	4.1322	104.40	1880.2	48
50	0.22811	25.729	477.48	0.00887	0.03887	18.557	4.3839	112.79	2093.2	50
60	0.16973	27.675	583.05	0.00613	0.03613	21.067	5.8916	163.05	3435.1	60
70	0.12630	29.123	676.08	0.00434	0.03434	23.214	7.9178	230.59	5353.1	70
80	0.09398	30.200	756.08	0.00311	0.03311	25.035	10.640	321.36	8045.4	80
90	0.06993	31.002	823.63	0.00226	0.03226	26.566	14.300	443.34	11778.	90
100	0.05203	31.598	879.85	0.00165	0.03165	27.844	19.218	607.28	16909.	100
120	0.02881	32.373	963.86	0.00089	0.03089	29.773	34.711	1123.7	33456.	120
180	0.00489	33.170	1076.3	0.00015	0.03015	32.448	204.50	6783.4	220115.	180
240	0.00083	33.305	1103.5	0.00002	0.03002	33.134	1204.8	40128.	>10**6	240
INF	0.00000	33.333	1111.1	0.0000	0.03000	33.333	INF	INF	INF	INF

i = 4%

	PRESENT SUM, P			UNIFORM SERIES, A			FUTURE SUM, F			
n	P/F	P/A	P/G	A/F	A/P	A/G	F/P	F/A	F/G	n
1	0.96154	0.9615	0.0000	1.00000	1.04000	0.0000	1.0400	1.0000	0.0000	1
2	0.92456	1.8860	0.9245	0.49020	0.53020	0.4902	1.0816	2.0400	1.0000	2
3	0.88900	2.7750	2.7025	0.32035	0.36035	0.9738	1.1248	3.1216	3.0400	3
4	0.85480	3.6299	5.2669	0.23549	0.27549	1.4510	1.1698	4.2464	6.1616	4
5	0.82193	4.4518	8.5546	0.18463	0.22463	1.9216	1.2166	5.4163	10.408	5
6	0.79031	5.2421	12.506	0.15076	0.19076	2.3857	1.2653	6.6329	15.824	6
7	0.75992	6.0020	17.065	0.12661	0.16661	2.8433	1.3159	7.8982	22.457	7
8	0.73069	6.7327	22.180	0.10853	0.14853	3.2944	1.3685	9.2142	30.355	8
9	0.70259	7.4353	27.801	0.09449	0.13449	3.7390	1.4233	10.582	39.569	9
10	0.67556	8.1109	33.881	0.08329	0.12329	4.1772	1.4802	12.006	50.152	10
11	0.64958	8.7604	40.377	0.07415	0.11415	4.6090	1.5394	13.486	62.158	11
12	0.62460	9.3850	47.247	0.06655	0.10655	5.0343	1.6010	15.025	75.645	12
13	0.60057	9.9856	54.454	0.06014	0.10014	5.4532	1.6650	16.626	90.670	13
14	0.57748	10.563	61.961	0.05467	0.09467	5.8658	1.7316	18.291	107.29	14
15	0.55526	11.118	69.735	0.04994	0.08994	6.2720	1.8009	20.023	125.59	15
16	0.53391	11.652	77.744	0.04582	0.08582	6.6720	1.8729	21.824	145.61	16
17	0.51337	12.165	85.958	0.04220	0.08220	7.0656	1.9479	23.697	167.43	17
18	0.49363	12.659	94.349	0.03899	0.07899	7.4530	2.0258	25.645	191.13	18
19	0.47464	13.133	102.89	0.03614	0.07614	7.8341	2.1068	27.671	216.78	19
20	0.45639	13.590	111.56	0.03358	0.07358	8.2091	2.1911	29.778	244.45	20
21	0.43883	14.029	120.34	0.03128	0.07128	8.5779	2.2787	31.969	274.23	21
22	0.42196	14.451	129.20	0.02920	0.06920	8.9406	2.3699	34.248	306.19	22
23	0.40573	14.856	138.12	0.02731	0.06731	9.2972	2.4647	36.617	340.44	23
24	0.39012	15.247	147.10	0.02559	0.06559	9.6479	2.5633	39.082	377.06	24
25	0.37512	15.622	156.10	0.02401	0.06401	9.9925	2.6658	41.645	416.14	25
26	0.36069	15.982	165.12	0.02257	0.06257	10.331	2.7724	44.311	457.79	26
28	0.33348	16.663	183.14	0.02001	0.06001	10.990	2.9987	49.967	549.19	28
30	0.30832	17.292	201.06	0.01783	0.05783	11.627	3.2434	56.084	652.12	30
35	0.25342	18.664	244.87	0.01358	0.05358	13.119	3.9460	73.652	966.30	35
36	0.24367	18.908	253.40	0.01289	0.05289	13.401	4.1039	77.598	1039.9	36
40	0.20829	19.792	286.53	0.01052	0.05052	14.476	4.8010	95.025	1375.6	40
45	0.17120	20.720	325.40	0.00826	0.04826	15.704	5.8411	121.02	1900.7	45
48	0.15219	21.195	347.24	0.00718	0.04718	16.383	6.5705	139.26	2281.5	48
50	0.14071	21.482	361.16	0.00655	0.04655	16.812	7.1066	152.66	2566.6	50
55	0.11566	22.108	393.68	0.00523	0.04523	17.807	8.6463	191.15	3403.9	55
60	0.09506	22.623	422.99	0.00420	0.04420	18.697	10.519	237.99	4449.7	60
70	0.06422	23.394	472.47	0.00275	0.04275	20.196	15.571	364.29	7357.2	70
80	0.04338	23.915	511.11	0.00181	0.04181	21.371	23.049	551.24	11781.	80
90	0.02931	24.267	540.73	0.00121	0.04121	22.282	34.119	827.98	18449.	90
100	0.01980	24.505	563.12	0.00081	0.04081	22.980	50.504	1237.6	28440.	100
120	0.00904	24.774	592.24	0.00036	0.04036	23.905	110.66	2741.5	65539.	120
180	0.00086	24.978	620.59	0.00003	0.04003	24.845	1164.1	29078.	722456.	180
INF	0.00000	25.000	625.00	0.0000	0.04000	25.000	INF	INF	INF	INF

Appendix A (Continued)

	$i=5\%$			$i=5\%$			$i=5\%$			
	PRESENT SUM, P			UNIFORM SERIES, A			FUTURE SUM, F			
n	P/F	P/A	P/G	A/F	A/P	A/G	F/P	F/A	F/G	n
1	0.95238	0.9523	0.0000	1.00000	1.05000	0.0000	1.0500	1.0000	0.0000	1
2	0.90703	1.8594	0.9070	0.48780	0.53780	0.4878	1.1025	2.0500	0.0000	2
3	0.86384	2.7232	2.6347	0.31721	0.36721	0.9674	1.1576	3.1525	3.0500	3
4	0.82270	3.5459	5.1028	0.23201	0.28201	1.4390	1.2155	4.3101	6.2025	4
5	0.78353	4.3294	8.2369	0.18097	0.23097	1.9025	1.2762	5.5256	10.512	5
6	0.74622	5.0756	11.968	0.14702	0.19702	2.3579	1.3401	6.8019	16.038	6
7	0.71068	5.7863	16.232	0.12282	0.17282	2.8052	1.4071	8.1420	22.840	7
8	0.67684	6.4632	20.970	0.10472	0.15472	3.2445	1.4774	9.5491	30.982	8
9	0.64461	7.1078	26.126	0.09069	0.14069	3.6657	1.5513	11.026	40.531	9
10	0.61391	7.7217	31.652	0.07950	0.12950	4.0990	1.6288	12.577	51.557	10
11	0.58468	8.3064	37.498	0.07039	0.12039	4.5144	1.7103	14.206	64.135	11
12	0.55684	8.8632	43.624	0.06283	0.11283	4.9219	1.7958	15.917	78.342	12
13	0.53032	9.3935	49.987	0.05646	0.10646	5.3215	1.8856	17.713	94.259	13
14	0.50507	9.8986	56.553	0.05102	0.10102	5.7132	1.9799	19.598	111.97	14
15	0.48102	10.379	63.288	0.04634	0.09634	6.0973	2.0789	21.578	131.57	15
16	0.45811	10.837	70.159	0.04227	0.09227	6.4736	2.1828	23.657	153.15	16
17	0.43630	11.274	77.140	0.03870	0.08870	6.8422	2.2920	25.840	176.80	17
18	0.41552	11.689	84.204	0.03555	0.08555	7.2033	2.4066	28.132	202.64	18
19	0.39573	12.085	91.327	0.03275	0.08275	7.5569	2.5269	30.539	230.78	19
20	0.37689	12.462	98.488	0.03024	0.08024	7.9029	2.6533	33.066	261.31	20
21	0.35894	12.821	105.66	0.02800	0.07800	8.2416	2.7859	35.719	294.38	21
22	0.34185	13.163	112.84	0.02597	0.07597	8.5729	2.9252	38.505	330.10	22
23	0.32557	13.488	120.00	0.02414	0.07414	8.8970	3.0715	41.430	368.61	23
24	0.31007	13.798	127.14	0.02247	0.07247	9.2139	3.2251	44.502	410.04	24
25	0.29530	14.093	134.22	0.02095	0.07095	9.5237	3.3863	47.727	454.54	25
26	0.28124	14.375	141.25	0.01956	0.06956	9.8265	3.5556	51.113	502.26	26
28	0.25509	14.898	155.11	0.01712	0.06712	10.411	3.9201	58.402	608.05	28
30	0.23138	15.372	168.62	0.01505	0.06505	10.969	4.3219	66.438	728.77	30
35	0.18129	16.374	200.58	0.01107	0.06107	12.249	5.5160	90.320	1106.4	35
36	0.17266	16.546	206.62	0.01043	0.06043	12.487	5.7918	95.836	1196.7	36
40	0.14205	17.159	229.54	0.00828	0.05828	13.377	7.0399	120.80	1616.0	40
45	0.11130	17.774	255.31	0.00626	0.05626	14.364	8.9850	159.70	2294.0	45
48	0.09614	18.077	269.24	0.00532	0.05532	14.894	10.401	188.02	2800.5	48
50	0.08720	18.255	277.91	0.00478	0.05478	15.223	11.467	209.34	3186.9	50
55	0.06833	18.633	297.51	0.00367	0.05367	15.966	14.635	272.71	4354.2	55
60	0.05354	18.929	314.34	0.00283	0.05283	16.606	18.679	353.58	5871.6	60
70	0.03287	19.342	340.84	0.00170	0.05170	17.621	30.426	588.52	10370.	70
80	0.02018	19.596	359.64	0.00103	0.05103	18.352	49.561	971.22	17824.	80
90	0.01239	19.752	372.74	0.00063	0.05063	18.871	80.730	1594.6	30092.	90
100	0.00760	19.847	381.74	0.00038	0.05038	19.233	131.50	2610.0	50200.	100
120	0.00287	19.942	391.97	0.00014	0.05014	19.655	348.91	6958.2	136765.	120
INF	0.00000	20.000	400.00	0.0000	0.05000	20.000	INF	INF	INF	INF

n	PRESENT SUM, P			UNIFORM SERIES, A			FUTURE SUM, F			n
	P/F	P/A	P/G	A/F	A/P	A/G	F/P	F/A	F/G	
1	0.94340	0.9434	0.0000	1.00000	1.06000	0.0000	1.0600	1.0000	0.0000	1
2	0.89000	1.8333	0.8900	0.48544	0.54544	0.4854	1.1236	2.0600	1.0000	2
3	0.83962	2.6730	2.5692	0.31411	0.37411	0.9611	1.1910	3.1836	3.0600	3
4	0.79209	3.4651	4.9455	0.22859	0.28859	1.4272	1.2624	4.3746	6.2436	4
5	0.74726	4.2123	7.9345	0.17740	0.23740	1.8836	1.3382	5.6370	10.518	5
6	0.70496	4.9173	11.459	0.14336	0.20336	2.3304	1.4185	6.9753	16.255	6
7	0.66506	5.5823	15.449	0.11914	0.17914	2.7675	1.5036	8.3938	23.230	7
8	0.62741	6.2097	19.841	0.10104	0.16104	3.1952	1.5938	9.8974	31.524	8
9	0.59190	6.8016	24.576	0.08702	0.14702	3.6133	1.6894	11.491	41.521	9
10	0.55839	7.3600	29.602	0.07587	0.13587	4.0220	1.7908	13.180	53.013	10
11	0.52679	7.8868	34.870	0.06679	0.12679	4.4212	1.8983	14.971	66.194	11
12	0.49697	8.3838	40.336	0.05928	0.11928	4.8112	2.0122	16.869	81.165	12
13	0.46884	8.8526	45.962	0.05296	0.11296	5.1919	2.1329	18.882	98.035	13
14	0.44230	9.2949	51.712	0.04758	0.10758	5.5635	2.2609	21.015	116.91	14
15	0.41727	9.7122	57.554	0.04296	0.10296	5.9259	2.3965	23.276	137.93	15
16	0.39365	10.105	63.459	0.03895	0.09895	6.2794	2.5403	25.672	161.20	16
17	0.37136	10.477	69.401	0.03544	0.09544	6.6239	2.6927	28.212	186.88	17
18	0.35034	10.827	75.356	0.03236	0.09236	6.9597	2.8543	30.905	215.09	18
19	0.33051	11.158	81.306	0.02962	0.08962	7.2867	3.0256	33.760	246.00	19
20	0.31180	11.469	87.230	0.02718	0.08718	7.6051	3.2071	36.785	279.76	20
21	0.29416	11.764	93.113	0.02500	0.08500	7.9150	3.3995	39.992	316.54	21
22	0.27751	12.041	98.941	0.02305	0.08305	8.2166	3.6035	43.392	356.53	22
23	0.26180	12.303	104.70	0.02128	0.08128	8.5099	3.8197	46.995	399.93	23
24	0.24698	12.550	110.38	0.01968	0.07968	8.7950	4.0489	50.815	446.92	24
25	0.23300	12.783	115.97	0.01823	0.07823	9.0722	4.2918	54.864	497.74	25
26	0.21981	13.003	121.46	0.01690	0.07690	9.3414	4.5493	59.156	552.60	26
27	0.20737	13.210	126.86	0.01570	0.07570	9.6029	4.8223	63.705	611.76	27
28	0.19563	13.406	132.14	0.01459	0.07459	9.8568	5.1116	68.528	675.46	28
29	0.18456	13.590	137.31	0.01358	0.07358	10.103	5.4183	73.639	743.99	29
30	0.17411	13.764	142.35	0.01265	0.07265	10.342	5.7434	79.058	817.63	30
35	0.13011	14.498	155.74	0.00897	0.06897	11.431	7.6860	111.43	1273.9	35
40	0.09722	15.046	185.95	0.00646	0.06646	12.359	10.285	154.76	1912.7	40
45	0.07265	15.455	203.11	0.00470	0.06470	13.141	13.764	212.74	2795.7	45
50	0.05429	15.761	217.45	0.00344	0.06344	13.796	18.420	290.33	4005.5	50
55	0.04057	15.990	229.32	0.00254	0.06254	14.341	24.650	394.17	5652.8	55
60	0.03031	16.161	239.04	0.00188	0.06188	14.790	32.987	533.12	7885.4	60
65	0.02263	16.289	246.94	0.00139	0.06139	15.160	44.145	719.08	10901.	65
70	0.01693	16.384	253.32	0.00103	0.06103	15.461	59.075	967.93	14965.	70
80	0.00945	16.509	262.54	0.00057	0.06057	15.903	105.79	1746.6	27776.	80
90	0.00528	16.578	268.39	0.00032	0.06032	16.189	189.46	3141.0	50851.	90
100	0.00295	16.617	272.04	0.00018	0.06018	16.371	339.30	5638.3	92306.	100
120	0.00092	16.651	275.68	0.00006	0.06006	16.556	1088.1	18119.	299997.	120
INF	0.00000	16.666	277.77	0.0000	0.06000	16.666	INF	INF	INF	INF

Appendix A *(Continued)*

	i = 7%			i = 7%			i = 7%			
	PRESENT SUM, P			UNIFORM SERIES, A			FUTURE SUM, F			
n	P/F	P/A	P/G	A/F	A/P	A/G	F/P	F/A	F/G	n
1	0.93458	0.9345	0.0000	1.00000	1.07000	0.0000	1.0700	1.0000	0.0000	1
2	0.87344	1.8080	0.8734	0.48309	0.55309	0.4830	1.1449	2.0700	1.0000	2
3	0.81630	2.6243	2.5060	0.31105	0.38105	0.9549	1.2250	3.2149	3.0700	3
4	0.76290	3.3872	4.7947	0.22523	0.29523	1.4155	1.3108	4.4399	6.2849	4
5	0.71299	4.1002	7.6466	0.17389	0.24389	1.8649	1.4025	5.7507	10.724	5
6	0.66634	4.7665	10.978	0.13980	0.20980	2.3032	1.5007	7.1532	16.475	6
7	0.62275	5.3892	14.714	0.11555	0.18555	2.7303	1.6057	8.6540	23.628	7
8	0.58201	5.9713	18.788	0.09747	0.16747	3.1465	1.7181	10.259	32.282	8
9	0.54393	6.5152	23.140	0.08349	0.15349	3.5517	1.8384	11.978	42.542	9
10	0.50835	7.0235	27.715	0.07238	0.14238	3.9460	1.9671	13.816	54.520	10
11	0.47509	7.4986	32.466	0.06336	0.13336	4.3296	2.1048	15.783	68.337	11
12	0.44401	7.9426	37.350	0.05090	0.12590	4.7025	2.2521	17.888	84.120	12
13	0.41496	8.3576	42.330	0.04965	0.11965	5.0648	2.4098	20.140	102.00	13
14	0.38782	8.7454	47.371	0.04434	0.11434	5.4167	2.5785	22.550	122.15	14
15	0.36245	9.1079	52.446	0.03979	0.10979	5.7582	2.7590	25.129	144.70	15
16	0.33873	9.4466	57.527	0.03586	0.10586	6.0896	2.9521	27.888	169.82	16
17	0.31657	9.7632	62.592	0.03243	0.10243	6.4110	3.1588	30.840	197.71	17
18	0.29586	10.059	67.621	0.02941	0.09941	6.7224	3.3799	33.999	228.55	18
19	0.27651	10.335	72.599	0.02675	0.09675	7.0241	3.6165	37.379	262.55	19
20	0.25842	10.594	77.509	0.02439	0.09439	7.3163	3.8696	40.995	299.93	20
21	0.24151	10.835	82.339	0.02229	0.09229	7.5990	4.1405	44.865	340.93	21
22	0.22571	11.061	87.079	0.02041	0.09041	7.8724	4.4304	49.005	385.79	22
23	0.21095	11.272	91.720	0.01871	0.08871	8.1368	4.7405	53.436	434.80	23
24	0.19715	11.469	96.254	0.01719	0.08719	8.3923	5.0723	58.176	488.23	24
25	0.18425	11.653	100.67	0.01581	0.08581	8.6391	5.4274	63.249	546.41	25
26	0.17220	11.825	104.98	0.01456	0.08456	8.8773	5.8073	68.676	609.66	26
27	0.16093	11.986	109.16	0.01343	0.08343	9.1072	6.2138	74.483	678.34	27
28	0.15040	12.137	113.22	0.01239	0.08239	9.3289	6.6488	80.697	752.82	28
29	0.14056	12.277	117.16	0.01145	0.08145	9.5427	7.1142	87.346	833.52	29
30	0.13137	12.409	120.97	0.01059	0.08059	9.7486	7.6122	94.460	920.86	30
35	0.09366	12.947	138.13	0.00723	0.07723	10.668	10.676	138.23	1474.8	35
40	0.06678	13.331	152.29	0.00501	0.07501	11.423	14.974	199.63	2280.5	40
45	0.04761	13.605	163.75	0.00350	0.07350	12.036	21.002	285.74	3439.2	45
50	0.03395	13.800	172.90	0.00246	0.07246	12.528	29.457	406.52	5093.2	50
55	0.02420	13.939	180.12	0.00174	0.07174	12.921	41.315	575.92	7441.8	55
60	0.01726	14.039	185.76	0.00123	0.07123	13.232	57.946	813.52	10764.	60
65	0.01230	14.109	190.14	0.00087	0.07087	13.476	81.272	1146.7	15453.	65
70	0.00877	14.160	193.51	0.00062	0.07062	13.666	113.98	1614.1	22059.	70
80	0.00446	14.222	198.07	0.00031	0.07031	13.927	224.23	3189.0	44415.	80
90	0.00227	14.253	200.70	0.00016	0.07016	14.081	441.10	6287.1	88531.	90
100	0.00115	14.269	202.20	0.00008	0.07008	14.170	867.71	12381.	175452.	100
120	0.00030	14.281	203.51	0.00002	0.07002	14.250	3357.7	47954.	683345.	120
INF	0.00000	14.285	204.08	0.0000	0.07000	14.285	INF	INF	INF	INF

i = 8%

n	P/F	P/A	P/G	A/F	A/P	A/G	F/P	F/A	F/G
1	0.92593	0.9259	0.0000	1.00000	1.08000	0.0000	1.0800	1.0000	0.0000
2	0.85734	1.7832	0.8573	0.48077	0.56077	0.4807	1.1664	2.0800	1.0000
3	0.79383	2.5771	2.4450	0.30803	0.38803	0.9487	1.2597	3.2464	3.0800
4	0.73503	3.3121	4.6500	0.22192	0.30192	1.4039	1.3604	4.5061	6.3264
5	0.68058	3.9927	7.3724	0.17046	0.25046	1.8464	1.4693	5.8666	10.832
6	0.63017	4.6228	10.523	0.13632	0.21632	2.2763	1.5868	7.3359	16.699
7	0.58349	5.2063	14.024	0.11207	0.19207	2.6936	1.7138	8.9228	24.035
8	0.54027	5.7466	17.806	0.09401	0.17401	3.0985	1.8509	10.636	32.957
9	0.50025	6.2468	21.808	0.08008	0.16008	3.4910	1.9990	12.487	43.594
10	0.46319	6.7100	25.976	0.06903	0.14903	3.8713	2.1589	14.486	56.082
11	0.42888	7.1389	30.265	0.06008	0.14008	4.2395	2.3316	16.645	70.558
12	0.39711	7.5360	34.633	0.05270	0.13270	4.5957	2.5181	18.977	87.214
13	0.36770	7.9037	39.046	0.04652	0.12652	4.9402	2.7196	21.495	106.19
14	0.34046	8.2442	43.472	0.04130	0.12130	5.2730	2.9371	24.214	127.68
15	0.31524	8.5594	47.885	0.03683	0.11683	5.5944	3.1721	27.152	151.90
16	0.29189	8.8513	52.264	0.03298	0.11298	5.9046	3.4259	30.324	179.05
17	0.27027	9.1216	56.588	0.02963	0.10963	6.2037	3.7000	33.750	209.37
18	0.25025	9.3718	60.842	0.02670	0.10670	6.4920	3.9960	37.450	243.12
19	0.23171	9.6036	65.013	0.02413	0.10413	6.7696	4.3157	41.446	280.57
20	0.21455	9.8181	69.089	0.02185	0.10185	7.0369	4.6609	45.762	322.02
21	0.19866	10.016	73.062	0.01983	0.09983	7.2940	5.0338	50.422	367.78
22	0.18394	10.200	76.925	0.01803	0.09803	7.5411	5.4365	55.456	418.20
23	0.17032	10.371	80.672	0.01642	0.09642	7.7786	5.8714	60.893	473.66
24	0.15770	10.528	84.299	0.01498	0.09498	8.0066	6.3411	66.764	534.55
25	0.14602	10.674	87.804	0.01368	0.09368	8.2253	6.8484	73.105	601.32
26	0.13520	10.810	91.184	0.01251	0.09251	8.4351	7.3963	79.954	674.43
27	0.12519	10.935	94.439	0.01145	0.09145	8.6362	7.9980	87.350	754.38
28	0.11591	11.051	97.568	0.01049	0.09049	8.8288	8.6271	95.338	841.73
29	0.10733	11.158	100.57	0.00962	0.08962	9.0132	9.3172	103.96	937.07
30	0.09938	11.257	103.45	0.00883	0.08883	9.1897	10.062	113.28	1041.0
35	0.06763	11.654	116.09	0.00580	0.08580	9.9610	14.785	172.31	1716.4
40	0.04603	11.924	126.04	0.00386	0.08386	10.569	21.724	259.05	2738.2
45	0.03133	12.108	133.73	0.00259	0.08259	11.044	31.920	386.50	4268.8
50	0.02132	12.233	139.59	0.00174	0.08174	11.410	46.901	573.77	6547.1
55	0.01451	12.318	144.00	0.00118	0.08118	11.690	68.913	848.92	9924.0
60	0.00988	12.376	147.30	0.00080	0.08080	11.901	101.25	1253.2	14915.
65	0.00672	12.416	149.73	0.00054	0.08054	12.060	148.78	1847.2	22278.
70	0.00457	12.442	151.53	0.00037	0.08037	12.178	218.60	2720.0	33126.
80	0.00212	12.473	153.80	0.00017	0.08017	12.330	471.95	5886.9	72586.
90	0.00098	12.487	154.99	0.00008	0.08008	12.411	1018.9	12723.	157924.
100	0.00045	12.494	155.61	0.00004	0.08004	12.454	2199.7	27484.	342306.
120	0.00010	12.498	156.08	0.00001	0.08001	12.488	10253.	128150.	>10**6
INF	0.00000	12.500	156.25	0.0000	0.08000	12.500	INF	INF	INF

PRESENT SUM, P — UNIFORM SERIES, A — FUTURE SUM, F

Appendix A (Continued)

	$i = 9\%$			$i = 9\%$			$i = 9\%$			
	PRESENT SUM, P			UNIFORM SERIES, A			FUTURE SUM, F			
n	P/F	P/A	P/G	A/F	A/P	A/G	F/P	F/A	F/G	n
1	0.91743	0.9174	0.0000	1.00000	1.09000	0.0000	1.0900	1.0000	0.0000	1
2	0.84168	1.7591	0.8416	0.47847	0.56505	0.4784	1.1881	2.0900	1.0000	2
3	0.77218	2.5312	2.3860	0.30505	0.39505	0.9426	1.2950	3.2781	3.0900	3
4	0.70843	3.2397	4.5113	0.21867	0.30867	1.3925	1.4115	4.5731	6.3681	4
5	0.64993	3.8896	7.1110	0.16709	0.25709	1.8282	1.5386	5.9847	10.941	5
6	0.59627	4.4859	10.092	0.13292	0.22292	2.2497	1.6771	7.5233	16.925	6
7	0.54703	5.0329	13.374	0.10869	0.19869	2.6574	1.8280	9.2004	24.449	7
8	0.50187	5.5348	16.887	0.09067	0.18067	3.0511	1.9925	11.028	33.649	8
9	0.46043	5.9952	20.571	0.07680	0.16680	3.4312	2.1718	13.021	44.678	9
10	0.42241	6.4176	24.372	0.06582	0.15582	3.7977	2.3673	15.192	57.699	10
11	0.38753	6.8051	28.248	0.05695	0.14695	4.1509	2.5804	17.560	72.892	11
12	0.35553	7.1607	32.159	0.04965	0.13965	4.4910	2.8126	20.140	90.452	12
13	0.32618	7.4869	36.073	0.04357	0.13357	4.8181	3.0658	22.953	110.59	13
14	0.29925	7.7861	39.963	0.03843	0.12843	5.1326	3.3417	26.019	133.54	14
15	0.27454	8.0606	43.806	0.03406	0.12406	5.4346	3.6424	29.360	159.56	15
16	0.25187	8.3125	47.584	0.03030	0.12030	5.7244	3.9703	33.003	188.92	16
17	0.23107	8.5436	51.282	0.02705	0.11705	6.0023	4.3276	36.973	221.93	17
18	0.21199	8.7556	54.886	0.02421	0.11421	6.2686	4.7171	41.301	258.90	18
19	0.19449	8.9501	58.386	0.02173	0.11173	6.5235	5.1416	46.018	300.20	19
20	0.17843	9.1285	61.777	0.01955	0.10955	6.7674	5.6044	51.160	346.22	20
21	0.16370	9.2922	65.050	0.01762	0.10762	7.0005	6.1088	56.764	397.38	21
22	0.15018	9.4424	68.204	0.01590	0.10590	7.2337	6.6586	62.873	454.14	22
23	0.13778	9.5802	71.235	0.01438	0.10438	7.4357	7.2578	69.531	517.02	23
24	0.12640	9.7066	74.143	0.01302	0.10302	7.6384	7.9110	76.789	586.55	24
25	0.11597	9.8225	76.926	0.01181	0.10181	7.8316	8.6230	84.700	663.34	25
26	0.10639	9.9289	79.586	0.01072	0.10072	8.0155	9.3991	93.324	748.04	26
27	0.09761	10.026	82.124	0.00973	0.09973	8.1906	10.245	102.72	841.36	27
28	0.08955	10.116	84.541	0.00885	0.09885	8.3571	11.167	112.96	944.09	28
29	0.08215	10.198	86.842	0.00806	0.09806	8.5153	12.172	124.13	1057.0	29
30	0.07537	10.273	89.028	0.00734	0.09734	8.6656	13.267	136.30	1181.1	30
35	0.04899	10.566	98.359	0.00464	0.09464	9.3082	20.414	215.71	2007.9	35
40	0.03184	10.757	105.37	0.00296	0.09296	9.7957	31.409	337.88	3309.8	40
45	0.02069	10.881	110.55	0.00190	0.09190	10.160	48.327	525.85	5342.8	45
50	0.01345	10.961	114.32	0.00123	0.09123	10.429	74.357	815.08	8500.9	50
55	0.00874	11.014	117.03	0.00079	0.09079	10.626	114.40	1260.0	13389.	55
60	0.00568	11.048	118.96	0.00051	0.09051	10.768	176.03	1944.7	20942.	60
65	0.00369	11.070	120.33	0.00033	0.09033	10.870	270.84	2998.2	32592.	65
70	0.00240	11.084	121.29	0.00022	0.09022	10.942	416.73	4619.2	50546.	70
80	0.00101	11.099	122.43	0.00009	0.09009	11.029	986.55	10950.	120784.	80
90	0.00043	11.106	122.97	0.00004	0.09004	11.072	2335.5	25939.	287213.	90
100	0.00018	11.109	123.23	0.00002	0.09002	11.093	5529.0	61422.	681363.	100
120	0.00003	11.110	123.41	0.00000	0.09000	11.107	30987.	344289.	>10**6	120
INF	0.00000	11.111	123.45	0.0000	0.09000	11.111	INF	INF	INF	INF

i = 10%

	PRESENT SUM, P			UNIFORM SERIES, A			FUTURE SUM, F			
n	P/F	P/A	P/G	A/F	A/P	A/G	F/P	F/A	F/G	n
1	0.90909	0.9090	0.0000	1.00000	1.10000	0.0000	1.1000	1.0000	0.0000	1
2	0.82645	1.7355	0.8264	0.47619	0.57619	0.4761	1.2100	2.1000	1.0000	2
3	0.75131	2.4868	2.3290	0.30211	0.40211	0.9365	1.3310	3.3100	3.1000	3
4	0.68301	3.1698	4.3781	0.21547	0.31547	1.3811	1.4641	4.6410	6.4100	4
5	0.62092	3.7907	6.8618	0.16380	0.26380	1.8101	1.6105	6.1051	11.051	5
6	0.56447	4.3552	9.6841	0.12961	0.22961	2.2235	1.7715	7.7156	17.156	6
7	0.51316	4.8684	12.763	0.10541	0.20541	2.6216	1.9487	9.4871	24.871	7
8	0.46651	5.3349	16.028	0.08744	0.18744	3.0044	2.1435	11.435	34.358	8
9	0.42410	5.7590	19.421	0.07364	0.17364	3.3723	2.3579	13.579	45.794	9
10	0.38554	6.1445	22.891	0.06275	0.16275	3.7254	2.5937	15.937	59.374	10
11	0.35049	6.4950	26.396	0.05396	0.15396	4.0640	2.8531	18.531	75.311	11
12	0.31863	6.8136	29.901	0.04676	0.14676	4.3884	3.1384	21.384	93.842	12
13	0.28966	7.1033	33.377	0.04078	0.14078	4.6987	3.4522	24.522	115.22	13
14	0.26333	7.3666	36.800	0.03575	0.13575	4.9955	3.7975	27.975	139.75	14
15	0.23939	7.6060	40.152	0.03147	0.13147	5.2789	4.1772	31.772	167.72	15
16	0.21763	7.8237	43.416	0.02782	0.12782	5.5493	4.5949	35.949	199.49	16
17	0.19784	8.0215	46.581	0.02466	0.12466	5.8071	5.0544	40.544	235.44	17
18	0.17986	8.2014	49.639	0.02193	0.12193	6.0525	5.5599	45.599	275.99	18
19	0.16351	8.3649	52.582	0.01955	0.11955	6.2861	6.1159	51.159	321.59	19
20	0.14864	8.5135	55.406	0.01746	0.11746	6.5080	6.7275	57.275	372.75	20
21	0.13513	8.6486	58.109	0.01562	0.11562	6.7188	7.4002	64.002	430.02	21
22	0.12285	8.7715	60.689	0.01401	0.11401	6.9188	8.1402	71.402	494.02	22
23	0.11168	8.8832	63.146	0.01257	0.11257	7.1084	8.9543	79.543	565.43	23
24	0.10153	8.9847	65.481	0.01130	0.11130	7.2880	9.8497	88.497	644.97	24
25	0.09230	9.0770	67.696	0.01017	0.11017	7.4579	10.834	98.347	733.47	25
26	0.08391	9.1609	69.794	0.00916	0.10916	7.6186	11.918	109.18	831.81	26
27	0.07628	9.2372	71.777	0.00826	0.10826	7.7704	13.110	121.10	940.99	27
28	0.06934	9.3065	73.649	0.00745	0.10745	7.9137	14.421	134.21	1062.1	28
29	0.06304	9.3696	75.414	0.00673	0.10673	8.0488	15.863	148.63	1196.3	29
30	0.05731	9.4269	77.076	0.00608	0.10608	8.1762	17.449	164.49	1344.9	30
35	0.03558	9.6441	83.987	0.00369	0.10369	8.7086	28.102	271.02	2360.2	35
40	0.02209	9.7790	88.952	0.00226	0.10226	9.0962	45.259	442.59	4025.9	40
45	0.01372	9.8628	92.454	0.00139	0.10139	9.3740	72.890	718.90	6739.0	45
50	0.00852	9.9148	94.888	0.00086	0.10086	9.5704	117.39	1163.9	11139.	50
55	0.00529	9.9471	96.561	0.00053	0.10053	9.7075	189.05	1880.5	18255.	55
60	0.00328	9.9671	97.701	0.00033	0.10033	9.8022	304.48	3034.8	29748.	60
65	0.00204	9.9796	98.470	0.00020	0.10020	9.8671	490.37	4893.7	48287.	65
70	0.00127	9.9873	98.987	0.00013	0.10013	9.9112	789.74	7887.4	78174.	70
80	0.00049	9.9951	99.560	0.00005	0.10005	9.9609	2048.4	20474.	203940.	80
90	0.00019	9.9981	99.811	0.00002	0.10002	9.9830	5313.	53120.	530302.	90
100	0.00007	9.9992	99.920	0.00001	0.10001	9.9927	13780.	137796.	>10**6	100
120	0.00001	9.9998	99.986	0.00000	0.10000	9.9987	92709.	927081	>10**6	120
INF	0.00000	10.000	130.00	0.0000	0.10000	10.000	INF	INF	INF	INF

Appendix A (Continued)

	i = 11%			i = 11%			i = 11%			
	PRESENT SUM, P			UNIFORM SERIES, A			FUTURE SUM, F			
n	P/F	P/A	P/G	A/F	A/P	A/G	F/P	F/A	F/G	n
1	0.90090	0.9009	0.0000	1.00000	1.11000	0.0000	1.1100	1.0000	0.0000	1
2	0.81162	1.7125	0.8116	0.47393	0.58393	0.4739	1.2321	2.1100	0.0000	2
3	0.73119	2.4437	2.2740	0.29921	0.40921	0.9305	1.3676	3.3421	3.1100	3
4	0.65873	3.1024	4.2502	0.21233	0.32233	1.3699	1.5180	4.7097	6.4521	4
5	0.59345	3.6959	6.6240	0.16057	0.27057	1.7922	1.6850	6.2278	11.161	5
6	0.53464	4.2305	9.2972	0.12638	0.23638	2.1976	1.8704	7.9128	17.389	6
7	0.48166	4.7122	12.187	0.10222	0.21222	2.5863	2.0761	9.7832	25.302	7
8	0.43393	5.1461	15.224	0.08432	0.19432	2.9584	2.3045	11.859	35.085	8
9	0.39092	5.5370	18.352	0.07060	0.18060	3.3144	2.5580	14.164	46.945	9
10	0.35218	5.8892	21.521	0.05980	0.16980	3.6544	2.8394	16.722	61.109	10
11	0.31728	6.2065	24.694	0.05112	0.16112	3.9788	3.1517	19.561	77.831	11
12	0.28584	6.4923	27.838	0.04403	0.15403	4.2879	3.4984	22.713	97.392	12
13	0.25751	6.7498	30.929	0.03815	0.14815	4.5821	3.8832	26.211	120.10	13
14	0.23199	6.9818	33.944	0.03323	0.14323	4.8618	4.3104	30.094	146.31	14
15	0.20900	7.1908	35.870	0.02907	0.13907	5.1274	4.7845	34.405	176.41	15
16	0.18829	7.3791	39.695	0.02552	0.13552	5.3793	5.3108	39.189	210.81	16
17	0.16963	7.5487	42.007	0.02247	0.13247	5.6180	5.8950	44.500	250.00	17
18	0.15282	7.7016	45.007	0.01984	0.12984	5.8438	6.5435	50.395	294.50	18
19	0.13768	7.8392	47.485	0.01756	0.12756	6.0573	7.2633	56.939	344.90	19
20	0.12403	7.9633	49.842	0.01558	0.12558	6.2589	8.0623	64.202	401.84	20
21	0.11174	8.0750	52.077	0.01384	0.12384	6.4491	8.9491	72.265	466.04	21
22	0.10067	8.1757	54.191	0.01231	0.12231	6.6282	9.9335	81.214	538.31	22
23	0.09069	8.2664	55.186	0.01097	0.12097	6.7969	11.026	91.147	619.52	23
24	0.08170	8.3481	55.065	0.00979	0.11979	6.9555	12.239	102.17	710.67	24
25	0.07361	8.4217	59.832	0.00874	0.11874	7.1044	13.585	114.41	812.84	25
26	0.06631	8.4880	61.490	0.00781	0.11781	7.2443	15.079	127.99	927.26	26
27	0.05974	8.5478	63.043	0.00699	0.11699	7.3753	16.738	143.07	1055.2	27
28	0.05382	8.6016	64.496	0.00626	0.11626	7.4981	18.579	159.81	1198.3	28
29	0.04849	8.6501	65.854	0.00561	0.11561	7.6131	20.623	178.39	1358.1	29
30	0.04368	8.6937	67.121	0.00502	0.11502	7.7205	22.892	199.02	1536.5	30
35	0.02592	8.8552	72.253	0.00293	0.11293	8.1594	38.574	341.59	2787.1	35
40	0.01538	8.9510	75.778	0.00172	0.11172	8.4659	65.000	581.82	4925.6	40
45	0.00913	9.0079	78.155	0.00101	0.11101	8.6762	109.53	986.63	8560.3	45
50	0.00542	9.0416	79.734	0.00060	0.11060	8.8185	184.56	1668.7	14716.	50
55	0.00322	9.0616	80.771	0.00035	0.11035	8.9134	311.00	2818.2	25120.	55
60	0.00191	9.0735	81.446	0.00021	0.11021	8.9762	524.05	4755.0	42682.	60
65	0.00113	9.0806	81.881	0.00012	0.11012	9.0172	883.06	8018.7	72307.	65
70	0.00067	9.0848	82.161	0.00007	0.11007	9.0438	1488.0	13518.	122258.	70
INF	0.00000	9.0909	82.644	0.0000	0.11000	9.0909	INF	INF	INF	INF

	$i=12\%$				$i=12\%$				$i=12\%$		
	PRESENT SUM, P				UNIFORM SERIES, A				FUTURE SUM, F		
n	P/F	P/A	P/G	A/F	A/P	A/G	F/P	F/A	F/G	n
1	0.89286	0.8928	0.0000	1.00000	1.12000	0.0000	1.1200	1.0000	0.0000	1
2	0.79719	1.6900	0.7971	0.47170	0.59170	0.4717	1.2544	2.1200	1.0000	2
3	0.71178	2.4018	2.2207	0.29635	0.41635	0.9246	1.4049	3.3744	3.1200	3
4	0.63552	3.0373	4.1273	0.20923	0.32923	1.3588	1.5735	4.7793	6.4944	4
5	0.56743	3.6047	6.3970	0.15741	0.27741	1.7745	1.7623	6.3528	11.273	5
6	0.50663	4.1114	8.9301	0.12323	0.24323	2.1720	1.9738	8.1151	17.626	6
7	0.45235	4.5637	11.644	0.09612	0.21912	2.5514	2.2106	10.089	25.741	7
8	0.40388	4.9676	14.471	0.08130	0.20130	2.9131	2.4759	12.299	35.830	8
9	0.36061	5.3282	17.356	0.06768	0.18768	3.2574	2.7730	14.775	48.130	9
10	0.32197	5.6502	20.254	0.05698	0.17698	3.5846	3.1058	17.548	62.906	10
11	0.28748	5.9377	23.128	0.04842	0.16842	3.8952	3.4785	20.654	80.454	11
12	0.25668	6.1943	25.952	0.04144	0.16144	4.1896	3.8959	24.133	101.10	12
13	0.22917	6.4235	28.702	0.03568	0.15568	4.4683	4.3634	28.029	125.24	13
14	0.20462	6.6281	31.362	0.03087	0.15087	4.7316	4.8871	32.392	153.27	14
15	0.18270	6.8108	33.920	0.02682	0.14682	4.9803	5.4735	37.279	185.66	15
16	0.16312	6.9739	36.367	0.02339	0.14339	5.2146	6.1303	42.753	222.94	16
17	0.14564	7.1196	38.697	0.02046	0.14046	5.4353	6.8660	48.883	265.69	17
18	0.13004	7.2496	40.908	0.01794	0.13794	5.6427	7.6899	55.749	314.58	18
19	0.11611	7.3657	42.997	0.01576	0.13576	5.8375	8.6127	63.439	370.33	19
20	0.10367	7.4694	44.967	0.01388	0.13388	6.0202	9.6462	72.052	433.77	20
21	0.09256	7.5620	46.818	0.01224	0.13224	6.1913	10.803	81.698	505.82	21
22	0.08264	7.6446	48.554	0.01081	0.13081	6.3514	12.100	92.502	587.52	22
23	0.07379	7.7184	50.177	0.00956	0.12956	6.5010	13.552	104.60	680.02	23
24	0.06588	7.7843	51.692	0.00846	0.12846	6.6406	15.178	118.15	784.62	24
25	0.05882	7.8431	53.104	0.00750	0.12750	6.7708	17.000	133.33	902.78	25
26	0.05252	7.8956	54.417	0.00665	0.12665	6.8921	19.040	150.33	1036.1	26
27	0.04689	7.9425	55.636	0.00590	0.12590	7.0049	21.324	169.37	1186.4	27
28	0.04187	7.9844	56.767	0.00524	0.12524	7.1097	23.883	190.69	1355.8	28
29	0.03738	8.0218	57.814	0.00466	0.12466	7.2071	26.749	214.58	1546.5	29
30	0.03338	8.0551	58.782	0.00414	0.12414	7.2974	29.959	241.33	1761.1	30
35	0.01894	8.1755	62.605	0.00232	0.12232	7.6576	52.799	431.66	3305.5	35
40	0.01075	8.2437	65.115	0.00130	0.12130	7.8987	93.051	767.09	6059.1	40
45	0.00610	8.2825	66.734	0.00074	0.12074	8.0572	163.98	1358.2	10943.	45
50	0.00346	8.3045	67.762	0.00042	0.12042	8.1597	289.00	2400.0	19583.	50
55	0.00196	8.3169	68.408	0.00024	0.12024	8.2251	509.32	4236.0	34841.	55
60	0.00111	8.3240	68.810	0.00013	0.12013	8.2664	897.59	7471.6	61763.	60
65	0.00063	8.3280	69.058	0.00008	0.12008	8.2922	1581.8	13173.	109241.	65
70	0.00036	8.3303	69.210	0.00004	0.12004	8.3082	2787.8	23223.	192944.	70
INF	0.00000	8.3333	69.444	0.0000	0.12000	8.3333	INF	INF	INF	INF

Appendix A *(Continued)*

| | i = 13% | | | i = 13% | | | i = 13% | | | |
| | PRESENT SUM, P | | | UNIFORM SERIES, A | | | FUTURE SUM, F | | | |
n	P/F	P/A	P/G	A/F	A/P	A/G	F/P	F/A	F/G	n
1	0.88496	0.8849	0.0000	1.00000	1.13000	0.0000	1.1300	1.0000	0.0000	1
2	0.78315	1.6681	0.7831	0.46948	0.59948	0.4694	1.2769	2.1300	1.0000	2
3	0.69305	2.3611	2.1692	0.29352	0.42352	0.9187	1.4429	3.4069	3.1300	3
4	0.61332	2.9744	4.0092	0.20619	0.33631	1.3478	1.6304	4.8498	6.5369	4
5	0.54276	3.5172	6.1802	0.15431	0.28431	1.7571	1.8424	6.4802	11.386	5
6	0.48032	3.9975	8.5818	0.12015	0.25015	2.1467	2.0819	8.3227	17.867	6
7	0.42506	4.4226	11.132	0.09611	0.22611	2.5171	2.3526	10.404	26.189	7
8	0.37616	4.7987	13.765	0.07839	0.20839	2.8685	2.6584	12.757	36.594	8
9	0.33288	5.1316	15.428	0.06487	0.19487	3.2013	3.0040	15.415	49.351	9
10	0.29459	5.4262	19.079	0.05429	0.18429	3.5161	3.3945	18.419	64.767	10
11	0.26070	5.6869	21.686	0.04584	0.17584	3.8134	3.8358	21.814	83.187	11
12	0.23071	5.9176	24.224	0.03899	0.16899	4.0935	4.3345	25.650	105.00	12
13	0.20416	6.1218	26.674	0.03335	0.16335	4.3572	4.8980	29.984	130.65	13
14	0.18068	6.3024	29.023	0.02867	0.15867	4.6050	5.5347	34.882	160.63	14
15	0.15989	6.4623	31.261	0.02474	0.15474	4.8374	6.2542	40.417	195.51	15
16	0.14150	6.6038	33.384	0.02143	0.15143	5.0552	7.0673	46.671	235.93	16
17	0.12522	6.7290	35.387	0.01861	0.14861	5.2589	7.9860	53.739	282.60	17
18	0.11081	6.8399	37.271	0.01620	0.14620	5.4491	9.0242	61.725	336.34	18
19	0.09806	6.9379	39.036	0.01413	0.14413	5.6265	10.197	70.749	398.07	19
20	0.08678	7.0247	40.685	0.01235	0.14235	5.7917	11.523	80.946	468.82	20
21	0.07680	7.1015	42.221	0.01081	0.14081	5.9453	13.021	92.469	549.76	21
22	0.06796	7.1695	43.648	0.00948	0.13948	6.0880	14.713	105.49	642.23	22
23	0.06014	7.2296	44.971	0.00832	0.13832	6.2204	16.626	120.20	747.73	23
24	0.05323	7.2828	46.196	0.00731	0.13731	6.3430	18.788	136.83	867.93	24
25	0.04710	7.3299	47.326	0.00643	0.13643	6.4565	21.230	155.62	1004.7	25
26	0.04168	7.3716	48.368	0.00565	0.13565	6.5614	23.990	176.85	1160.3	26
27	0.03689	7.4085	49.327	0.00498	0.13498	6.6581	27.109	200.84	1337.2	27
28	0.03264	7.4412	50.209	0.00439	0.13439	6.7474	30.633	227.95	1538.0	28
29	0.02889	7.4700	51.017	0.00387	0.13387	6.8296	34.615	258.58	1766.0	29
30	0.02557	7.4956	51.759	0.00341	0.13341	6.9052	39.115	293.19	2024.6	30
35	0.01388	7.5855	54.614	0.00183	0.13183	7.1998	72.068	546.68	3936.0	35
40	0.00753	7.6343	56.408	0.00099	0.13099	7.3887	132.78	1013.7	7490.0	40
45	0.00409	7.6608	57.514	0.00053	0.13053	7.5076	244.64	1874.1	14070.	45
50	0.00222	7.6752	58.187	0.00029	0.13029	7.5811	450.73	3459.5	26227.	50
55	0.00120	7.6830	58.590	0.00016	0.13016	7.6260	830.45	6380.4	48656.	55
60	0.00065	7.6872	58.831	0.00009	0.13009	7.6530	1530.0	11761.	90015.	60
65	0.00035	7.6895	58.973	0.00005	0.13005	7.6692	2819.0	21677.	166247.	65
70	0.00019	7.6908	59.056	0.00003	0.13003	7.6788	5193.8	39945.	306732.	70
INF	0.00000	7.6923	59.171	0.0000	0.13000	7.6923	INF	INF	INF	INF

i = 14%

n	PRESENT SUM, P			UNIFORM SERIES, A			FUTURE SUM, F			n
	P/F	P/A	P/G	A/F	A/P	A/G	F/P	F/A	F/G	
1	0.87719	0.8771	0.0000	1.00000	1.14000	0.0000	1.1400	1.0000	0.0000	1
2	0.76947	1.5466	0.7694	0.46729	0.60729	0.4672	1.2996	2.1400	1.0000	2
3	0.67497	2.3216	2.1194	0.29073	0.43073	0.9129	1.4815	3.4396	3.1400	3
4	0.59208	2.9137	3.8956	0.20320	0.34320	1.3370	1.6889	4.9211	6.5796	4
5	0.51937	3.4330	5.9731	0.15128	0.29128	1.7398	1.9254	6.6101	11.500	5
6	0.45559	3.8886	8.2510	0.11716	0.25716	2.1218	2.1949	8.5355	18.110	6
7	0.39964	4.2883	10.648	0.09319	0.23319	2.4832	2.5022	10.730	26.646	7
8	0.35056	4.6388	13.102	0.07557	0.21557	2.8245	2.8525	13.232	37.376	8
9	0.30751	4.9463	15.562	0.06217	0.20217	3.1463	3.2519	16.085	50.609	9
10	0.26974	5.2161	17.990	0.05171	0.19171	3.4490	3.7072	19.337	66.695	10
11	0.23662	5.4527	20.356	0.04339	0.18339	3.7333	4.2262	23.044	86.032	11
12	0.20756	5.6602	22.639	0.03667	0.17667	3.9997	4.8179	27.270	109.07	12
13	0.18207	5.8423	24.824	0.03116	0.17116	4.2490	5.4924	32.088	136.34	13
14	0.15971	6.0020	26.900	0.02661	0.16661	4.4819	6.2613	37.581	168.43	14
15	0.14010	6.1421	28.862	0.02281	0.16281	4.6990	7.1379	43.842	206.01	15
16	0.12289	6.2650	30.705	0.01962	0.15962	4.9011	8.1372	50.980	249.86	16
17	0.10780	6.3728	32.430	0.01692	0.15692	5.0888	9.2764	59.117	300.84	17
18	0.09456	6.4674	34.038	0.01462	0.15462	5.2629	10.575	68.394	359.95	18
19	0.08295	6.5503	35.531	0.01266	0.15266	5.4242	12.055	78.969	428.35	19
20	0.07276	6.6231	36.913	0.01099	0.15099	5.5734	13.743	91.024	507.32	20
21	0.06383	6.6869	38.190	0.00954	0.14954	5.7111	15.667	104.76	598.34	21
22	0.05599	6.7429	39.365	0.00830	0.14830	5.8380	17.861	120.43	703.11	22
23	0.04911	6.7920	40.446	0.00723	0.14723	5.9549	20.361	138.29	823.55	23
24	0.04308	6.8351	41.437	0.00630	0.14630	6.0623	23.212	158.65	961.84	24
25	0.03779	6.8729	42.344	0.00550	0.14550	6.1610	26.461	181.87	1120.5	25
26	0.03315	6.9060	43.172	0.00480	0.14480	6.2514	30.166	208.33	1302.3	26
27	0.02908	6.9351	43.928	0.00419	0.14419	6.3337	34.389	238.49	1510.7	27
28	0.02551	6.9606	44.617	0.00366	0.14366	6.4099	39.204	272.88	1749.2	28
29	0.02237	6.9830	45.244	0.00320	0.14320	6.4791	44.693	312.09	2022.1	29
30	0.01963	7.0026	45.813	0.00280	0.14280	6.5422	50.950	356.78	2334.1	30
35	0.01019	7.0700	47.951	0.00144	0.14144	6.7824	98.100	693.57	4704.0	35
40	0.00529	7.1050	49.237	0.00075	0.14075	6.9299	188.88	1342.0	9300.1	40
45	0.00275	7.1232	49.996	0.00039	0.14039	7.0187	363.67	2590.5	18182.	45
50	0.00143	7.1326	50.437	0.00020	0.14020	7.0713	700.23	4994.5	35318.	50
55	0.00074	7.1375	50.691	0.00010	0.14010	7.1020	1348.2	9623.1	68343.	55
60	0.00039	7.1401	50.835	0.00005	0.14005	7.1197	2595.9	18535.	131965.	60
65	0.00020	7.1414	50.917	0.00003	0.14003	7.1298	4998.2	35694.	254496.	65
70	0.00010	7.1421	50.963	0.00001	0.14001	7.1355	9623.6	68733.	490451.	70
INF	0.00000	7.1428	51.020	0.0000	0.14000	7.1428	INF	INF	INF	INF

Appendix A (Continued)

	i = 15%			i = 15%			i = 15%			
	PRESENT SUM, P			UNIFORM SERIES, A			FUTURE SUM, F			
n	P/F	P/A	P/G	A/F	A/P	A/G	F/P	F/A	F/G	n
1	0.86957	0.8695	0.0000	1.00000	1.15000	0.0000	1.1500	1.0000	0.0000	1
2	0.75614	1.6257	0.7561	0.46512	0.61512	0.4651	1.3225	2.1500	1.0000	2
3	0.65752	2.2832	2.0711	0.28798	0.43798	0.9071	1.5208	3.4725	3.1500	3
4	0.57175	2.8549	3.7864	0.20027	0.35027	1.3262	1.7490	4.9933	6.6225	4
5	0.49718	3.3521	5.7751	0.14832	0.29832	1.7228	2.0113	6.7423	11.615	5
6	0.43233	3.7844	7.9367	0.11424	0.26424	2.0971	2.3130	8.7537	18.358	6
7	0.37594	4.1604	10.192	0.09036	0.24036	2.2498	2.6600	11.066	27.112	7
8	0.32690	4.4873	12.480	0.07285	0.22285	2.7813	3.0590	13.726	38.178	8
9	0.28426	4.7715	14.754	0.05957	0.20957	3.0922	3.5178	16.785	51.905	9
10	0.24718	5.0187	16.979	0.04925	0.19925	3.3832	4.0455	20.303	68.691	10
11	0.21494	5.2337	19.128	0.04107	0.19107	3.6549	4.6523	24.349	88.995	11
12	0.18691	5.4206	21.184	0.03448	0.18448	3.9082	5.3502	29.001	113.34	12
13	0.16253	5.5831	23.135	0.02911	0.17911	4.1437	6.1527	34.351	142.34	13
14	0.14133	5.7244	24.972	0.02469	0.17469	4.3624	7.0757	40.504	176.69	14
15	0.12289	5.8473	26.693	0.02102	0.17102	4.5649	8.1370	47.580	217.20	15
16	0.10686	5.9542	28.296	0.01795	0.16795	4.7522	9.3576	55.717	264.78	16
17	0.09293	6.0471	29.782	0.01537	0.16537	4.9250	10.761	65.075	320.50	17
18	0.08081	6.1279	31.156	0.01319	0.16319	5.0843	12.375	75.836	385.57	18
19	0.07027	6.1982	32.421	0.01134	0.16134	5.2307	14.231	88.211	461.41	19
20	0.06110	6.2593	33.582	0.00976	0.15976	5.3651	16.366	102.44	549.62	20
21	0.05313	6.3124	34.644	0.00842	0.15842	5.4883	18.821	118.81	652.06	21
22	0.04620	6.3586	35.615	0.00727	0.15727	5.6010	21.644	137.63	770.87	22
23	0.04017	6.3988	36.498	0.00628	0.15628	5.7039	24.891	159.27	908.50	23
24	0.03493	6.4337	37.302	0.00543	0.15543	5.7978	28.625	184.16	1067.7	24
25	0.03038	6.4641	38.031	0.00470	0.15470	5.8834	32.919	212.79	1251.9	25
26	0.02642	6.4905	38.691	0.00407	0.15407	5.9612	37.856	245.71	1464.7	26
27	0.02297	6.5135	39.289	0.00353	0.15353	6.0319	43.535	283.56	1710.4	27
28	0.01997	6.5308	39.828	0.00306	0.15306	6.0960	50.065	327.10	1994.0	28
29	0.01737	6.5508	40.314	0.00255	0.15265	6.1540	57.575	377.17	2321.1	29
30	0.01510	6.5659	40.752	0.00230	0.15230	6.2066	66.211	434.74	2698.3	30
35	0.00751	6.6166	42.358	0.00113	0.15113	6.4018	133.17	881.17	5641.1	35
40	0.00373	6.6417	43.283	0.00056	0.15056	6.5167	267.86	1779.0	11593.	40
45	0.00186	6.6542	43.805	0.00028	0.15028	6.5829	538.76	3585.1	23600.	45
50	0.00092	6.6605	44.095	0.00014	0.15014	6.6204	1083.6	7217.7	47784.	50
55	0.00046	6.6636	44.255	0.00007	0.15007	6.6414	2179.6	14524.	96461.	55
60	0.00023	6.6651	44.343	0.00003	0.15003	6.6529	4384.0	29220.	194400.	60
65	0.00011	6.6659	44.390	0.00002	0.15002	6.6592	8817.7	58778.	391424.	65
INF	0.00000	6.6666	44.444	0.0000	0.15000	6.6666	INF	INF	INF	INF

		$i=20\%$			$i=20\%$			$i=20\%$			
	PRESENT SUM, P			UNIFORM SERIES, A			FUTURE SUM, F				
n	P/F	P/A	P/G	A/F	A/P	A/G	F/P	F/A	F/G	n
1	0.83333	0.8333	0.0000	1.00000	1.20000	0.0000	1.2000	1.0000	0.0000	1
2	0.69444	1.5277	0.6944	0.45455	0.65455	0.4545	1.4400	2.2000	1.0000	2
3	0.57870	2.1064	1.8518	0.27473	0.47473	0.8791	1.7280	3.6400	3.2000	3
4	0.48225	2.5887	3.2986	0.18629	0.38629	1.2742	2.0736	5.3680	6.8400	4
5	0.40188	2.9906	4.9061	0.13438	0.33438	1.6405	2.4883	7.4416	12.208	5
6	0.33490	3.3255	6.5806	0.10071	0.30071	1.9788	2.9859	9.9299	19.649	6
7	0.27908	3.6045	8.2551	0.07742	0.27742	2.2901	3.5831	12.915	29.579	7
8	0.23257	3.8371	9.8830	0.06061	0.26061	2.5756	4.2998	16.499	42.495	8
9	0.19381	4.0309	11.433	0.04808	0.24808	2.8364	5.1597	20.798	58.994	9
10	0.16151	4.1924	12.887	0.03852	0.23852	3.0738	6.1917	25.958	79.793	10
11	0.13459	4.3270	14.233	0.03110	0.23110	3.2892	7.4300	32.150	105.75	11
12	0.11216	4.4392	15.466	0.02526	0.22526	3.4841	8.9161	39.580	137.90	12
13	0.09346	4.5326	16.588	0.02062	0.22062	3.6597	10.699	48.496	177.48	13
14	0.07789	4.6105	17.600	0.01689	0.21689	3.8174	12.839	59.195	225.98	14
15	0.06491	4.6754	18.509	0.01388	0.21388	3.9588	15.407	72.035	285.17	15
16	0.05409	4.7295	19.320	0.01144	0.21144	4.0851	18.488	87.442	357.21	16
17	0.04507	4.7746	20.041	0.00944	0.20944	4.1975	22.186	105.93	444.65	17
18	0.03756	4.8121	20.680	0.00781	0.20781	4.2975	26.623	128.11	550.58	18
19	0.03130	4.8435	21.243	0.00646	0.20646	4.3860	31.948	154.74	678.70	19
20	0.02608	4.8695	21.739	0.00536	0.20536	4.4643	38.337	186.68	833.44	20
21	0.02174	4.8913	22.174	0.00444	0.20444	4.5333	46.005	225.02	1020.1	21
22	0.01811	4.9094	22.554	0.00369	0.20369	4.5941	55.206	271.03	1245.1	22
23	0.01509	4.9245	22.886	0.00307	0.20307	4.6475	66.247	326.23	1516.1	23
24	0.01258	4.9371	23.176	0.00255	0.20255	4.6942	79.496	392.48	1842.4	24
25	0.01048	4.9475	23.427	0.00212	0.20212	4.7351	95.396	471.98	2234.9	25
26	0.00874	4.9563	23.646	0.00176	0.20176	4.7708	114.47	567.37	2706.8	26
27	0.00728	4.9636	23.835	0.00147	0.20147	4.8020	137.37	681.85	3274.2	27
28	0.00607	4.9696	23.999	0.00122	0.20122	4.8291	164.84	819.22	3956.1	28
29	0.00506	4.9747	24.140	0.00102	0.20102	4.8526	197.81	984.06	4775.3	29
30	0.00421	4.9789	24.262	0.00085	0.20085	4.8730	237.37	1181.8	5759.4	30
35	0.00169	4.9915	24.661	0.00034	0.20034	4.9406	590.66	2948.3	14566.	35
40	0.00068	4.9966	24.846	0.00014	0.20014	4.9727	1469.7	7343.8	36519.	40
45	0.00027	4.9986	24.931	0.00005	0.20005	4.9876	3657.2	18281.	91181.	45
50	0.00011	4.9994	24.969	0.00002	0.20002	4.9945	9100.4	45497.	227236.	50
55	0.00004	4.9997	24.986	0.00001	0.20001	4.9975	22644.	113219.	565820.	55
60	0.00002	4.9999	24.994	0.00000	0.20000	4.9989	56347.	281733	>10**6	60
INF	0.00000	5.0000	25.000	0.0000	0.20000	5.0000	INF	INF	INF	INF

Appendix A *(Continued)*

| | i=25% | | | i=25% | | | i=25% | | | |
| | PRESENT SUM, P | | | UNIFORM SERIES, A | | | FUTURE SUM, F | | | |
n	P/F	P/A	P/G	A/F	A/P	A/G	F/P	F/A	F/G	n
1	0.80000	0.8000	0.0000	1.00000	1.25000	0.0000	1.2500	1.0000	0.0000	1
2	0.64000	1.4400	0.6400	0.44444	0.69444	0.4444	1.5625	2.2500	1.0000	2
3	0.51200	1.9520	1.6640	0.26230	0.51230	0.8524	1.9531	3.8125	3.2500	3
4	0.40960	2.3616	2.8928	0.17344	0.42344	1.2249	2.4414	5.7656	7.0625	4
5	0.32768	2.6892	4.2035	0.12185	0.37185	1.5630	3.0517	8.2070	12.828	5
6	0.26214	2.9514	5.5142	0.08882	0.33882	1.8683	3.8147	11.258	21.035	6
7	0.20972	3.1611	6.7725	0.06634	0.31634	2.1424	4.7683	15.073	32.293	7
8	0.16777	3.3289	7.9469	0.05040	0.30040	2.3872	5.9604	19.841	47.367	8
9	0.13422	3.4631	9.0206	0.03876	0.28876	2.6047	7.4505	25.802	67.209	9
10	0.10737	3.5705	9.9870	0.03007	0.28007	2.7971	9.3132	33.252	93.011	10
11	0.08590	3.6564	10.846	0.02349	0.27349	2.9663	11.641	42.566	126.26	11
12	0.06872	3.7251	11.602	0.01845	0.26845	3.1145	14.551	54.207	168.83	12
13	0.05498	3.7801	12.261	0.01454	0.26454	3.2437	18.189	68.759	223.03	13
14	0.04398	3.8240	12.833	0.01150	0.26150	3.3559	22.737	86.949	291.79	14
15	0.03518	3.8592	13.326	0.00912	0.25912	3.4529	28.421	109.68	378.74	15
16	0.02815	3.8874	13.748	0.00724	0.25724	3.5366	35.527	138.10	488.43	16
17	0.02252	3.9099	14.108	0.00576	0.25576	3.6083	44.408	173.63	626.54	17
18	0.01801	3.9279	14.414	0.00459	0.25459	3.6697	55.511	218.04	800.17	18
19	0.01441	3.9423	14.674	0.00366	0.25366	3.7221	69.388	273.55	1018.2	19
20	0.01153	3.9538	14.893	0.00292	0.25292	3.7667	86.736	342.94	1291.7	20
21	0.00922	3.9631	15.077	0.00233	0.25233	3.8045	108.42	429.68	1634.7	21
22	0.00738	3.9704	15.232	0.00186	0.25186	3.8364	135.52	538.10	2064.4	22
23	0.00590	3.9763	15.362	0.00148	0.25148	3.8634	169.40	673.62	2602.5	23
24	0.00472	3.9811	15.471	0.00119	0.25119	3.8861	211.75	843.03	3276.1	24
25	0.00378	3.9848	15.561	0.00095	0.25095	3.9051	264.69	1054.7	4119.1	25
26	0.00302	3.9879	15.637	0.00076	0.25076	3.9211	330.87	1319.4	5173.9	26
27	0.00242	3.9903	15.700	0.00061	0.25061	3.9345	413.59	1650.3	6493.4	27
28	0.00193	3.9922	15.752	0.00048	0.25048	3.9457	516.98	2063.9	8143.8	28
29	0.00155	3.9938	15.795	0.00039	0.25039	3.9550	646.23	2580.9	10207.	29
30	0.00124	3.9950	15.831	0.00031	0.25031	3.9628	807.79	3227.1	12788.	30
35	0.00041	3.9983	15.936	0.00010	0.25010	3.9858	2465.1	9856.7	39287.	35
40	0.00013	3.9994	15.976	0.00003	0.25003	3.9946	7523.1	30088.	120195.	40
45	0.00004	3.9998	15.991	0.00001	0.25001	3.9980	22958.	91831.	367146.	45
INF	0.00000	4.0000	16.000	0.0000	0.25000	4.0000	INF	INF	INF	INF

	i = 30%				i = 30%			i = 30%			
	PRESENT SUM, P			UNIFORM SERIES, A				FUTURE SUM, F			
n	P/F	P/A	P/G	A/F	A/P	A/G	F/P	F/A	F/G	n
1	0.76923	0.7692	0.0000	1.00000	1.30000	0.0000	1.3000	1.0000	0.0000	1
2	0.59172	1.3609	0.5917	0.43478	0.73478	0.4347	1.6900	2.3000	1.0000	2
3	0.45517	1.8161	1.5020	0.25063	0.55063	0.8270	2.1970	3.9900	3.3000	3
4	0.35013	2.1662	2.5524	0.16163	0.46163	1.1782	2.8561	6.1870	7.2900	4
5	0.26933	2.4355	3.6297	0.11058	0.41058	1.4903	3.7129	9.0431	13.477	5
6	0.20718	2.6427	4.6656	0.07839	0.37839	1.7654	4.8268	12.756	22.520	6
7	0.15937	2.8021	5.6218	0.05687	0.35687	2.0062	6.2748	17.582	35.276	7
8	0.12259	2.9247	6.4799	0.04192	0.34192	2.2155	8.1573	23.857	52.859	8
9	0.09430	3.0190	7.2343	0.03124	0.33124	2.3962	10.604	32.015	76.716	9
10	0.07254	3.0915	7.8871	0.02346	0.32346	2.5512	13.785	42.619	108.73	10
11	0.05580	3.1473	8.4451	0.01773	0.31773	2.6832	17.921	56.405	151.35	11
12	0.04292	3.1902	8.9173	0.01345	0.31345	2.7951	23.298	74.327	207.75	12
13	0.03302	3.2232	9.3135	0.01024	0.31024	2.8894	30.287	97.625	282.08	13
14	0.02540	3.2486	9.6436	0.00782	0.30782	2.9685	39.373	127.91	379.70	14
15	0.01954	3.2682	9.9172	0.00598	0.30598	3.0344	51.185	167.28	507.62	15
16	0.01503	3.2832	10.142	0.00458	0.30458	3.0892	66.541	218.47	674.90	16
17	0.01156	3.2948	10.327	0.00351	0.30351	3.1345	86.504	285.01	893.38	17
18	0.00889	3.3036	10.478	0.00269	0.30269	3.1718	112.45	371.51	1178.3	18
19	0.00684	3.3105	10.601	0.00207	0.30207	3.2024	146.19	483.97	1549.9	19
20	0.00526	3.3157	10.701	0.00159	0.30159	3.2275	190.05	630.16	2033.8	20
21	0.00405	3.3198	10.782	0.00122	0.30122	3.2479	247.06	820.21	2664.0	21
22	0.00311	3.3229	10.848	0.00094	0.30094	3.2646	321.18	1067.2	3484.2	22
23	0.00239	3.3253	10.900	0.00072	0.30072	3.2781	417.53	1388.4	4551.5	23
24	0.00184	3.3271	10.943	0.00055	0.30055	3.2890	542.80	1806.0	5940.0	24
25	0.00142	3.3286	10.977	0.00043	0.30043	3.2978	705.64	2348.8	7746.0	25
26	0.00109	3.3297	11.004	0.00033	0.30033	3.3049	917.33	3054.4	10094.	26
27	0.00084	3.3305	11.026	0.00025	0.30025	3.3106	1192.5	3971.7	13149.	27
28	0.00065	3.3311	11.043	0.00019	0.30019	3.3152	1550.2	5164.3	17121.	28
29	0.00050	3.3316	11.057	0.00015	0.30015	3.3189	2015.0	6714.6	22285.	29
30	0.00038	3.3320	11.068	0.00011	0.30011	3.3218	2620.0	8729.9	29000.	30
35	0.00010	3.3329	11.098	0.00003	0.30003	3.3297	9727.8	32422.	107960.	35
INF	0.00000	3.3333	11.111	0.0000	0.30000	3.3333	INF	INF	INF	INF

Appendix A *(Continued)*

	$i=40\%$ PRESENT SUM, P			$i=40\%$ UNIFORM SERIES, A			$i=40\%$ FUTURE SUM, F			
n	P/F	P/A	P/G	A/F	A/P	A/G	F/P	F/A	F/G	n
1	0.71429	0.7142	0.0000	1.00000	1.40000	0.0000	1.4000	1.0000	0.0000	1
2	0.51020	1.2244	0.5102	0.41667	0.81667	0.4166	1.9600	2.4000	1.0000	2
3	0.36443	1.5889	1.2390	0.22936	0.62936	0.7798	2.7440	4.3600	3.4000	3
4	0.26031	1.8492	2.0199	0.14077	0.54077	1.0923	3.8416	7.1040	7.7600	4
5	0.18593	2.0351	2.7637	0.09136	0.49136	1.3579	5.3782	10.945	14.864	5
6	0.13281	2.1679	3.4277	0.06126	0.46126	1.5811	7.5295	16.323	25.809	6
7	0.09486	2.2628	3.9969	0.04192	0.44192	1.7663	10.541	23.853	42.133	7
8	0.06776	2.3306	4.4712	0.02907	0.42907	1.9185	14.757	34.394	65.986	8
9	0.04840	2.3790	4.8584	0.02034	0.42034	2.0422	20.661	49.152	100.38	9
10	0.03457	2.4135	5.1696	0.01432	0.41432	2.1419	28.925	69.813	149.53	10
11	0.02469	2.4382	5.4165	0.01013	0.41013	2.2214	40.495	98.739	219.34	11
12	0.01764	2.4559	5.6106	0.00718	0.40718	2.2845	56.693	139.23	318.08	12
13	0.01260	2.4685	5.7617	0.00510	0.40510	2.3341	79.371	195.92	457.32	13
14	0.00900	2.4775	5.8787	0.00363	0.40363	2.3728	111.12	275.30	653.25	14
15	0.00643	2.4839	5.9687	0.00259	0.40259	2.4029	155.56	386.42	928.55	15
16	0.00459	2.4885	6.0376	0.00185	0.40185	2.4262	217.79	541.98	1314.9	16
17	0.00328	2.4918	6.0901	0.00132	0.40132	2.4440	304.91	759.78	1856.9	17
18	0.00234	2.4941	6.1299	0.00094	0.40094	2.4577	426.87	1064.7	2616.7	18
19	0.00167	2.4958	6.1600	0.00067	0.40067	2.4681	597.63	1491.5	3681.4	19
20	0.00120	2.4970	6.1827	0.00048	0.40048	2.4760	836.68	2089.2	5173.0	20
21	0.00085	2.4978	6.1998	0.00034	0.40034	2.4820	1171.3	2925.8	7262.2	21
22	0.00061	2.4984	6.2126	0.00024	0.40024	2.4865	1639.9	4097.2	10188.	22
23	0.00044	2.4989	6.2222	0.00017	0.40017	2.4899	2295.8	5737.1	14285.	23
24	0.00031	2.4992	6.2293	0.00012	0.40012	2.4925	3214.2	8033.0	20022.	24
25	0.00022	2.4994	6.2347	0.00009	0.40009	2.4944	4499.8	11247.	28055.	25
INF	0.00000	2.5000	6.2500	0.0000	0.40000	2.5000	INF	INF	INF	INF

| | i = 50% | | | i = 50% | | | i = 50% | | | |
| | PRESENT SUM, P | | | UNIFORM SERIES, A | | | FUTURE SUM, F | | | |
n	P/F	P/A	P/G	A/F	A/P	A/G	F/P	F/A	F/G	n
1	0.66667	0.6666	0.0000	1.00000	1.50000	0.0000	1.5000	1.0000	0.0000	1
2	0.44444	1.1111	0.4444	0.40000	0.90000	0.4000	2.2500	2.5000	1.0000	2
3	0.29630	1.4074	1.0370	0.21053	0.71053	0.7368	3.3750	4.7500	3.5000	3
4	0.19753	1.6049	1.6296	0.12308	0.62308	1.0153	5.0625	8.1250	8.2500	4
5	0.13169	1.7366	2.1563	0.07583	0.57583	1.2417	7.5937	13.187	16.375	5
6	0.08779	1.8244	2.5953	0.04812	0.54812	1.4225	11.390	20.781	29.562	6
7	0.05853	1.8829	2.9465	0.03108	0.53108	1.5648	17.085	32.171	50.343	7
8	0.03902	1.9219	3.2196	0.02030	0.52030	1.6751	25.628	49.257	82.515	8
9	0.02601	1.9479	3.4277	0.01335	0.51335	1.7596	38.443	74.886	131.77	9
10	0.01734	1.9653	3.5838	0.00882	0.50882	1.8235	57.665	113.33	206.66	10
11	0.01156	1.9768	3.6994	0.00585	0.50585	1.8713	86.497	170.99	319.99	11
12	0.00771	1.9845	3.7841	0.00388	0.50388	1.9067	129.74	257.49	490.98	12
13	0.00514	1.9897	3.8458	0.00258	0.50258	1.9328	194.62	387.23	748.47	13
14	0.00343	1.9931	3.8903	0.00172	0.50172	1.9518	291.92	581.85	1135.7	14
15	0.00228	1.9954	3.9223	0.00114	0.50114	1.9656	437.89	873.78	1717.5	15
16	0.00152	1.9969	3.9451	0.00076	0.50076	1.9756	656.84	1311.6	2591.3	16
17	0.00101	1.9979	3.9614	0.00051	0.50051	1.9827	985.26	1968.5	3903.0	17
18	0.00068	1.9986	3.9729	0.00034	0.50034	1.9878	1477.8	2953.7	5871.5	18
19	0.00045	1.9991	3.9810	0.00023	0.50023	1.9914	2216.8	4431.6	8825.3	19
20	0.00030	1.9994	3.9867	0.00015	0.50015	1.9939	3325.2	6648.5	13257.	20
21	0.00020	1.9996	3.9907	0.00010	0.50010	1.9957	4987.8	9973.7	19905.	21
22	0.00013	1.9997	3.9935	0.00007	0.50007	1.9970	7481.8	14961.	29879.	22
23	0.00009	1.9998	3.9955	0.00004	0.50004	1.9979	11222.	22443.	44841.	23
24	0.00006	1.9998	3.9969	0.00003	0.50003	1.9985	16834.	33666.	67284.	24
25	0.00004	1.9999	3.9978	0.00002	0.50002	1.9990	25251.	50500.	100951.	25
INF	0.00000	2.0000	4.0000	0.0000	0.50000	2.0000	INF	INF	INF	INF

Appendix B SUMMARY OF EQUATIONS AND CASH FLOW DIAGRAMS

DESCRIPTION & NOTATION	EQUATION	CASH FLOW DIAGRAM	AS $n \to \infty$
Future worth (lump sum), F, of a present amount (lump sum), P $F = P(F/P, i, n)$ (n = number of time periods)	$F = P(1+i)^n$ $F = Pe^{ni}$ (for continuous compounding)		$F/P \to \infty$
Future worth (lump sum), F, of a periodic uniform series, A $F = A(F/A, i, n)$ (n = number of end of period payments)	$F = A\dfrac{(1+i)^n - 1}{i}$		$F/A \to \infty$
Present worth (lump sum), P, of a periodic series, A $P = A(P/A, i, n)$ (n = number of end of period payments)	$P = A\left[\dfrac{(1+i)^n - 1}{i(1+i)^n}\right]$		$P/A \to 1/i$
Future worth, F, of an arithmetic gradient series, G $F = G(F/G, i, n)$ (n = number of payments plus 1)	$F = \dfrac{G}{i}\left[\dfrac{(1+i)^n - 1}{i} - n\right]$		$F/G \to \infty$

Present worth, P, of an arithmetic gradient series, G

$P = G(P/G.i.n)$

(n = number of payments plus 1)

$$P = \frac{G}{i}\left[\frac{(1+i)^n - 1}{i(1+i)^n} - \frac{n}{(1+i)^n}\right]$$

$P/G \to 1/i^2$

Periodic uniform series, A equivalent of arithmetic gradient series, G

$A = G(A/G,i,n)$

$$A = G\left[\frac{1}{i} - \frac{n}{(1+i)^n - 1}\right]$$

$A/G \to 1/i$

Present worth, P, of a geometric gradient series

(n = number of end of period payments)

1. When $r > i$, then $w = \dfrac{1+r}{1+i} - 1$, and $P = \dfrac{C}{(1+i)}(F/A, w, n)$

$$= \frac{C}{(1+i)}\left[\frac{(1+w)^n - 1}{w}\right]$$

$P/C \to \infty$

2. When $r < i$, then $w = \dfrac{1+i}{1+r} - 1$, and $P = \dfrac{C}{(1+r)}(P/A, w, n)$

$$= \frac{C}{(1+r)}\left[\frac{(1+w)^n - 1}{w(1+w)^n}\right]$$

$P/C \to \dfrac{1}{(1+r)w}$

3. When $r = i$, then $P = \dfrac{Cn}{(1+r)} = \dfrac{Cn}{(1+i)}$

$P/C \to \infty$

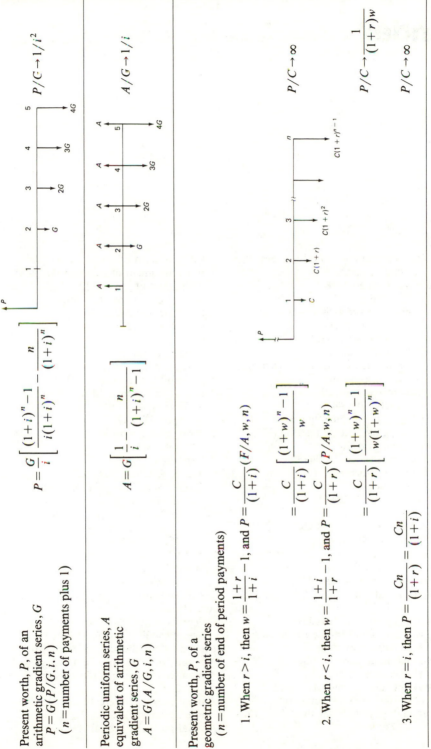

Index

86 87 88 89 90   10 9 8 7 6 5